MW00843381

Covered Bridges in the Southeastern United States

Covered Bridges in the Southeastern United States

A Comprehensive Illustrated Catalog

by Warren H. White

McFarland & Company, Inc., Publishers
Jefferson, North Carolina, and London

Dedicated to my wife, Pat, for her forbearance
as I wandered around the southeastern states
for 166 days and nights over several trips
during two and one-half years,
researching each covered bridge in this book.

Happily, we are still married.

Library of Congress Cataloguing-in-Publication Data

White, Warren H., 1935–
Covered bridges in the southeastern United States :
a comprehensive illustrated catalog / by Warren H. White.
p. cm.
Includes index.

ISBN 0-7864-1536-3 (illustrated case binding : 50# and 70# alkaline papers) 

1. Covered Bridges—Southern States. I. Title.
TG23.W48 2003 388.1'32—dc21 2002156432

British Library cataloguing data are available

Cover photograph: Baker Park Bridge, Carroll Creek, Frederick, Maryland, 1965.

Manufactured in the United States of America

McFarland & Company, Inc., Publishers
Box 611, Jefferson, North Carolina 28640
www.mcfarlandpub.com

Contents

Preface

While passing through Georgia in 1999, I photographed some authentic covered bridges. Shortly thereafter, during research on those bridges, I found apparent discrepancies, returned to the bridges to check the information, and concluded that discrepancies do exist. I wondered whether this was typical of available covered bridge data. Unfortunately, the answer is yes. While initial errors may not be common, data passed down over time becomes altered by renovations, relocations, renaming of roads and waterways, and so on. Discrepancies in the record are an inevitable result.

This book is an attempt to rectify the situation and, in so doing, to present a listing of all authentic (usually historic) and all non-authentic (usually modern and privately owned) covered bridges existing in the southeastern United States in one volume. To achieve the objectives, I visited each bridge to obtain accurate statistics defined in a consistent manner—encompassing measurements, bridge construction details and photographs. For consistency, length measurements were obtained from portal to portal at the point where the portal meets the decking along the downstream side, and width measurements were obtained at each portal from the exterior of the siding to the exterior of the siding, also at the point where the portal meets the decking; the widest dimension was used.

Included in each listing is the most commonly used bridge name, alternate bridge names, measurements, name of waterway or other underneath passage, construction date and builder (where known), truss or bridge type, construction details, World Guide Covered Bridge Number, and concise directions to the bridge.

Not listed in this book are historic covered bridges that no longer exist. Many modern covered bridges have also been destroyed and are therefore not listed. Although considerable effort has been made to locate and include every privately owned covered bridge, regrettably some remote covered bridges have undoubtedly been missed.

I hope this book will promote interest in covered bridges, a small but important part of our national heritage, and lead to additional efforts to preserve covered bridges for the enjoyment of future generations. I also hope to inspire the reader to personally visit many of the remaining historic covered bridges, to marvel at the craftsmanship of the bridge builders.

I wish to acknowledge Richard T. Donovan, without whose vast covered bridge knowledge and assistance this effort would never have materialized; and Howard Rogers, for his assistance with the bridges in North and South Carolina.

Introduction

This reference book lists all extant covered bridges known to the author in the southeastern United States including the states of Alabama, Delaware, Florida, Georgia, Kentucky, Maryland, Mississippi, North Carolina, South Carolina, Tennessee, Virginia and West Virginia. (Mississippi is included as a formality because of its location in the Southeast, but it has no covered bridges known to exist today.) The covered bridges are presented in alphabetical order by state, each state subdivided alphabetically by county and the bridges within a county listed alphabetically by the bridge name, utilizing the most commonly or locally used name where more than one name exists. Alternate bridge names are noted in the text and listed in the index.

To be included in this book, a bridge must have a covered portion at least ten feet long. It must also be a true bridge or originally built and intended as a true bridge, as in the case of some relocated bridges that have been placed on display over dry ground. A true bridge provides for the conveyance of persons or vehicles over an obstacle, typically a waterway, and is not attached to buildings solely for access to the building or between buildings.

World Guide Covered Bridge (WGCB) numbers assigned by the National Society for the Preservation of Covered Bridges (NSPCB) are referenced for all listed bridges. Assigned numbers are coded, such as AL-05-12 or GA-115-a. The first two characters are the postal abbreviation for the state (formerly digits starting at 01 represented the states in alphabetical order). The second two digits represent the county in alphabetical order, starting at 01 for each state. The third pair of characters represents the bridge's place in the NSPCB tally of that county's bridges, starting at 01 for authentic truss type bridges or a letter for non-authentic type bridges, an upper case letter for motor vehicle bridges and a lower case letter for footbridges.

The covered bridges are categorized into four classes: authentic historic, authentic modern, non-authentic historic and non-authentic modern. Truss bridges are classed as authentic. Stringer bridges and post supported roofs over a roadway, provided a waterway passes beneath the roadway, are considered non-authentic. Post supported roofs over a roadway that has no waterway passing beneath and never had a waterway passing beneath are not included in this book. Bridges at least fifty years old are considered historic, except in the case of stringer bridges where the cover was added at a later date; then the age is considered from the date of the cover.

In addition to compass points, orientation of a bridge is referred to as the upstream or downstream side, with the portals referred to as the right downstream or left downstream portal, when viewed from the downstream side. All compass points are given viewed from the downstream side with the first compass point being the left downstream portal; e.g., southwest/northeast would have the left downstream portal at the southwest end.

Length dimensions are recorded for the downstream side, from portal to portal at the point where the portal meets the floor. Width dimensions are the wider of measurements taken at each portal, from the exterior of the siding to the exterior of the siding at the point where the portal meets the floor. The portal opening dimensions are approximate, as the dimensions vary between portals, the shorter of the dimensions being used.

The appearance of covered bridges may vary from that described in this book due to changes made during ongoing renovations and reconstruction. Renovations and reconstruction are happening more and more frequently as the public becomes aware of the need to preserve covered bridges.

Text for the historic bridges include the county seat, which the reader desiring more historical information about the bridge may find a useful starting point for research. For those interested in statistics, Goddard Bridge in Fleming County, Kentucky, is the oldest covered bridge in the southeastern states; Swann Bridge in Blount County, Alabama, is the longest covered bridge in the southeastern states; and Hortons Mill Bridge, also in Blount County, Alabama, is the highest covered bridge above a waterway in the United States.

The information included in this book was gathered mainly during the author's visits to the bridge sites. It is supplemented by historical references, interviews and various other sources. Many times contradictions existed. The information is the most accurate based on available data, observations and, finally, the author's experience and judgment. Notice of any errors or new information, including new bridges, would be sincerely appreciated in order to maintain the utmost accuracy in this book. Please send it to the author in care of the publisher.

Alabama

Hundreds of covered bridges were erected throughout the State of Alabama during the nineteenth and early twentieth centuries. Of these historic structures, two hundred remained in 1930, forty-six remained in 1958, fifteen remained in 1973, twelve remained in 2000 and ten remain in 2002. In retrospect, a regretful attrition of our national heritage, but, hopefully, the remaining ten will endure for future generations to enjoy and reflect upon the life of earlier times. The Civil War took its toll on Alabama's covered bridges, followed by floods and, in recent times, arson (at least ten since the 1960s). Alabama lost Oakachoy Bridge (arson) and Lidy Walker Bridge (collapsed) in 2001; however, an attempt to re-erect Lidy Walker Bridge is in the planning.

Of the twenty-five covered bridges in Alabama, twelve are authentic and thirteen are non-authentic, ten of the twenty-five being historic. These twenty-five bridges were constructed between circa 1850 and 1999, representing seventeen of the sixty-seven counties in Alabama.

Alabama's historic covered bridges, built between 1850 and 1935, span placid streams to deep, roaring gorges, providing picturesque settings in which to reflect upon the simplicity of life and the hardships of those times.

Overland Road Covered Bridge

Bibb County

Brief Statistics

Type: Non-authentic modern covered bridge. World Guide Covered Bridge Number: AL-04-A. Originally built in 1970s by Howard Harkins, dismantled and rebuilt at present site in March 1994 by Ted Burnett. One-span bridge crossing Furnace Branch. Stringer bridge 47.6 feet long by 16.7 feet wide, with approximately 9.5-foot wide by 10.5-foot high portal openings. Alternate Name: Montevallo Bridge. Located in Brierfield Ironworks Historical State Park.

Overland Road Bridge was the former World Guide Number AL-59-A built in the 1970s by Howard Harkins, a three-span stringer bridge crossing Shoal Creek on Overland Road. Constructed with a 49-foot center span and two 15-foot end spans, for a total length of 79 feet, the bridge was originally located in Montevallo, in Shelby County. It was also called Montevallo Bridge. Retaining its original name, the bridge was dismantled and relocated to nearby Brierfield Ironworks Historical State Park in Bibb County, where the center span was rebuilt in March 1994 by Ted Burnett, a local wood carver, salvaging the steel I-beams, flooring and tin roof. The two 15-foot end spans were joined and rebuilt on flat ground, forming a 30-foot long picnic pavilion. The State of Alabama now owns the 47.6-foot long bridge and the two end spans.

The single-span stringer bridge is supported by three longitudinal steel I-beam stringers set directly on the stream banks. The decking consists of transverse planking with three-plank-wide wheel treads; the sides are open with a three-board railing along each side; the gables have natural vertical boarding with battens; and the roof is dull red painted tin. Seven wood posts, slanting outward toward the eaves, support the roof. An external walkway covered by the roof overhang is on the upstream or south-southeast side. *OVERLAND ROAD/COVERED BRIDGE* signs are mounted above both entrances.

The City of Montevallo donated Overland Road Bridge to Brierfield Ironworks Historical State Park. Spanning Furnace Branch, a wet weather stream that flows into the nearby Little Cahaba River, thence into the Cahaba River before entering the Alabama River at Cahaba in Dallas County, the bridge is open to State Park visitor motor traffic. Aligned east-northeast/west-southwest, the bridge is in a wooded area near a shooting range at the north edge of the park. Brierfield Ironworks Historical State Park preserves many historical buildings, including Bibb County Iron Company's brick iron furnace built in 1862, purchased in 1863 by the Confederacy and expanded to increase production, raided by Federal troops on March 31, 1865, and set afire, reopened after the Civil War, then permanently closed in 1894 due to superior furnaces in Birmingham.

Directions: From I-65 exit 228, go west on SR 25 to Montevallo, continuing 6.4 miles past the junction with SR 119 to Frederick Pass Road; go left to Furnace Parkway on the left and follow to Brierfield Ironworks Historical State Park. The bridge is in the park.

Easley Covered Bridge

Blount County

Brief Statistics

Type: Authentic historic covered bridge. World Guide Covered Bridge Number: AL-05-12. Built in 1927 by Zelma Tidwell. One-span bridge crossing Dub Branch of Calvert Prong. Town lattice truss bridge 82.3 feet long by 14.4 feet wide, with approximately 9.5-foot wide by 8.8-foot high portal openings. Alternate Names: Old Easley Bridge, Old Easley Road Bridge, Rosa Bridge. Located northeast of Easley on Old Easley Road.

Built in 1927 by Zelma Tidwell, Easley Bridge is one of seven Town lattice truss bridges remaining in Alabama and shares with Hortons Mill Bridge and Swann Bridge the distinction of being the only historic covered bridges remaining in Alabama still carrying motor traffic, all three located in Blount County! Easley Bridge, named after the nearby community of Easley, has also been called Rosa Bridge, after the nearby Rosa Community, Old Easley Road Bridge after the road transiting the bridge, and Old Easley Bridge.

The single-span bridge rests on poured concrete abutments with additional support from five sets of wood-timber stanchions added beneath the bridge in 1979. The Town lattice truss, fastened with bolts—the nuts attached on the outside to prevent vandalism or thievery by souvenir hunters—has been strengthened by the addition of six vertical iron tension rods along each side between the upper chord and the lower chord. Transverse planking with multi-plank-wide wheel treads makes up the flooring. The sides have red-painted, vertical, tin siding, leaving the upper 52 inches open and exposing the Town lattice truss, the tin replacing the previous siding during repairs in 1967. The gables have natural, weathered vertical boarding with battens and *EASLEY BRIDGE* signs mounted above each entrance. The roof and the gables extend beyond the decking, sheltering the entrances from the weather. Chin braces are just inside each portal. In 1967, a bright tin roof replaced the original wood shingles, weathered now with small areas of rust. The 82.3-foot long bridge has developed a downstream tilt.

Blount County-owned Easley Bridge, aligned east-southeast/west-northwest, spans Dub Branch that flows into nearby Calvert Prong, then enters the Little Warrior River at Little Warrior before joining the waters of Locust Fork on its way to the Black Warrior River west of Short Creek in Jefferson County. Dub Branch is a small stream, cascading over a rock bottom at low flow, heavily wooded with dense undergrowth at the bridge site.

One of three historic covered bridges remaining in Blount County, Easley Bridge was listed on the National Register of Historic Places on August 20, 1981.

Directions: From Oneonta, the county seat, at the intersection of US 231/SR 53 and SR 75, go 3.4 miles west on US 231/SR 53 to Pine Grove Road on the left (at church); go 1.1 miles to the tee at Old Easley Road. The bridge is on the right.

Hortons Mill Covered Bridge

Blount County

Brief Statistics

Type: Authentic historic covered bridge. World Guide Covered Bridge Number: AL-05-07. Built in 1935 by Zelma Tidwell. Two-span bridge crossing Calvert Prong. Town lattice truss bridge 203.4 feet long by 14.1 feet wide, with approximately 11.1-foot wide by 8.5-foot high portal openings. Alternate Name: None known. Located north of Oneonta on unnamed road.

Hortons Mill Bridge was built in 1935 by Zelma Tidwell to replace a previous bridge, built in 1895, about 0.75 mile downstream. It provided access to Talmedge M. Horton's mill, after which the present bridge was named. The Alabama Historical Commission-owned bridge is one of seven Town lattice truss bridges remaining in Alabama and shares with Easley Bridge and Swann Bridge the distinction of being the only historic covered bridges remaining in Alabama still carrying motor traffic. At approximately 70 feet above Calvert Prong below, Hortons Mill Bridge has the highest span above a waterway for a covered bridge in the United States.

Abutments of poured concrete, anchored on ledge-rock embankments, and one poured concrete pier, assisted by a two-utility-pole-wide stanchion added after 1983 on the southwest or left downstream end, support the bridge. The Town lattice truss, fastened with bolts—the nuts attached on the outside to prevent vandalism or thiev-

Lofty 203.4-foot long Hortons Mill Bridge is approximately 70 feet above Calvert Prong, making it the highest covered bridge in the United States.

ery—has been strengthened along each side with fifteen vertical iron tension rods per span between the upper chord and the lower chord. The sagging floor is comprised of transverse planking with five-plank-wide wheel treads; the sides and gables have natural vertical boarding with battens, the upper half of the sides open, exposing the Town lattice truss. The roof is covered with tin that replaced the wood shingles sometime after 1983. The roof and the gables extend beyond the decking to provide shelter from the elements for the entrances. The bridge has steel pipe and I-beam buttresses along both sides at each abutment and the pier.

Aligned southwest/northeast, the two-span bridge crosses Calvert Prong of the Little Warrior River that flows into Locust Fork, whose waters enter the Black Warrior River west of Short Creek in Jefferson County. The 203.4-foot long bridge spans a heavily wooded, deep and wide, rocky gorge with boulder-strewn Calvert Prong 70 feet below, presenting a very picturesque sight.

One of three historic covered bridges remaining in Blount County, Hortons Mill Bridge was the first bridge in Alabama placed on the National Register of Historic Places, being listed on December 29, 1970.

Directions: From Oneonta, the county seat, at the intersection of US 231/SR 53 and SR 75, go north 5.8 miles on SR 75 to the bridge on the left. Parking is available just past the bridge, on the left side of SR 75.

London Park Covered Bridge

BLOUNT COUNTY

Brief Statistics

Type: Non-authentic modern covered footbridge. World Guide Covered Bridge Number: AL-05-a. Built in 1979 by K.E. and Helen Henderson. One-span footbridge crossing highway culvert outflow. Stringer footbridge 11.5 feet long by 5.2 feet wide, with approximately 4.3-foot wide by 6.6-foot high portal openings. Alternate Name: None known. Located in Cleveland off US 231 east.

London Park Bridge, built in the summer of 1979 by K.E. and Helen Henderson as a scenic addition to their front yard, was given its name by the Hendersons' grandchildren. The single-span bridge aligns west/east

paralleling US Highway 231 as it crosses the outflow from the highway culvert. At 11.5 feet long, London Park Bridge is the shortest non-authentic covered footbridge in Alabama.

The 11.5-foot long stringer footbridge has a transverse planked floor supported by two wood posts under each portal. The sides have a decorative lattice on the upper half with rust-covered corrugated tin on the lower half. The gables are open. The tin roof is supported by four posts along each side and stabilized by half-height 2-inch by 4-inch wood buttress braces at each side of both portals. The roof extends beyond the entrances for weather protection.

Sadly, London Park Bridge is falling apart from neglect and the overgrown vegetation that now obscures the footbridge from the highway although less than twenty feet away. The approach ramps have collapsed, the tin siding is falling off, and the bridge is in rapid decay due to the encompassing vegetation depriving the bridge of the drying rays of the sun.

Directions: From Cleveland at the junction of SR 160 and US 231/SR 53, go east 0.9 mile on US 231/SR 53 to the bridge hidden in dense vegetation on the right, directly across from Blount County Area Vocational Center. Private property: request permission to visit.

Swann Covered Bridge

BLOUNT COUNTY

Brief Statistics

Type: Authentic historic covered bridge. World Guide Covered Bridge Number: AL-05-05. Built in 1933 by Zelma Tidwell. Three-span bridge crossing Locust Fork. Town lattice truss bridge 304.5 feet long by 14.3 feet wide, with approximately 9.6-foot wide by 9.0-foot high portal openings. Alternate Name: Joy Bridge. Located west of Cleveland on Swann Bridge Road.

At 304.5 feet long, Swann Bridge is the longest authentic covered bridge remaining in Alabama and in the entire southeastern United States. Swann Bridge, also known as Joy Bridge, was built in 1933 near the Swann Farm by Zelma Tidwell to provide access to the nearby community of Joy. Swann Bridge derived its name from the Swann family, and Joy Bridge was named after the nearby Joy community. The Blount County-owned bridge is one of seven Town lattice truss bridges remaining in Alabama.

Swann Bridge rests on poured concrete abutments and two poured concrete piers. The Town lattice truss is fastened with bolts—the nuts attached on the outside to prevent vandalism or thievery by souvenir hunters—and is strengthened by seven vertical iron tension rods in each span. The wavy flooring is comprised of longitudinal planking with four- to six-plank-wide wheel treads. The siding is gleaming corrugated tin, with the Town lattice truss exposed for the upper two feet two inches in the first and third spans and exposed for the upper five feet seven inches in the center span. The gables have natural vertical boarding with battens and *SWANN BRIDGE* signs mounted above each entrance. The current rust-streaked tin roof was installed in 1979. The roof and the gables extend beyond the decking, sheltering the entrances from inclement weather. Chin braces are inside each portal and at the junction of each span. A plank guardrail extends along each side of the interior for the length of the bridge.

The three-span bridge, aligned south-southeast/ north-northwest, crosses Locust Fork, which flows into the Black Warrior River west of Short Creek in Jefferson County. Locust Fork is a usually placid, wide, sandy-bottomed stream at the bridge, but rock strewn, narrow and turbulent upstream and downstream of the bridge. The banks are heavily wooded, with expansive sand on the south-southeast or left downstream end.

Swann Bridge was listed on the National Register of Historic Places on August 20, 1981, one of three historic covered bridges remaining in Blount County. The county seat is Oneonta.

Directions: From Cleveland, go north on US 231/SR 53 to the junction with SR 79; go left 0.7 mile on SR 79 to Swann Bridge Road on the right, following 1.0 mile to the bridge.

Coldwater Covered Bridge

CALHOUN COUNTY

Brief Statistics

Type: Authentic historic covered bridge. World Guide Covered Bridge Number: AL-08-01. Built circa 1850 by unknown builder, relocated to present site in 1990. One-span bridge crossing outflow of Oxford Lake. Modified kingpost truss bridge 63.1 feet long by 13.3 feet wide, with approximately 10.7-foot wide by 10.6-foot high portal openings. Alternate Name: Coldwater Creek Bridge. Located in Oxford at Oxford Lake Civic Center.

Originally spanning Coldwater Creek near Coldwater on the Calhoun-Talladega county line (therefrom deriving the names Coldwater Bridge and Coldwater Creek

***Opposite:* Perched high above rock-bottomed Locust Fork, 304.5-foot long Swann Bridge is the longest covered bridge in the southeastern United States.**

Bridge), Coldwater Bridge dates from around 1850, possibly as early as 1834. Its builder is unknown. The single-span bridge was relocated to its present site over the Oxford Lake outflow, at the Oxford Lake Civic Center, on August 17, 1990. A sign attesting to the relocation of the bridge and its donation to the City of Oxford by the Calhoun County Commission is affixed over the entrance on the west-northwest or left downstream portal. The Calhoun County bridge is the oldest authentic covered bridge remaining in Alabama and one of only two kingpost truss bridges remaining in Alabama, the other being modern Jimmy Lay Bridge.

Now closed to motor traffic, Coldwater Bridge has a six-panel multiple kingpost truss, modified by using iron rods in place of the vertical truss timbers, and rests on concrete abutments faced with upright railroad ties. Transverse planking topped with three-plank-wide wheel treads comprises the floor; old weathered, horizontal lapped boards cover the portal gables and sides, with the upper two feet of the sides open, exposing the truss; cedar shingles cover the roof. The west-northwest portal has a noticeable lean upstream.

Aligned west-northwest/east-southeast, Coldwater Bridge is located in a park-like area at the Oxford Lake Civic Center, fully exposed at the edge of Oxford Lake over the outflow stream. A walking path traverses the 63.1-foot long bridge.

Listed on the National Register of Historic Places on April 11, 1973, Coldwater Bridge is the only historic covered bridge remaining in Calhoun County. Anniston is the county seat of Calhoun County.

Directions: From south of Anniston in Oxford on SR 21, going north from I-20, take the first right on Recreation Drive to the Oxford Lake Civic Center. The bridge is near the south side of Oxford Lake.

Clarkson Covered Bridge

CULLMAN COUNTY

Brief Statistics

Type: Authentic historic covered bridge. World Guide Covered Bridge Number: AL-22-01. Built in 1904 by unknown builder. Four-span bridge crossing Crooked Creek. Town lattice-multiple kingpost combination truss bridge 249.7 feet long by 12.5 feet wide, with approximately 9.5-foot wide by 8.7-foot high portal openings. Alternate Names: Clarkson-Legg Bridge, Legg Bridge. Located west of Cullman on bypassed section of CR 53 in public park.

Bypassed and closed to motor traffic, Clarkson Bridge, named after the former nearby community of Clarkson, was originally built in 1904, severely damaged by a 1921 flood and rebuilt in 1922. The materials for the original construction were donated by James Wordlaw Legg, a local mail carrier for the Clarkson Post Office, hence the bridge was also known as Clarkson-Legg Bridge or Legg Bridge. Following years of neglect, the Cullman County-owned bridge underwent, as an American Revolution Bicentennial Project, an extensive renovation in 1975 by Ivan Williams. The four-span bridge shares the distinction with modern Bob Saunders Family Bridge of being, at 12.5 feet wide, the narrowest authentic covered bridge remaining in Alabama. Clarkson Bridge spans Crooked Creek at the site where a Civil War skirmish, known as the Battle of Hog Mountain, was fought on April 30, 1863.

Clarkson Bridge is the only combination Town lattice and multiple kingpost truss in Alabama. The northwest/southeast-aligned bridge rests on mortared cut stone abutments and three massive mortared cut stone piers. A transverse planked floor with three-plank-wide wheel treads is enclosed by a shiny corrugated tin roof and dark weathered, horizontal lapped boarding on the portal gables and lower sides, the upper 4.3 feet of the sides open, exposing the trusses. To shelter the entrances from the weather, the roof and the gables extend beyond the decking. There is a double-plank-high guard railing on both sides of the interior for the length of the bridge.

Clarkson Bridge is located in a small park that includes the Clarkson Grist Mill, with its millpond, Dogtrot Log Cabin, shaded picnic grounds and hiking trails. An informational sign is at the left of the right downstream portal. The 249.7-foot long bridge spans a gorge, through which flows Crooked Creek 45 feet below, on its way to Lewis Smith Lake near Crane Hill. The new concrete bypass bridge is on the downstream side, and tall trees shading the picnic area are on the upstream side.

A sign over the right downstream portal signifies that the bridge was listed on the National Register of Historic Places on June 25, 1974. Clarkson Bridge is one of two historic covered bridges remaining in Cullman County.

Directions: From Cullman, the county seat, proceed west on US 278, 7.3 miles past the intersection with I-65 to CR 1117 on the right; go 0.6 mile to CR 1013 on the left; go 0.5 mile to the bridge, on the right side of the road, in a small park.

Opposite: **Built in Coldwater circa 1850, pre–Civil War Coldwater Bridge displays a noticeable lean in the west-northwest portal.**

Lidy Walker Covered Bridge

Cullman County

Brief Statistics

Type: Authentic historic covered bridge. World Guide Covered Bridge Number: AL-22-12. Built in 1926 by unknown builder. One-span bridge crossing outlet to Lidy's Lake. Town lattice truss 50.5 feet long by 12.7 feet wide, with approximately 10.7-foot wide portal openings (gables are open with no roof beams at portals for measuring height). Alternate Name: Big Branch Bridge. Located in Berlin off CR 747.

Lidy Walker Bridge was originally known as Big Branch Bridge, located east of Berlin in Blount County, approximately three miles from the Cullman County line. Big Branch Bridge had been bypassed by a new concrete bridge and was in danger of being torn down. W.I. "Lidy" Walker was able to save the bridge by purchasing it from Blount County for $50.00 and moving it fourteen miles to its present location. Big Branch Bridge was built in 1926 by an unknown builder, most likely Zelma Tidwell, as the bridge shares the features of Tidwell-built Easley Bridge, Hortons Mill Bridge, Nectar Bridge and Swann Bridge, namely, the Town lattice truss fastened by bolts with the nuts on the outside of the truss members, the chin braces, and the roof extensions over the portals.

The privately owned bridge is set on mortared stone abutments, with two support posts added under the bridge at the west-southwest or left downstream end. The 50.5-foot long bridge is one of seven historic bridges remaining in Alabama with the Town lattice truss. Diagonal planking with wheel treads forms the floor; gray weathered, horizontal lapped siding covers the sides, with no siding under the eaves or in the gables; rusted tin completes the roof. Tin angle strips outline the entrances and gable fasciae.

The single-span bridge, aligned west-southwest/east-northeast and closed to motor traffic, crosses the outlet to Lidy's Lake. It is situated in open pastureland, leaning upstream and in need of repair, near the wood margin. Lidy Walker Bridge is one of two historic covered bridges remaining in Cullman County.

Directions: From Cullman, at the junction of US 278 and SR 69, go east 2.7 miles on US 278 to Walker's Corner USA Store on the left, in Berlin. Request permission to visit the bridge at the store, where they most likely will provide an escort to the bridge.

Author's Note: In early August 2001, the guy wires stabilizing Lidy Walker Bridge failed, causing the sides and roof to collapse upon the decking, which remained in place. As the owner is going to attempt to re-erect the sides and roof, the author is retaining the listing of the bridge in this reference book.

Valley Creek Park Covered Bridge

Dallas County

Brief Statistics

Type: Non-authentic modern covered footbridge. World Guide Covered Bridge Number: AL-24-a. Built in 1989 by City of Selma. Three-span footbridge crossing Valley Creek. Stringer footbridge 82.3 feet long by 8.9 feet wide, with approximately 7.1-foot wide by 8.9-foot high portal openings. Alternate Name: None known. Located in Selma in Valley Creek Park.

The City of Selma built Valley Creek Park Bridge, in 1989, in Valley Creek Park, near the 1865 Battle of Selma re-enactment site. The stringer footbridge has three reinforced concrete slab stringers, set on abutments, constructed from concrete and rubble placed behind two steel I-beam piles capped with a reinforced concrete crossbeam, and two piers. The piers are constructed from a pair of steel I-beam piles, which are cross-braced and capped with a reinforced concrete crossbeam. The sides have natural vertical boarding above and below a central opening along the length of the bridge and natural vertical boarding in the gables. The low peaked roof is supported along each side by twenty-one 2-inch by 4-inch posts, each post stabilized with a small external buttress-type brace at the floor level. A transverse-planked floor and a bright, galvanized, corrugated sheet-metal roof completes the bridge.

The three-span bridge crosses swiftly flowing Valley Creek that empties into the nearby Alabama River. Aligned northwest/southeast, the 82.3-foot long bridge is exposed at the portal ends, but heavily foliated woods along the creek prevent a full side view until early winter.

Directions: From Selma, at the junction of US 80/SR 8 (Broad Street) and SR 22 (Dallas Avenue), go west 1.1 miles on SR 22 to Bloch Park on the left, just past the Old Live Oak Cemetery, also on the left. Enter Bloch Park, bearing left to two mortared-rock columns at each side of the entrance to Valley Creek Park. The bridge is 0.2 mile ahead.

Old Union Covered Bridge

DeKalb County

Brief Statistics

Type: Non-authentic modern covered bridge. World Guide Covered Bridge Number: AL-25-A. Built in 1980 by Jack E. Jones, Sr. Three-span bridge crossing West Fork Little River. Stringer bridge 41.9 feet long by 10.1 feet wide, with approximately 9.0-foot wide by 8.6-foot high portal openings. Alternate Names: Cloudmont Bridge, Old

Triple span Old Union Bridge was completed in 1980 utilizing the cover portion from an old stringer bridge originally located at Union Crossing, a ford near Lincoln.

Union Crossing Bridge. Located south-southwest of Mentone at Cloudmont Ski and Golf Resort.

Old Union Bridge was constructed from an old stringer covered bridge, originally located at Union Crossing, a ford near Lincoln, Talladega County, where local people crossed Otter Creek to get to New Union Church. The bridge was moved, in 1972, to its present site. The 101.8-foot long stringer portion of the bridge was started in 1969, and completed in 1970, by Jack E. Jones, Jr., Tom Kerby and Charles Edwards, with the relocated 41.9-foot long cover portion installed in 1980 by Jack E. Jones, Sr., in his son's memory. Old Union Bridge was named after its origin, Union Crossing. The bridge is also known as Cloudmont Bridge, after its current location, and Old Union Crossing Bridge. At 10.1 feet wide, the privately owned bridge is the narrowest non-authentic motor vehicle covered bridge in Alabama.

The stringer bridge is supported by two large steel I-beam stringers resting on rock and gravel river banks and two piers constructed with two steel I-beam uprights, cross-braced and set in poured concrete foundations. Transverse planked flooring with four-plank-wide wheel treads runs the length of the bridge. The sides are open for the upper six feet, with gray weathered, vertical boarding on the lower sides and gables, the lower ends of the gable boards angle cut on both edges of the board. Six wood posts along each side support the steeply-peaked rusted tin roof. *OLD UNION BRIDGE* signs are mounted above each entrance.

The three-span bridge crosses West Fork Little River that joins with East Fork Little River in the Little River Wildlife Management Area, to form the Little River that flows into Weiss Lake. Aligned west/east, Old Union Bridge is picturesquely set among heavily wooded mountain slopes and high over a swift-flowing boulder-strewn mountain river, in summer its rusty roof standing out vividly from the surrounding verdant foliage.

Directions: From west of Mentone at I-59 exit 231 on SR 40, go east 5.0 miles, following the signs to Cloudmont Ski and Golf Resort, going right up the mountain. Request permission, at the golf shop office, to visit the bridge. The bridge is on the access road to Shady Grove Dude Ranch.

Mother's Covered Bridge

Elmore County

Brief Statistics

Type: Non-authentic modern covered bridge. World Guide Covered Bridge Number: AL-26-A. Built in 1999 by Jim Lee, Gee Eason and James Calhoun. One-span bridge crossing Coosada Creek. Stringer bridge 40.1 feet long by 12.3 feet wide, with approximately 11.0-foot wide by 10.1-foot high portal openings. Alternate Name: None known. Located in Millbrook at the end of Willow Tree Drive.

Originally built in the summer of 1979 by Jim Lee, Gee Eason and James Calhoun, Mother's Bridge was blown down by a tornado in July 2000. Jim Lee, Gee Eason and James Calhoun rebuilt the bridge with salvaged materials later that fall. Named by Jim Lee in honor of his mother Evelyn—who insisted the bridge honor mothers all over the world—Mother's Bridge has brass plaques mounted inside the bridge subscribed to by children from around the world in honor of their mothers.

The 40.1-foot long bridge is supported by four steel I-beam stringers set on the creek embankments. The stringer bridge has a transverse-planked floor made from swamp white oak lumber; gray weathered, horizontal lapped siding on the sides and gables, with a trapezoidal open lattice area in the upper center of the sides; and old gray wood shingles, from a 120-year old plantation near Wetumpka, on the roof. The low peaked roof is supported by 2-inch by 6-inch studding along each side. The roof and gables extend beyond the end posts to provide shelter from the elements for the entrances. A *MOTHER'S* sign, carved in an old board, is mounted above the left downstream entrance. Wood guardrails at each end of the bridge channel the traffic into the single traffic lane.

Owned by Jim Lee, the single-span bridge crosses Coosada Creek, which flows into the Alabama River about three miles southeast of Coosada. Open to private motor traffic and aligned south/north, Mother's Bridge is situated at the end of a short, home-lined street leading to the entrance to an expansive open area that the bridge shares with an adjoining working grist mill and nearby log cabin, both constructed by Jim Lee and his friends.

Directions: From Millbrook on SR 143 (Main Street), go north to the United States Post Office on the left, proceeding an additional 0.9 mile north to Linda Ann Drive on the right; go 0.2 mile to the four-way stop sign; go left 0.2 mile on Grace Street (becomes Willow Brook Drive) to Willow Tree Drive on the left; go 0.1 mile to the bridge, visible at the end of the road. Private property: request permission from the owner at 5500 Willow Bend Drive, further down Willow Brook Drive.

Gargus Covered Bridge

Etowah County

Brief Statistics

Type: Non-authentic modern covered bridge. World Guide Covered Bridge Number: AL-28-A. Built in 1966 by Chester L. Gargus. Four-span bridge crossing Gargus Bass Lake. Stringer bridge 22.2 feet long by 12.3 feet wide, with approximately 11.4-foot wide by 8.8-foot high portal openings. Alternate Name: None known. Located in Gallant off Gallant Road.

The oldest non-authentic covered bridge in Alabama, Gargus Bridge, was built, in 1966, by Chester L. Gargus, in the landscaped front yard of his residence, over a cove to a small pond named Gargus Bass Lake. The privately owned bridge, tilted in a 1997 tornado and subsequently repaired, is now owned by his daughter and son-in-law, Joann and Charles Bryson.

The four-span bridge is set on concrete abutments, with additional support from three concrete pillars along each side. The stringer bridge has a transverse planked floor; diagonal boarding on the sides, with a large, lighted, gray weathered wood cross on the pond side; open portals; and a shiny tin roof, showing some rust. The low peaked roof is supported by 2-inch by 4-inch studding and extends beyond the entrances, providing them extra shelter from the elements. Open to private motor traffic on the circular driveway, the 22.2-foot long red-painted bridge aligns north/south, producing the best photographic opportunities in the morning.

Directions: From Gallant Post Office, go north on Gallant Road about 0.25 mile, past the church on the right, to Bryson residence on the right. The bridge is visible in the front yard. Private property: request permission to visit.

Gilliland Covered Bridge

Etowah County

Brief Statistics

Type: Authentic historic covered bridge. World Guide Covered Bridge Number: AL-28-02. Built originally in 1899 by Jesse Gilliland, relocated and rebuilt in 1967. Former-one-span bridge crossing small pond at overflow spillway. Originally a Town lattice truss now resembles a Howe truss bridge 80.8 feet long by 13.3 feet wide, with approximately 12.4-foot wide by 8.9-foot high portal openings. Alternate Names: Reece City Bridge, Old Reece City Bridge. Located in Gadsden at Noccalula Falls Park.

The reconstructed Gilliland Bridge has vertical iron tension rods between cross X timbers that resemble a Howe truss.

Jesse Gilliland originally built Gilliland Bridge in 1899 over Big Wills Creek on the Gilliland Plantation near Reece City, Etowah County, where it was also known as Reece City Bridge and Old Reece City Bridge. In 1967, Gilliland Bridge was donated to the City of Gadsden, dismantled, and taken by truck to Noccalula Falls Park, where it was rebuilt over the overflow spillway of a small pond. During the reconstruction, the single-span Town lattice truss bridge became an 80.8-foot long non-functional truss bridge, retaining the upper and lower chords and the iron tension rods connecting the upper chords to transverse beams below the lower chords. Discarding the Town lattice truss members, cross X beams were added between the tension rods, creating a resemblance to a Howe truss. The concrete abutments and five concrete pillars along each side placed under alternate transverse beams support the lower chords. Transverse planks comprise the floor; gray weathered, foot-wide diagonal boards cover the sides, which have nine square window openings each. The portals are open, and gray redwood shingles cover the roof. The steeply peaked roof extends beyond the entrances to give the interior additional shelter from the weather. A *GILLILAND'S/COVERED BRIDGE* sign is mounted on the roof beam inside the east-northeast entrance.

Closed to motor traffic, the east-northeast/west-southwest aligned bridge is in a pioneer village setting, spanning a fishpond within a city-owned park. Gilliland Bridge is the last of the historic covered bridges that served Etowah County.

Directions: From Gadsden at I-59 exit 183 (US 431/US 278/SR 74/SR 1), go east into Gadsden, the county seat, to the junction with SR 211 (Noccalula Road) on the left; follow SR 211 to Noccalula Falls Park and Campground on the left. The bridge is in the park. Admission.

Gabe's Covered Bridge

HOUSTON COUNTY

Brief Statistics

Type: Non-authentic modern covered bridge. World Guide Covered Bridge Number: AL-35-A. Built in 1994 by unknown builder. One-span bridge crossing wet weather stream. Stringer bridge 48.5 feet long by 20.0 feet wide, with approximately 12.4-foot wide by 10.4-foot high portal openings. Alternate Name: None known. Located north of Cottonwood off CR 33 at Gabe's Covered Bridge Restaurant.

An unknown builder constructed Gabe's Bridge in 1994 at the entrance to Gabe's Covered Bridge Restaurant. The stringer bridge has external walkways along each side for the convenience of patrons. Gabe's Bridge, at 48.5 feet long, is the longest single-span non-authentic covered bridge carrying motor traffic in Alabama.

The privately owned bridge has two steel I-beam stringers set on large rocks at the embankments of a wet weather stream. The floor has transverse planking with triple-plank-wide wheel treads. The sides are open, with handrails between the seven wood-post roof supports along each side, which separate the traffic lane from the external walkways. The outside railings of the walkways having fallen off due to a lack of maintenance. Natural horizontal lapped siding encloses the gables and green-painted aluminum covers the roof.

The west-northwest/east-southeast aligned bridge is situated close to County Road 33 on the lengthy, heavily wooded driveway into Gabe's Covered Bridge Restaurant, obscured behind thick woods. The waterway forms a large pool on the downstream side which mirrors the bridge, with its bright green roof against a deep blue sky. Since the restaurant closed in 1998, a large falling tree branch has damaged the upstream side of the roof, and neglect is taking its toll.

Directions: From Campbellton, go north on US 231, north of Madrid, to Smithville Road on the right (to Hodgesville); go right 1.7 miles and Smithville Road becomes CR 33; continue 3.3 miles to the bridge, visible on the left. During the foliage season, dense woods conceal the restaurant. Private property.

Mountain Oaks Estates Covered Bridge

Jefferson County

Brief Statistics

Type: Non-authentic modern covered bridge. World Guide Covered Bridge Number: AL-37-A. Built in 1970 by unknown builder. Post supported roof over roadway crossing Hackberry Creek. Bridge 25.8 feet long by 33.3 feet wide, with approximately 27.1-foot wide by 14.6-foot high portal openings. Alternate Name: None known. Located in Volusia near Hoover in Mountain Oaks Estates development on Mountain Oaks Drive.

Built in 1970 by an unknown builder, Mountain Oaks Estates Bridge, at 25.8 feet long, is the shortest non-authentic covered bridge carrying motor traffic in Alabama. An unusual covered bridge that, at 33.3 feet wide, is wider than it is long, the Mountain Oaks Estates-owned bridge also has the distinction of being the widest non-authentic covered bridge in Alabama.

Mountain Oaks Estates Bridge is actually a post-supported roof over a roadway that is over a conduit. Four wood posts along each side support the roof of the south/north-aligned bridge. The bridge is over asphalt-paved Mountain Oaks Drive where wet weather Hackberry Creek flows beneath the roadway through a large galvanized pipe. Each side of the roadbed is restrained by a mortared natural stone facing beside and above the large pipe, lending a picturesque touch to the bridge. Natural, weathered vertical boarding with battens cover the sides, each of which have three large, vertical rectangular window openings. The same siding also covers the portals, which have gas-lit carriage lamps at each side of the entrances. Gray weathered, wood shingles clad the low peaked roof.

Directions: From I-65 in Hoover, take exit 252, US 31 West, to the bottom of the ramp, going straight across US 31 on Columbiana Road to the traffic light at Tyler Road; go left to Hackberry Lane; go left 1.3 miles to Blue Ridge Parkway; go left 0.3 mile to Mountain Oaks Drive; go left to the bridge.

Tannehill Valley Covered Bridge

Jefferson County

Brief Statistics

Type: Non-authentic modern covered bridge. World Guide Covered Bridge Number: AL-37-B. Built in 1972 by unknown builder. One-span bridge crossing Tannehill Mill Stream. Stringer bridge 41.8 feet long by 20.0 feet wide, with approximately 19.9-foot wide by 13.9-foot high portal openings. Alternate Name: None known. Located in Kimbrell at entrance to Tannehill Valley Estates development.

An unknown builder built Tannehill Valley Bridge, in 1972, at the entrance to Tannehill Valley Estates development. The 41.8-foot long stringer bridge is supported by seven steel I-beam stringers, four large and three smaller I-beams, set on steel posts and steel barrier abutments, railroad ties serving as earth retaining walls at the sides. The flooring is transverse planking with no wheel treads. Natural vertical boarding covers the sides and portals, a three-foot high opening extends down each side on the exterior of the walkways that are on both sides, separated

Viewed from the unusually shaped east-southeast portal, Tannehill Valley Bridge, built in 1972, has walkways on both sides of the traffic lane.

from the single traffic lane by the roof supports and a hand railing between the supports. Twelve wood posts along each side of the walkways support the wood-shingled roof. Four external wood braces running from the deck level to the roof overhang along each side stabilize the roof. A *TANNEHILL VALLEY* sign, yellow letters on a brown background, is above the left downstream entrance. The portals and vehicle entrance flanked by the walkway entrances and roof braces, give Tannehill Valley Bridge an unusual and distinctive appearance.

The single-span bridge crosses Tannehill Mill Stream, a tributary to nearby Mill Creek that flows into Shades Creek before entering the Cahaba River north of Marvel. Aligned east-southeast/west-northwest, Tannehill Valley Bridge is open to traffic passing in and out of the development. The bridge is exposed at County Road 18, but the stream is heavily wooded on both sides of the bridge and extending up the hill behind the bridge, concealing the development.

Directions: From west of Bucksville at I-59 exit 100, go east on SR 216 through Bucksville to the tee at CR 97, near the entrance to Tannehill Ironworks Historical State Park on the right; go left on CR 97 (becomes CR 18 at the county line) 3.5 miles to the bridge at the entrance to Tannehill Valley Estates development on the right.

Askew Covered Bridge

Lee County

Brief Statistics

Type: Non-authentic modern covered footbridge. World Guide Covered Bridge Number: AL-41-a. Built in 1968 by William K. Askew. One-span footbridge crossing unnamed stream. Stringer footbridge 24.4 feet long by 4.3 feet wide, with approximately 3.1-foot wide by 6.9-foot high portal openings. Alternate Name: None known. Located in Auburn at 543 Sanders Street.

William K. Askew, a former member of the Southern Covered Bridge Society, in 1968, constructed the Askew Covered Footbridge across a small, unnamed stream in his front yard. Mr. Askew used to take his children to play at the nearby historic Askew Covered Bridge prior to its destruction. Seeking to continue his children's playing enjoyment, Mr. Askew was inspired to build a covered bridge in his front yard, incorporating four diamond window openings on each side, the lower two appropriately placed at a child's level. Askew Bridge holds the distinctions of being the oldest (1968) non-authentic covered footbridge, the longest (24.4 feet) single-span non-authentic covered footbridge and the narrowest (4.3 feet) non-authentic covered footbridge remaining in Alabama.

Two stringers, 2-inch by 10-inch lumber placed on edge, resting on stone abutments, support the private bridge. Transverse 2-inch by 8-inch wood planks compose the floor; natural vertical boarding covers the sides and the portals; and brown asphalt shingles cover the steeply peaked roof. There is an electric light mounted atop a wood post that illuminates the bridge at the left downstream portal or street end. A pine tree, felled by Hurricane Opal on October 5, 1995, damaged the bridge, since repaired, but necessitated the placement of 2-inch by 4-inch wood braces at each side of each portal to correct skewing.

The Lee County footbridge, aligned west-southwest/east-northeast, nestles amid large trees and thick foliage created by the numerous camellias and other shrubbery in the front yard. The pathway from the street to the house traverses the bridge.

Directions: From Auburn, at the intersection of SR 14 and SR 147 (College Street North), go north 0.3 mile to Drake Street on the left; proceed to Sanders Street, the first right. The bridge is on the left at No. 543. Private property: request permission to visit.

Salem-Shotwell Covered Bridge

Lee County

Brief Statistics

Type: Authentic historic covered bridge. World Guide Covered Bridge Number: AL-41-04. Built in 1900 by Otto Puls. One-span bridge crossing Wacoochee Creek. Town lattice truss bridge 76.3 feet long by 14.6 feet wide, with approximately 12.2-foot wide by 10.1-foot high portal openings. Alternate Names: Pea Ridge Bridge, Salem Bridge, Shotwell Bridge. Located northeast of Salem on CR 252 (Lee Road).

Salem-Shotwell Bridge, also known as Salem Bridge and Shotwell Bridge after the nearby communities, and sometimes called Pea Ridge Bridge, was built in 1900 by Otto Puls. The Lee County bridge is one of seven Town lattice truss bridges remaining in Alabama and the only historic covered bridge remaining in Lee County.

Now closed to motor traffic, Salem-Shotwell Bridge rests on mortared natural stone abutments, with additional support from a mortared natural stone pier near the left downstream abutment and two mortared natural stone pillars near the right downstream abutment. The 76.3-foot long bridge has a Town lattice truss, fastened with white oak treenails (wooden pegs), and chin braces inside the left downstream portal; those inside the right downstream portal have been removed. Transverse planking with two-plank-wide wheel treads comprises the floor;

red-painted vertical boarding with battens covers the sides and portals; and dark weathered, wood shingles cover the roof. The roof and the gables extend beyond the truss end posts forming shelter to the entrance. A *SALEM-SHOTWELL* sign, white letters on a red background, is above the entrance on the right downstream end.

Aligned east-southeast/west-northwest, the single-span bridge crosses Wacoochee Creek that flows eastward into the Chattahoochee River. The sand and mud bottomed creek is heavily wooded along its banks and in the near vicinity of the bridge. County Road 252, Lee Road, is paved to the west-northwest end of the bridge and continues as a dirt road from the other end. Northwest of the bridge, just up US 280, on the left side, is a local history mural, painted on the side of a building by Ans Steenmener, that depicts Salem-Shotwell Bridge. *See Salem Shotwell Covered Bridge in the color photograph section: C-1.*

Directions: From Opelika, the county seat, at the intersection of I-85 (exit 62) and US 431/SR 1, go east 8.7 miles on US 431/US 280/SR 1 to CR 254 on the left; go 1.3 miles to CR 252 (Lee Road) on the right; follow 0.3 mile to the bridge.

Cambron Covered Bridge

MADISON COUNTY

Brief Statistics

Type: Non-authentic modern covered footbridge. World Guide Covered Bridge Number: AL-45-a. Built in 1974 by Madison County. Eight-span footbridge crossing cove of Sky Lake. Stringer footbridge 90.3 feet long by 10.3 feet wide, with approximately 9.0-foot wide by 8.9-foot high portal openings. Alternate Name: None known. Located southeast of Huntsville atop Green Mountain at Madison County Nature Trail Park.

Cambron Bridge derived its name in memory of Joe E. Cambron, who was the Madison County Bridge Foreman from 1958 until his death, in 1974. Madison County built the stringer footbridge, in 1974, along the Nature Trail over a cove of Sky Lake. The Madison County-owned bridge has the dual distinctions of being the longest covered footbridge, at 90.3 feet, and the widest covered footbridge, at 10.3 feet, in Alabama.

Cambron Bridge has eight spans supported by seven sets of double wood-utility-pole pillars and wood-utility-pole abutments. Uncovered ramps extend from the abutments to the first set of pillars at each end. The south ramp has rail fencing at both sides, and the north ramp has handrails at both sides. The flooring has transverse planking; the sides are covered with gray weathered, diagonal boards, with four rectangular window openings along each side; the portals are open; and the steeply peaked roof has weathered cedar shingles. The roof, supported by 2-inch by 4-inch studding, extends beyond the end studs a short distance over the ramps. *CAMBRON/COVERED BRIDGE* signs are mounted above both entrances.

Aligned south/north, Cambron Bridge is in a heavily wooded area of the Nature Trail, spanning a cove of Sky Lake, home of nesting Canada Geese. Nestled among the verdant foliage around the cove, the bridge presents a beautiful sight from across the lake. The Madison County Nature Trail Park is open daily, free to the public, and offers trails, pavilions, a chapel, picnic tables and restrooms.

Directions: From Huntsville, go south on US 231 (South Memorial Parkway) to Weatherly Road; go left to Bailey Cove Road; go right to Green Mountain Road; go left to South Shawdee Road; go right to Madison County Nature Trail Park, on the right.

Poole's Covered Bridge

PIKE COUNTY

Brief Statistics

Type: Non-authentic modern covered bridge. World Guide Covered Bridge Number: AL-55-A. Built in 1998 by Wyndel Eiland. Two-span bridge crossing small pond. Stringer bridge 60.1 feet long by 12.4 feet wide, with approximately 10.6-foot wide by 14.3-foot high portal openings. Alternate Name: None known. Located in Troy at Pike Pioneer Museum.

Poole's Bridge was built in 1998 by Wyndel Eiland, a local resident, for the Pike Pioneer Museum. The 60.1-foot long stringer bridge, with a decorative Town lattice truss, was named after Grover Poole, a local elder who logged in the area for many years, using his horses to remove the timber.

Three steel I-beam stringers, resting on three mortared natural-stone-faced poured concrete piers, support Poole's Bridge. The bridge has approach ramps at each end. Transverse planking with four-plank-wide wheel treads forms the floor that extends to the end of the approach ramps. The sides have natural vertical boarding to rail height, with the upper portion open, exposing the decorative Town lattice truss. The gables also have natural vertical boarding, and the roof has black asphalt shingles. The steeply peaked roof and the gables extend beyond the portal end posts, providing a weather shelter for the entrances. *Poole's Bridge* signs are mounted above each entrance. Hand railings extend the length of the approach ramps.

The two-span bridge crosses a small pond created by

damming the adjacent US Highway 231 runoff conduits. Aligned northwest/southeast, the bridge is situated in Pike Pioneer Museum, with US Highway 231 on the upstream side and dense woods on the downstream side. An old log cabin chapel, at the southeast end, can be nicely photographed through the bridge portals. The museum features many other old structures, making a visit very worthwhile.

Directions: From Troy, at the intersection of US 231/SR 53 and US 29/SR 15, go north 1.3 miles on US 231/SR 53 to Pike Pioneer Museum, on the left. The bridge abuts the highway.

Bob Saunders Family Covered Bridge

Shelby County

Brief Statistics

Type: Authentic modern covered bridge. World Guide Covered Bridge Number: AL-59-01. Built in 1988 by unknown builder. One-span bridge crossing spillway of Lake Lauralee. Long truss bridge 50.8 feet long by 12.5 feet wide, with approximately 10.4-foot wide by 9.5-foot high portal openings. Alternate Name: Twin Pines Bridge. Located west of Sterrett at Twin Pines Conference Center Resort.

Bob Saunders Family Bridge, named after the original owner, was built in 1988 by an unknown builder, utilizing a Long truss for support, thereby creating an authentic covered bridge in modern times. To achieve an old appearance for the modern bridge, the natural, weathered vertical boarding on the sides and portals was acquired from a house located in Chance in Clarke County that was built circa 1888, and the rusted metal on the roof came from another old house. The Shelby County bridge, spanning the spillway of Lake Lauralee in a single span, is also known as Twin Pines Bridge, due to its location at Twin Pines Conference Center Resort. At 12.5 feet wide, this private bridge shares with Clarkson Bridge, in Cullman County, the distinction of being the narrowest authentic covered bridge remaining in Alabama.

Bob Saunders Family Bridge has a five-panel Long truss, constructed from 4-inch by 8-inch wood timbers, secured by bolts through galvanized steel plates at all intersections and joints, set on poured concrete abutments. The flooring has transverse planking with three-plank-wide wheel treads. The sides are open between the railing and the boxed roof overhang, both covered with aged vertical boarding, as are the portals. The roof is covered with rusty, corrugated galvanized steel.

The south-southwest/north-northeast aligned bridge is at the far side of Lake Lauralee, carrying the perimeter road over the spillway, surrounded on three sides by thick woods and the lake on the remaining side. The spillway flows northerly into nearby Bear Creek that joins Kelly Creek, before entering the Coosa River near Vincent. The 50.8-foot long bridge makes a lovely scene when viewed from across the lake.

Directions: From Sterrett, take SR 25 north to CR 45; go left 1.6 miles to Twin Pines Road, on the left; follow to Twin Pines Conference Center Resort. The bridge is on the lake perimeter roadway at the far side of the lake. Private property: request permission to visit, at the main lobby of the resort.

Pumpkin Hollow Covered Bridge

Shelby County

Brief Statistics

Type: Non-authentic modern covered bridge. World Guide Covered Bridge Number: AL-59-C. Built in 1992 by unknown builder. Two-span bridge crossing Bear Creek. Stringer bridge 68.0 feet long by 18.2 feet wide, with approximately 13.5-foot wide by 13.2-foot high portal openings. Alternate Name: None known. Located northeast of Sterrett in New Pumpkin Hollow Lake development.

Built in 1992, by an unknown builder, Pumpkin Hollow Bridge is located in a private gated community of summer homes called New Pumpkin Hollow Lake. Owned by New Pumpkin Hollow Lake development, Pumpkin Hollow Bridge, at 68.0 feet long, is the longest non-authentic covered bridge carrying motor traffic in Alabama.

The stringer bridge with a decorative Town lattice truss has two reinforced concrete slab stringers, one set between each poured concrete abutment and the poured concrete central pier. Transverse planked flooring with triple-plank-wide wheel treads is laid on the concrete slabs. Natural vertical boarding with battens covers the sides to rail height, with the decorative Town lattice exposed above the railing to the roof. Natural vertical boarding with battens also covers the gables, the right downstream gable having *PUMPKIN HOLLOW* and *CLEARANCE 13' 0"* signs mounted above the entrance. The gleaming tin roof completes the structure, the roof and gables with a pronounced extension beyond the portal end posts, providing a weather shelter for the entrances.

The two-span bridge, aligned east-southeast/west-northwest, crosses Bear Creek, a tributary to Kelly Creek that flows into the Coosa River near Vincent. The bridge is on the undeveloped portion of the entry road, situated in a heavily wooded area with dense undergrowth that has been cleared near the bridge.

Directions: From Sterrett, take SR 25 north to CR 55, on the right; go to the New Pumpkin Hollow Lake development, on the right at No. 18274, after the Church of God of Prophecy, also on the right. There are no development signs. This is a private gated community. Access permission from the caretaker is required. Call box is at the gate.

Alamuchee Covered Bridge

SUMTER COUNTY

Brief Statistics

Type: Authentic historic covered bridge. World Guide Covered Bridge Number: AL-60-01. Built in 1861 by Lorenzo Don Norville and James Thomas Praytor, relocated to present site in 1970. One-span bridge crossing a small lake. Town lattice truss bridge 82.3 feet long by 19.4 feet wide, with approximately 16.6- foot wide by 12.2-foot high portal openings. Alternate Name: Bellamy Bridge. Located in Livingston on University of West Alabama-Livingston campus.

Designed by Captain William Alexander Campbell Jones, a civil engineer who came to Livingston with the railroad in 1853, Alamuchee Bridge was built by Lorenzo Don Norville and James Thomas Praytor in 1861, originally over the Sucarnoochee River on US Highway 11, south of Livingston. During the Civil War, Confederate General Nathan Bedford Forrest used this covered bridge to march his troops into Mississippi. In 1924, the bridge was moved by John Ross, to make way for a new bridge, to a location five miles away in Bellamy, on old Bellamy-Livingston Road over Alamuchee Creek, where it remained in use until 1958. While at this location, the bridge was also known as Bellamy Bridge. Alamuchee Bridge was acquired by the University of West Alabama and relocated by Clayton Marcus, in 1970, to its present site on the University campus in Livingston, the county seat of Sumter County. During restoration, the roof was rebuilt, adding portal overhangs with supports. Alamuchee Bridge, at 19.4 feet wide, has the distinction of being the widest authentic covered bridge remaining in Alabama. The privately owned bridge is not open to motor traffic.

The single-span Alamuchee Bridge has neither abutments nor piers, but is set on five equally-spaced mortared natural stone pillars under each side. The bridge, aligned on a south-southeast/north-northwest orientation, has a transverse-planked floor with a plywood walkway inside the east-northeast side. The truss is a Town lattice fastened with treenails, one of the seven remaining Town lattice truss bridges in Alabama. Two electric carriage lamps, one toward each end, are mounted on the truss members inside the bridge, on the west-northwest side. The exterior of the picturesque bridge has gray weathered, vertical boarding with battens, two diamond-shaped window openings, cut through the siding, opposing on each side, and a dark gray weathered, wood-shingled roof.

The last historic covered bridge remaining in Sumter County is situated in a beautiful, fully exposed setting on a landscaped University campus. The 82.3-foot long bridge divides a lake, the larger west-northwest side with a small central island. *See Alamuchee Covered Bridge in the color photograph section: C-1.*

Directions: From Livingston, at the junction of US 11/SR 7 and SR 28, go north 0.6 mile on US 11/SR 7 to the University of West Alabama, on the left. The bridge is located on the University grounds, behind the dormitories.

Kymulga Covered Bridge

TALLADEGA COUNTY

Brief Statistics

Type: Authentic historic covered bridge. World Guide Covered Bridge Number: AL-61-01. Built circa 1860 by unknown builder. One-span bridge crossing Talladega Creek. Howe truss bridge 104.9 feet long by 14.5 feet wide, with approximately 10.4-foot wide by 10.6-foot high portal openings. Alternate Name: None known. Located northeast of Childersburg in Kymulga Grist Mill Park.

Built around 1860, by an unknown builder, Kymulga Bridge, named after the nearby community of Kymulga, is still at its original location, now part of Kymulga Grist Mill Park and near the operating Kymulga Grist Mill that was built around 1862. Kymulga Bridge originally carried Glovers Ferry Road across Talladega Creek, but the approach ramps were removed during World War II to restrict access into the Alabama Ordnance Plant and were not replaced until the 1990s. Kymulga Bridge is the last of the Howe truss bridges remaining in Alabama, although Waldo Bridge is a Howe-Queenpost combination truss bridge.

Kymulga Bridge has a unique ten-panel Howe truss, modified with one of the two timbers in each X panel doubled, sandwiching the single timber, and two vertical iron tension rods between each panel, giving the bridge added strength for the long span. The 104.9-foot long bridge rests on two dry chip-cut stone piers that have had later repairs using mortar. Upright utility poles support the ramps, as they were intended for foot traffic only, the bridge having been closed to motor traffic since the early 1940s. The flooring is comprised of transverse planking atop longitudinal planking, with no wheel treads. The sides, flanks of the portals, and roof are covered with old

Pre–Civil War Kymulga Bridge has three latticed windows along each side cut into the old galvanized metal siding.

galvanized steel, the roof showing considerable rust, enhancing the historic appeal of the bridge. There are three small, opposing, latticed window openings along each side. Horizontal lapped boarding covers the gables, with a *KYMULGA BRIDGE* sign hanging from the right downstream or southwest portal.

The single-span bridge crosses Talladega Creek that flows into the Coosa River, north of Childersburg. Aligned northeast/southwest, Kymulga Bridge is in a relatively open, maintained park surrounded by scattered mature trees, including the largest Sugarberry (also called Hackberry) tree in Alabama. The park also has a large population of White Oak trees, in addition to nature trails and barbecue facilities for groups and families, and, of course, the old working Kymulga Grist Mill that houses a country store.

Kymulga Bridge, owned by Childersburg Heritage Society and one of only two historic covered bridges remaining in Talladega County, was listed, along with the Kymulga Grist Mill, on the National Register of Historic Places on October 29, 1976. Talladega is the county seat of Talladega County.

Directions: From Childersburg, at the junction of US 231/ US280 and SR 76, go 1.8 miles east on SR 76 to CR 180, on the left; go 0.1 mile to Kymulga Grist Mill Road (still CR 180), on the right; follow for 3.7 miles to Kymulga Grist Mill Park, on the left. The bridge is in the park.

Waldo Covered Bridge

TALLADEGA COUNTY

Brief Statistics

Type: Authentic historic covered bridge. World Guide Covered Bridge Number: AL-61-02. Built circa 1858 by unknown builder. One-span bridge crossing Talladega Creek. Howe-queenpost combination truss bridge approximately 115 feet long by approximately 13.5 feet wide (inaccessible to measure). Alternate Names: Riddle Mill Bridge, Socapatoy Bridge, Waldo-Riddle Bridge. Located southeast of Talladega in Waldo at The Old Mill Restaurant.

Built around 1858, by an unknown builder, Waldo Bridge obtained its name from the nearby small community of Waldo. The Talladega County bridge was also known as Riddle Mill Bridge (after the Riddle Grist Mill—now The Old Mill Restaurant—formerly operated by the Riddle brothers at the bridge site), Socapatoy Bridge (as the bridge is on the former Riddle Mill Road that followed the Socapatoy Indian Trail), and Waldo-Riddle Bridge. At approximately 115 feet long, Waldo Bridge is the longest single-span covered bridge in Alabama and one of only two historic covered bridges remaining in Talladega County.

Waldo Bridge has an unusual Howe-queenpost combination truss resting on two piers, constructed of mortared cut stone on top of dry natural stone, the dry natural stone portion probably from an earlier bridge. The queenpost truss is a double truss, constructed with double horizontal beams and end braces. The inside truss has eight timber braces under the horizontal beam, four on each side of the center, slanting downward away from the center. The outside truss has three, possibly four, braces under the horizontal beam, slanting in the reverse direction. The purported Howe truss is between the queenpost and the siding, not allowing the author to discern the details, as the bridge interior is inaccessible.

The flooring construction is transverse planking with three-plank-wide wheel treads on top of longitudinal planking.

The upper portion of the sides has rusted corrugated tin over the gray weathered, vertical boarding with battens; rusted corrugated tin skirts protect the lower chord timbers. The gables are closed, with dark weathered, vertical boarding, upon which *BRIDGE CONDEMNED* signs were mounted by the Talladega County Road Department, in some bygone time. The roof covering is rusted corrugated tin. The roof and the gables extend beyond the portal end posts, forming a weather shelter for the entrances. The approach ramps were long ago removed, rendering the interior of the bridge inaccessible, due to the high piers.

Situated across an open grassy area, amid a backdrop of verdant foliage and behind The Old Mill Restaurant, Waldo Bridge perches high on its piers, looking down at Talladega Creek, a wide and shallow rock-bottomed stream that flows into the Coosa River north of Childersburg. Aligned northeast/southwest, the privately owned bridge is secluded in summertime by the tree-lined creek on the upstream side.

Directions: From Talladega, take SR 77 southeast for about 4 miles to The Old Mill Restaurant, on the left. The bridge is behind the restaurant, at the end of the parking lot. Private property.

Jimmy Lay Covered Bridge

WALKER COUNTY

Brief Statistics

Type: Authentic modern covered bridge. World Guide Covered Bridge Number: AL-64-01. Built circa 1984 by unknown builder. One-span bridge crossing tributary to Sloan Creek. Kingpost truss bridge 21.0 feet long by 19.3 feet wide, with approximately 17.1-foot wide by 9.8-foot high portal openings. Alternate Name: None known. Located in Empire off Empire Road.

Jimmy Lay Bridge was built circa 1984 by an unknown builder, contracted by the original owner, Jimmy Lay. The 21.0-foot long bridge is the shortest authentic covered bridge in Alabama and, for that matter, the entire southeastern United States. There are two kingpost truss bridges remaining in Alabama, this privately owned bridge and the historic Coldwater Bridge.

Jimmy Lay Bridge has a single kingpost truss set on mortared natural stone abutments, with a transverse-planked deck capped with three-plank-wide wheel treads. The sides and gables are covered with gray weathered, horizontal lapped boards. The middle three feet, eight inches along the length of each side is open, exposing the two-panel kingpost truss. Weathered wood shingles cover the roof that, with the gables, extends beyond the truss end posts, forming a weather shelter for the entrances. Wooden benches are attached to the interior of each side.

The single-span bridge, aligned northwest/southeast, crosses a tributary to Sloan Creek, whose waters eventually enter the Black Warrior River. Set back from Empire Road, in an open pasture, against a backdrop of stately pine trees, Jimmy Lay Bridge offers a splendid scene to passing motorists. Currently owned by Harold Hamrick, the bridge carries only an occasional private motor vehicle these days.

Directions: From US 78, in Sumiton, go north on Bryan Road (CR 22) to the first traffic light; go left on Main Street to the second traffic light; go right on Empire Road 7.3 miles to the bridge, visible in a pasture, on the left. Private property: request permission to visit.

Delaware

Delaware never was blessed with many covered bridges, and the few historic covered bridges the state had were all located in the northernmost county of New Castle. In 1937, there were over thirty-five covered bridges, but, by 1954, these historic structures numbered only four; today, only two remain, and these two still carry motor traffic.

Of the six covered bridges in Delaware today, two are authentic historic covered bridges, and four are non-authentic modern covered bridges. Four covered bridges are located in New Castle County, and one each in Kent and Sussex Counties. Ironically, the historic bridges are of similar age—built in 1870—and the modern bridges were built between 1960 and 1987. A Town lattice truss supports both of Delaware's authentic covered bridges.

Fortunately, Delaware is prudently maintaining their two historic covered bridges.

Loockerman Landing Covered Bridge

KENT COUNTY

Brief Statistics

Type: Non-authentic modern covered bridge. World Guide Covered Bridge Number: DE-01-A. Built in 1987 by unknown builder. Three-span bridge crossing spillway to Mill Pond. Stringer bridge 33.7 feet long by 11.8 feet wide, with approximately 10.3-foot wide by 8.5-foot high portal openings. Alternate Name: Delaware Agricultural Museum and Village Bridge. Located in Dover at 866 North Dupont Highway (US 13).

A local contractor built Loockerman Landing Bridge in 1987 for the Delaware Agricultural Museum and Village. Deriving its name from the local historic area, Loockerman Landing, the bridge was constructed over the spillway of Mill Pond, whose waters run down the mill race to operate adjacent Silver Lake Mill and, as with the waters of the spillway, flow into Silver Lake. Loockerman Landing Bridge has also been called Delaware Agricultural Museum and Village Bridge.

The 33.7-foot long stringer bridge is supported by four stringers, consisting of triple 2-inch by 8-inch pressure-treated timbers bolted together that replaced the original three stringers in 2001. The new stringers are set on poured concrete abutments and two piers, each pier constructed with four steel screw jacks—one under each stringer—that are placed on the poured concrete retaining walls of the spillway. The screw jacks were installed in 2001, replacing the wood posts that supported the original three stringers. The flooring is constructed with transverse 2-inch by 8-inch planks attached to the stringers and replaced the original flooring in 2001, when the new stringers were installed. Gray weathered vertical boarding with battens covers the sides to rail height, the rail capped with a 1-inch by 10-inch board. Weathered cedar shingles cover the steeply peaked roof that is supported by four 6-inch by 6-inch wood posts. Decorative treenails or wooden pegs are used in the timbers.

The three-span bridge is fully exposed, with Silver Lake Mill near the south-southeast end and the Blacksmith/Wheelwright Shop near the north-northwest end. Mill Pond is at the east-northeast side. Silver Lake is on the downstream side and is bordered by trees and bushes. Three-rail split-rail fencing funnels pedestrian and maintenance vehicle traffic along the gravel pathway and into the bridge. The bridge fits in especially well with the other structures in the village of Loockerman Landing. *See Loockerman Landing Covered Bridge in the color photograph section: C-2.*

Directions: From Dover, go north on North Dupont Highway (US 13) to No. 866, on the left. Admission.

Ashland Covered Bridge

NEW CASTLE COUNTY

Brief Statistics

Type: Authentic historic covered bridge. World Guide Covered Bridge Number: DE-02-02. Built circa 1870 by unknown builder. One-span bridge crossing Red Clay Creek. Town lattice truss bridge 51.9 feet long by 17.6 feet wide, with approximately 14.4-foot wide by 12.5-foot high portal openings. Alternate Name: None known. Located in Ashland on Brackenville Road.

Ashland Bridge, built circa 1870 by an unknown builder, adjoins the Ashland Nature Center that offers parking, a visitor center, picnic facilities and hiking trails. Children exploring the Ashland Nature Center heard the rumbling clickety-clack of loose and uneven deck planking on Ashland Bridge, as vehicles passed through, and gave the bridge the nickname "Thunder Bridge." Ashland Bridge underwent a major reconstruction in 1982.

The 51.9-foot long bridge has a Town lattice truss set on mortared rough-cut stone abutments. Ashland Bridge, named after the local community of Ashland, has four steel I-beam stringers that were added circa 1983 to help the aging truss support the everyday load of passing vehicles. The bridge timbers are fastened with 1½-inch diameter wooden pegs, called treenails. The old bridge has a transverse-planked floor that, with the truss and the interior of the sides, is painted a light gray that effectively lightens the interior of the bridge. Vertical boarding on the sides, horizontal boarding on the gables, and gray weathered cedar shingles on the roof enclose the bridge, which is painted red, with white trim on the portals. Bridge clearance signs, *11' 3"*, are mounted above each entrance. Mortared rough-cut stone guard walls at each end funnel the traffic into the single-lane bridge.

The single-span bridge carries tree-lined Brackenville Road across Red Clay Creek which flows into White Clay Creek at Stanton, before entering the Christina River. The placid creek is heavily wooded along the embankments and in the south-southwest upstream corner; the other three corners contain open fields with scattered mature trees. The bridge is aligned in a south-southwest/north-northeast orientation.

One of two historic covered bridges remaining in New Castle County, Ashland Bridge was listed on the National Register of Historic Places on March 20, 1973. *See Ashland Covered Bridge in the color photograph section: C-2.*

Directions: From Wilmington, the county seat, take SR 48 west to Centerville Road; go right to Barley Mill Road; go left to the tee at Brackenville Road; go right 0.2 mile to the bridge.

Covered Bridge Farms Covered Bridge

NEW CASTLE COUNTY

Brief Statistics

Type: Non-authentic modern covered bridge. World Guide Covered Bridge Number: DE-02-B. Built in 1961 by unknown builder. Post supported roof over roadway crossing Christina River. Bridge 39.5 feet long by 22.4 feet wide, with approximately 19.8-foot wide by 13.4-foot high portal openings. Alternate Name: None known. Located in Newark in Covered Bridge Farms Subdivision on Covered Bridge Lane.

An unknown builder built Covered Bridge Farms Bridge, in 1961, in Covered Bridge Farms Subdivision. The 39.5-foot long bridge is actually a post-supported roof over an asphalt roadway, under which the Christina River flows through a concrete culvert on its way to the Delaware River, at Wilmington. Covered Bridge Farms Bridge is the longest non-authentic covered bridge in Delaware.

The brown asphalt-shingled roof is supported by eleven 4-inch by 6-inch wood posts along each side, with double posts at the portals. The sides are enclosed with vertical tongue and groove boarding that is open, at eye level and also under the eaves, the length of the sides, the lower portion being boxed in and extending out both portals, forming guard walls that funnel traffic into the two-lane bridge. These guard walls were added at a later date, most likely when the bridge was re-sided. The gables are also enclosed with vertical tongue and groove boarding.

Open to motor traffic, the red painted bridge aligns west-northwest/east-southeast. Densely wooded at both sides, the bridge is open at both ends, with a residence in the right downstream corner and a pond in the left downstream corner.

Directions: From Newark, at the intersection of SR 273 and SR 896, go north on SR 896, then immediately left on New London Road to Covered Bridge Lane, on the left (there is a sign for Covered Bridge Farms on the right); follow 0.8 mile to the bridge. From Maryland, on SR 273, go to Wedgewood Road, on the left (0.2 mile west of the Delaware state line); go 0.4 mile to Covered Bridge Lane. The bridge is on the right.

Westminster Covered Bridge

NEW CASTLE COUNTY

Brief Statistics

Type: Non-authentic modern covered bridge. World Guide Covered Bridge Number: DE-02-A. Built in 1960 by Emilio Capaldi. One-span bridge crossing Hyde Run.

The Town lattice truss of the 1870 Wooddale Bridge is fastened together with wooden pegs. ***See Wooddale Covered Bridge in the color photograph section: C-3.***

Stringer bridge 32.5 feet long by 22.5 feet wide, with approximately 18.2-foot wide by 11.2-foot high portal openings. Alternate Name: None known. Located north of Faulkland in Westminster Subdivision on Ambleside Road.

Built in 1960 by Emilio Capaldi, at a cost of $17,000, Westminster Bridge is in Westminster Subdivision. Open to motor traffic, the 32.5-foot long two-lane stringer bridge is supported by seven steel I-beam stringers resting on abutments that are constructed from mortared stone set on poured concrete bases. Asphalt paves the floor, vertical boarding encloses the sides and portals, and gray weathered, cedar shingles cover the roof. The white-trimmed red-painted bridge is open under the eaves. Entry guardrails constructed from wood posts, boxed in with vertical facsimile tongue and groove paneling, are at each portal.

The single-span bridge crosses Hyde Run, a tributary to nearby Red Clay Creek, which flows into White Clay Creek at Stanton, before entering the Christina River. Hyde Run is densely wooded downstream of the west-southwest/east-northeast aligned bridge and relatively open upstream. There are homes situated in all four corners. The 22.5-foot wide bridge is the widest non-authentic covered bridge in Delaware.

Directions: From Faulkland, go north on SR 41 to Hercules Road; go right to Westminster Subdivision, on the left. Enter the subdivision, following Cheltenham Road to Ambleside Drive, on the right; follow to the bridge.

Wooddale Covered Bridge

NEW CASTLE COUNTY

Brief Statistics

Type: Authentic historic covered bridge. World Guide Covered Bridge Number: DE-02-04. Built in 1870 by unknown builder. One-span bridge crossing Red Clay Creek. Town lattice truss bridge 67.3 feet long by 17.1 feet wide, with approximately 13.4-foot wide by 12.0-foot high portal openings. Alternate Name: None known. Located in Wooddale on Foxhill Lane.

At 67.3 feet long, Wooddale Bridge is the longer of the two authentic, historic covered bridges remaining in

Delaware, both still carrying vehicle traffic. Wooddale Bridge was built in 1870 by an unknown builder, as was nearby Ashland Bridge, the construction similarities leading one to believe both bridges were built by the same builder. The old bridge derives its name from the local community of Wooddale.

Wooddale Bridge has a Town lattice truss, fastened with treenails, set on mortared rough-cut stone abutments, with rock-slab-capped poured concrete guard walls, at each portal, that channel the traffic into the single-lane bridge. Two large steel I-beam stringers were recently added under the bridge to augment its strength. The floor is comprised of diagonal planking; the sides have vertical boarding, with a square window opening, exposing the white painted truss on each side; the gables have horizontal boarding, with a *CLEARANCE 12 FEET* sign above each entrance; and the roof has dark gray weathered, cedar shingles. There are electric lights installed inside the red painted bridge with white trim on the portals. For safety, a *BLOW HORN* sign is mounted above the left downstream or north entrance, as the south exit is at a blind tee with Rolling Mill Road.

The single-span bridge crosses Red Clay Creek, which flows into White Clay Creek at Stanton, before joining the Christina River. Surrounded by dense woods, Wooddale Bridge rests on the steep banks of a wide, gravel and rock bed creek that has many boulders on the downstream side.

Listed on the National Register of Historic Places on April 11, 1973, Wooddale Bridge is one of only two historic covered bridges remaining in New Castle County. *See Wooddale Covered Bridge in the color photograph section: C-3.*

Directions: From Wilmington, the county seat, take SR 48 west to Centerville Road; go right about 0.7 mile to Rolling Mill Road, on the left; go 0.3 mile to the bridge, on the right.

Hopkins Covered Bridge Farm Covered Bridge

SUSSEX COUNTY

Brief Statistics

Type: Non-authentic modern covered bridge. World Guide Covered Bridge Number: DE-03-A. Built in 1981 by Alden S. Hopkins, Jr. One-span bridge crossing wet weather stream. Stringer bridge 21.6 feet long by 11.2 feet wide, with approximately 9.2-foot wide by 11.1-foot high portal openings. Alternate Name: None known. Located southwest of Lewes at 30227 Fisher Road (CR 262).

Hopkins Covered Bridge Farm Bridge was started in 1980 and completed the following year, by Alden S. Hopkins, Jr., at Hopkins Covered Bridge Farm, over a wet weather stream that flows into a man-made pond. Six 12-inch by 12-inch oak timber stringers set on the stream banks support the 21.6-foot long stringer bridge. Transverse 2-inch thick planks of 12-inch to 23-inch widths, attached directly to the stringers, constitute the floor. An 11.4-foot ramp at the west portal and a 9.5-foot ramp at the east portal have longitudinal 2-inch thick planking extending from the bridge down to the ground. The bridge has no abutments. Gray weathered, vertical, 1-inch by 12-inch poplar boards cover the sides and the portals; the gables with a circular arch above the entrances. Gray weathered red cedar shakes cover the roof which is supported by seven 2-inch by 12-inch timber posts along each side. The bridge boasts very unusual rafters that display the craftsmanship of the builder, for each rafter carries the continuous circular curvature of the gables through the bridge. This was accomplished by attaching, with an interlocking mortise and tenon joint, a cross-brace under the upper part of the rafter, with the circular arch of the gables sawed out of the lower part of the rafters and the cross-brace, in a continuous arc. The bridge has a decorative 2-inch by 6-inch Burr arch along each side, mounted across three horizontal 6-inch by 6-inch timbers, one spaced above the other.

The single-span bridge has open agricultural fields at both ends and a thick wood line along the banks of the stream at both sides, the nearest trees forming a canopy over the bridge. A small man-made pond, encircled by a split-rail fence, is in the downstream corner, at the east end. Due to the deteriorated condition of the ramps and bridge flooring, the bridge is closed to vehicle traffic.

Directions: From Lewes, at the junction of SR 1 and US 9, go west 0.1 mile on US 9 to SR 23; go left 2.4 miles to Fisher Road (CR 262), on the right; go 0.9 mile to Hopkins Covered Bridge Farm, on the right at No. 30227. Private property: request permission to visit.

Florida

Florida never had any historic covered bridges and presently has only one authentic covered bridge, the kingpost truss footbridge named Clover Leaf Farms Bridge, in Hernando County. Nevertheless, Florida has some impressive and interesting modern covered bridges.

In addition to the authentic covered bridge, Florida has twenty-one non-authentic covered bridges; six that are a post supported roof over a roadway and fifteen that are stringer bridges. Nine of the twenty-one covered bridges are footbridges.

Florida's twenty-two covered bridges were built between 1964 and 1997, representing fifteen of the sixty-seven counties, and all of the covered bridges are located in Central and South Florida. An interesting aside, only one covered bridge is privately owned at a residence; most bridges are in subdivisions.

Covered Bridge Apartments Covered Bridge

ALACHUA COUNTY

Brief Statistics

Type: Non-authentic modern covered bridge. World Guide Covered Bridge Number: FL-01-A. Built in 1973 by unknown builder. Four-span bridge crossing Hogtown Creek. Stringer bridge 51.5 feet long by 27.0 feet wide, with approximately 23.9-foot wide by 13.6-foot high portal openings. Alternate Name: None known. Located in Gainesville at 2401 NW 23rd Avenue.

Covered Bridge Apartments Bridge was built in 1973 by an unknown builder on the entrance road to Covered Bridge Apartments in Gainesville. This is a six-span stringer bridge, totaling 78.5 feet long, with only the four center spans covered, for a length of 51.5 feet. Dense woods run along both banks of the creek and, on the southwest end, up to 23rd Avenue. Trees and maintained grass areas adjoin the northeast end, with a bench placed in the upstream corner. The two-lane bridge spans sandy-bottomed Hogtown Creek that flows northwestward.

The stringer bridge is supported by thirteen timber stringers, resting on poured concrete abutments and five piers, the covered portion resting only on the five piers. Each pier is constructed with four wood utility pole piles, cross-braced and capped with a timber crossbeam that extends out each side of the bridge by about four feet; the second pier from the northeast end, on the downstream side, has two wood utility poles replaced by steel pipes. The flooring consists of transverse planking covered with asphalt pavement. The sides are open, with the upper quarter covered with brown weathered wood shingles; the lower quarter has handrails running the length of the bridge. The gables are covered with facsimile vertical board paneling painted cream; a tent-shaped wood sign inscribed *COVERED BRIDGE/APARTMENTS*, under a painting of trees flanking a stream, is mounted on the right downstream or southwest gable. Asphalt shingles cover the low peaked roof, which is supported along each side by five 6-inch by 10-inch timber posts aligned on the piers, with a double-wood-plank buttress brace positioned from each roof support post to the pier crossbeam extension. Creosoted timbers are used in the bridge construction. Electric lighting illuminates the interior, and globe lights atop black-painted metal posts are placed just beyond the ends of the handrails.

Directions: From Gainesville, on US 441 (NW 13th Street) heading north, go left on NW 23rd Avenue 0.4 mile to No. 2401 at Covered Bridge Apartments, on the right. The bridge is down the entrance road.

Coral Springs Hills Covered Bridge

BROWARD COUNTY

Brief Statistics

Type: Non-authentic modern covered bridge. World Guide Covered Bridge Number: FL-06-A. Built in 1964 by unknown builder. One-span bridge crossing canal. Stringer bridge 40.0 feet long by 25.2 feet wide, with approximately 19.4-foot wide by 11.8-foot high portal openings. Alternate Name: None known. Located in Coral Springs on NW 95th Avenue.

Coral Springs Hills Bridge was built in 1964 by an unknown builder, on NW 95th Avenue, spanning a canal. The two-lane bridge is the oldest non-authentic covered motor vehicle bridge in Florida. At 40.0 feet long, the bridge is also the longest single-span non-authentic covered motor vehicle bridge in Florida.

The stringer bridge is supported by a concrete slab on nine reinforced concrete stringers, set on poured concrete abutments, with poured concrete wing walls. The concrete slab forms the floor in the bridge and abuts the asphalt pavement of the roadway. Vertical boarding covers the sides to rail height, leaving a four-foot high opening the length of the bridge and exposing the roof support posts with braces that simulate an eight-panel multiple kingpost truss. Above this opening, plywood paneling covers the side up to the eave. On this paneling, a long advertisement, on each side, runs the length of the bridge, the downstream side advertising *Bull of the Woods Chewing Tobacco* and the upstream side advertising *Peach Sweet Snuff*. Vertical boarding also covers the portals of the brown painted bridge, a painted wood sign over the north-northeast or right downstream entrance stating *WELCOME TO/CORAL SPRINGS* and a painted wood sign over the south-southwest entrance stating *HURRY BACK TO/CORAL SPRINGS*. Asphalt shingles cover the low peaked roof that is supported along each side by nine wood posts, constructed with triple 2-inch by 8-inch lumber. A raised concrete sidewalk runs the length of the bridge interior, on the upstream side.

Dense foliage extends along both canal banks. Homes are in three corners; the fourth corner, in the downstream north-northeast sector, has a pond that is fed by the canal. White-painted board-rail fencing guides traffic in and out of the bridge.

Directions: From Coral Springs, at the intersection of SR 834 (Sample Road) and US 441, go north to Wiles Road, on the left; go 3.2 miles to NW 95th Avenue (the first left past SR 817, University Drive); go left 0.1 mile to the bridge.

Wozniuk Covered Bridge

BROWARD COUNTY

Brief Statistics

Type: Non-authentic modern covered footbridge. World Guide Covered Bridge Number: FL-06-b. Built in 1973 by Bill Wozniuk. One-span footbridge crossing pond. Stringer footbridge 18.0 feet long by 5.1 feet wide, with approximately 3.9-foot wide by 6.7-foot high portal openings. Alternate Name: None known. Located in Davie at 7301 Orange Drive.

Bill Wozniuk built Wozniuk Bridge in 1973, in the side yard of the residence, over the edge of a small pond. The bridge is fully exposed on the pond or south side, a Royal Palm tree at the west end and overgrown landscape foliage on the north side. The bridge is currently being used for storage.

The 18.0-foot long stringer footbridge is supported by three 4-inch by 8-inch timber stringers, set on poured concrete bases at each end. The flooring consists of transverse 2-inch by 6-inch planks. Facsimile tongue and groove paneling covers the sides and the gables, the sides having three large, opposing, rectangular window openings. Old dark weathered wood shingles cover the roof, which is supported along each side by four wood posts. The exterior and the interior of the single-span bridge is painted red, the exterior paint faded and severely weathered away.

Directions: From Cooper City, at the intersection of SR 817 (University Drive) and SR 818 (Griffin Road), go north on SR 817, across the South New River Canal, and immediately go right on Orange Drive (SW 45th Street) for 0.5 mile to SW 73rd Terrace, on the left. The bridge is at No. 7301 Orange Drive. The bridge is visible just off Orange Drive. Private property: request permission to visit.

Oxbow Golf Club Covered Bridge

HENDRY COUNTY

Brief Statistics

Type: Non-authentic modern covered bridge. World Guide Covered Bridge Number: FL-26-A. Built in 1971 by unknown builder. Two-span bridge crossing canal. Stringer bridge 40.5 feet long by 12.9 feet wide, with approximately 10.9-foot wide by 9.5-foot high portal openings. Alternate Name: None known. Located east of LaBelle off SR 80 at Port LaBelle Inn and Country Club.

The original golf course developer built Oxbow Golf Club Bridge, in 1971, at Port LaBelle Inn and Country

Club (Oxbow Country Club) on the River Course, over a wide canal that feeds into the nearby Caloosahatchee River. The two-span bridge is on an asphalt-paved roadway that accommodates course maintenance vehicles and golf carts.

Two reinforced concrete slab stringers, set on the abutments and one pier, support the 40.5-foot long stringer bridge. The abutments and pier are constructed with a very large reinforced concrete beam, capping three upright square concrete pillars, the abutments backfilled behind a retaining wall, restrained by the pillars. The floor has asphalt pavement on top of the concrete slabs. The sides and gables are open, a banister-type handrail running down each side of the north/south-aligned bridge and flaring out twenty feet beyond the portals, uniquely assuming full side height at the portal posts and tapering down to handrail height twenty feet out. Gray asphalt shingles cover the roof of the bridge, which is painted green on the exterior, the interior and the handrails. Supported along each side by five wood posts, the roof overhangs the portals, tapering out to a point at the peak. Lightning rods are mounted on the peak at each end.

Directions: From LaBelle, at the intersection of US 29 and SR 80, go east 2.9 miles on SR 80 to Port LaBelle Inn and Country Club (Oxbow Country Club), on the left. Follow the long driveway 1.3 miles to the Inn.

Clover Leaf Farms Covered Bridge

HERNANDO COUNTY

Brief Statistics

Type: Authentic modern covered footbridge. World Guide Covered Bridge Number: FL-27-01. Built in 1979 by unknown builder. One-span footbridge crossing lakes interconnect. Kingpost truss footbridge 23.3 feet long by 7.7 feet wide, with approximately 6.9-foot wide by 6.9-foot high portal openings. Alternate Name: None known. Located in Brooksville in Clover Leaf Farms Manufactured Homes Park off Twingate Avenue.

Clover Leaf Farms Bridge was built in 1979 by an unknown builder, in Clover Leaf Farms Manufactured Homes Park, over an interconnect between two man-made lakes. Clover Leaf Farms Bridge is the only authentic covered bridge in Florida. The north/south-aligned bridge is fully exposed along the lamppost lit, asphalt-paved walkway between the two lakes, one on each side of the bridge.

A four-panel double kingpost truss, set on beige-painted, mortared cement block abutments, with wing walls, supports the 23.3-foot long footbridge. The truss is constructed with 2-inch by 6-inch truss members bolted to the upper and lower chords, which are comprised of double 2-inch by 6-inch timbers encasing the truss members. The portal end posts are comprised of triple 2-inch by 6-inch timbers that are bolted to the upper and lower chords. The flooring consists of transverse 2-inch by 6-inch planking. The sides are covered with a thin 1-inch wide lattice, sandwiched between the truss and end post members, and 1-inch by 6-inch boards on the exterior side. The gables are covered with plywood paneling that have widely spaced battens. The bridge is painted dark green on the exterior and interior, except the lattice, which is painted beige. Lichen-covered, weathered wood shingles cover the roof of the single-span bridge. White-lettered, green-painted, cloverleaf-shaped, fiberboard signs, declaring *GOLF/CARTS/BICYCLES/AND PEDESTRIANS/ONLY* are mounted on each portal. Poured concrete ramps are at each portal. *See Clover Leaf Farms Covered Bridge in the color photograph section: C-4.*

Directions: From Brooksville, at the intersection of US 98 and US 41 (North Broad Street), go north 0.9 mile on US 41 to Clover Leaf Farms Manufactured Homes Park, on the right. Proceed into the park on Twingate Avenue to the clubhouse and the pool, on the left. The bridge is visible along the pathway, on the left, beyond the pool and between the lakes that are behind the pool.

Covered Bridge Retirement Community Covered Bridge

HIGHLANDS COUNTY

Brief Statistics

Type: Non-authentic modern covered bridge. World Guide Covered Bridge Number: FL-28-A #2. Built in 1988 by unknown builder. Post supported roof over roadway crossing Jack Creek. Bridge 48.0 feet long by 22.1 feet wide, with approximately 19.5-foot wide by 13.7-foot high portal openings. Alternate Name: Lehigh Acres Bridge. Located north of Lake Placid on Patterson Street in Covered Bridge Retirement Community Subdivision.

Covered Bridge Retirement Community Bridge was first built in 1975 by an unknown builder, contracted by Mitchel Miller, on Patterson Street, at the entrance to Covered Bridge Retirement Community. This covered bridge, also formerly called Lehigh Acres Bridge, was torn down in 1988 due to severe termite damage and an errant sod truck striking the bridge. Later in 1988, the bridge was rebuilt as a post supported roof over a roadway, over culverts that allow the waters of Jack Creek to pass beneath. This second bridge was built by an unknown builder. In 1989, the roadbed collapsed, causing a second closure of the bridge, but repairs quickly reopened the bridge.

The 48.0-foot long post supported roof is over an asphalt paved roadway, which is over three large, round, galvanized-steel culverts encased in concrete from the creek bed up to the roadway and extending from bank-to-bank. The sides are open, with handrails extending down each side, constructed from 1-inch by 6-inch boarding for the upper and lower rails, fastened to the inside of the roof support posts with 1-inch by 6-inch board braces between the rails, alternating up and down between the roof support posts. The gables of the brown painted bridge are enclosed with horizontal lapped siding, each gable with an electric lamp and large yellow cutout letters mounted above the entrances, the letters spelling *COVERED BRIDGE*. Brown asphalt shingles cover the low peaked roof of the bridge; seven 4-inch by 4-inch wood posts, Y braced under the eaves and set on a raised concrete base along each side, support the roof. A chain-link, motorized gate restricts motor vehicle access at the west-southwest entrance.

The two-lane bridge reposes, fully exposed, above Jack Creek, which joins nearby Lake Carrie with Lake June in Winter. Peachtree Road is at the west-southwest end of the east-northeast/west-southwest aligned bridge, the manufactured home community at the other end. Several mature trees surround the bridge, the picturesque upstream side being more open.

Directions: From Lake Placid, go north on US 27 to Lake June Road (CR 621), on the left; go 3.8 miles to the country club, on the right, going right 1.6 miles on Golfview Drive (becomes Peachtree Road) to Covered Bridge Retirement Community and Patterson Street, on the right. The gated bridge is on Patterson Street, at the entrance to the Community.

The unusual hip-roofed Riverbend Housing Development Bridge has 2.3-foot diameter portholes offering splendid views of Indian Creek.

Riverbend Housing Development Covered Bridge

LEE COUNTY

Brief Statistics

Type: Non-authentic modern covered bridge. World Guide Covered Bridge Number: FL-36-A. Built in 1981 by

Built in 1981, Golf Lakes Bridge, at 15.9 feet long, is the shortest non-authentic covered footbridge in Florida.

Thomas Hoolihan. Two-span bridge crossing Indian Creek. Stringer bridge 42.8 feet long by 11.8 feet wide, with approximately 9.8-foot wide by 10.7-foot high portal openings. Alternate Name: None known. Located in North Fort Myers on Riverbend Boulevard in Riverbend Housing Development.

Thomas Hoolihan, the developer, built Riverbend Housing Development Bridge, in 1981, on Riverbend Boulevard in Riverbend Housing Development. The two-span bridge crosses moderately wide and deep Indian Creek, which flows into the nearby Caloosahatchee River.

The east-southeast/west-northwest aligned bridge is fully exposed, with the short No. 1 tee and green on the upstream side and homes on the downstream side, the downstream side of the creek thickly foliated near the bridge. The 11.8-foot wide bridge is the narrowest covered motor vehicle bridge in Florida.

The 42.8-foot long stringer bridge is supported by two steel I-beam stringers, supplemented with timber stringers on the west-northwest span and two double 8-inch by 12-inch timber stringers on the east-southeast span. The bridge rests on the abutments and the central pier, with an additional pier, centered under the east-southeast span, lending added support for the four timber stringers. The west-northwest abutment consists of a vertical plank barrier restraining steel and earth fill; the east-southeast abutment consists of a poured concrete barrier restraining wood and earth fill. The two piers are constructed with two wood utility poles, set on round concrete pillars. Asphalt pavement on top of transverse planking comprises the floor. Vertical boarding covers the sides that, at a later date, had the centrally located square window openings closed in and two 2.3-foot round portholes added to each side. Short vertical boards cover the ends under the eaves, above the entrances, as the bridge has a low hip roof that is covered on the four slopes with shiny tin. The roof is supported along each side by five wood posts, cross-braced between, and an extra wood post at each portal. The exterior of the bridge is painted green, with white trim around the portholes; the interior is painted white. The bridge has interior electric lights.

Directions: From Bayshore, at the intersection of I-75 and SR 78 (Bayshore Road), go west 3.5 miles on SR 78 to Indian Creek Drive, on the left; follow 0.5 mile, and the road becomes Riverbend Boulevard; follow 0.2 mile to the bridge.

Golf Lakes Covered Bridge

MANATEE COUNTY

Brief Statistics

Type: Non-authentic modern covered footbridge. World Guide Covered Bridge Number: FL-41-a. Built in 1981 by unknown builder. Two-span footbridge crossing lakes interconnect. Stringer footbridge 15.9 feet long by 7.0 feet wide, with approximately 5.4-foot wide by 7.3-foot high portal openings. Alternate Name: None known. Located in Bradenton at 5050 East Fifth Street in Golf Lakes Estates.

Golf Lakes Bridge was built in 1981 by an unknown builder, over a wide interconnect between two man-made lakes in Golf Lakes Estates. The south/north-aligned bridge is fully exposed, surrounded by a maintained lawn, with the small limestone-rock banked lake on the east side and the limestone-rock banked interconnect on the west side.

A tall windmill-operated water pump and post-mounted pelican sculpture are near the north end. The 15.9-foot long bridge is the shortest non-authentic covered footbridge in Florida.

The two-span footbridge is supported by two sets of double 2-inch by 8-inch wood stringers, set on three piers, each pier constructed with two double 4-inch by 4-inch wood posts, with two 2-inch by 8-inch wood beams fastened to the sides of the posts at the top. Ramps extend from the ground up to the portals at each end, the ramps supported by two 2-inch by 8-inch wood stringers. A 13.9-foot long ramp at the south end and a 13.3-foot long ramp at the north end, added to the 15.9-foot long covered section, give the bridge a total length of 43.1 feet. The flooring consists of transverse 2-inch by 8-inch planks placed on the stringers and extending along the ramps. Gray weathered vertical boarding covers the sides. Weathered paneling, with most of the light colored paint washed off, covers the gables, which have a built-in multiple-hole bird box on each gable. Shiny galvanized metal covers the low peaked roof, which is supported along each side by three 2-inch by 4-inch wood posts. Plywood covers the interior ceiling, with an electric light mounted in the center. The ramps have gray-painted steel pipe handrails and an approximately 15-inch high, white painted lattice along each side.

Directions: From Bradenton, on SR 70 (53rd Avenue), east of US 41, take East 5th Street north to No. 5050, on the left (clubhouse for Golf Lakes Estates—no sign). The bridge is near the north end of the clubhouse.

Rustic Hills Covered Bridge

MARTIN COUNTY

Brief Statistics

Type: Non-authentic modern covered bridge. World Guide Covered Bridge Number: FL-43-A. Built in 1972 by unknown builder. Three-span bridge crossing Crooked Creek. Stringer bridge 57.7 feet long by 16.8 feet wide, with approximately 12.4-foot wide by 9.6-foot high portal openings. Alternate Name: None known. Located in Palm City at Rustic Hills Subdivision on Covered Bridge Road.

Rustic Hills Bridge was built, in 1972, by an unknown builder on Covered Bridge Road in Rustic Hills Subdivision. The three-span bridge crosses wide and deep Crooked Creek flowing into nearby Bessey Creek, which enters the North Fork St. Lucie River at Lighthouse Point. Homes surround the north/south-aligned bridge, several with boat docks, with an open lot in the south downstream corner. At 57.7 feet long, Rustic Hills Bridge is the longest non-authentic covered motor vehicle bridge in Florida.

The stringer bridge is supported by three concrete slab stringers, set on the abutments and two piers, the abutments and piers each constructed with a large reinforced concrete beam capping four square concrete pillars. The concrete slab stringers form the floor. The sides are open, with a wood handrail running between the roof support posts along each side. Gray weathered vertical boarding covers the gables, with a lattice height-safety barrier, added at a later date, hanging from the top of each entrance and a *CAUTION/EIGHT FOOT/CLEARANCE* sign at each portal. Gray asphalt shingles cover the low peaked roof, which is supported along each side by ten 4-inch by 4-inch wood posts, with Y braces at the eaves.

Directions: From Palm City, at the Florida Turnpike and SR 714, go east 1.1 miles on SR 714 to SW High Meadow Road; go left 1.0 mile to an intersection, crossing through on Covered Bridge Road 0.1 mile to the bridge in Rustic Hills Subdivision. Parking is available in the empty lot on the left, before the bridge.

Flipper's School Covered Bridge

MONROE COUNTY

Brief Statistics

Type: Non-authentic modern covered footbridge. World Guide Covered Bridge Number: FL-44-a. Built in 1972 by unknown builder. One-span footbridge set on dry ground. Stringer footbridge 35.1 feet long by 6.6 feet wide, with approximately 5.8-foot wide by 7.7-foot high portal

The 57.7-foot long Rustic Hills Bridge is the longest non-authentic covered motor vehicle bridge in Florida.

openings. Alternate Name: None known. Located on Grassy Key at US 1 mile marker No. 59.

Flipper's School Bridge was built, in 1972, by an unknown builder, on a pathway over a display pond at Flipper's Sea School, the name since changed to Dolphin Research Center. The single-span footbridge was moved, circa 1987, to a nearby location within the Dolphin Research Center and is currently used as a storage building, resting on its three wood utility pole stringers set on dry ground. The flooring consists of transverse 2-inch by 4-inch planking. Originally open sided with handrails, the bridge has been enclosed with facsimile vertical board paneling on the sides and the portals, two doors with screen doors added on one side and one shed door added on the end. Originally covered with wood shingles, the low peaked roof is now covered with black asphalt shingles. The bridge is now painted white, with blue trimmed fasciae. At 35.1 feet long, Flipper's School Bridge is the longest non-authentic covered footbridge in Florida.

Directions: From Key Largo, take US 1 south toward Key West to the Dolphin Research Center, on the right, at mile marker No. 59. Admission.

Covered Bridge Subdivision Covered Bridge

ORANGE COUNTY

Brief Statistics

Type: Non-authentic modern covered bridge. World Guide Covered Bridge Number: FL-48-C. Built in 1988 by Lennar Development Co. Post supported roof over roadway crossing ponds interconnect. Bridge 41.0 feet long by 39.1 feet wide, with approximately 34.6-foot wide by 17.1-foot high portal openings. Alternate Name: None known. Located in Orlando on Sagebrush Place in Covered Bridge Subdivision at Curry Ford Woods.

Lennar Development Company built Covered Bridge Subdivision Bridge, in 1988, at the entrance to Covered Bridge Subdivision at Curry Ford Woods, on Sagebrush Place, the entrance road. The two-lane bridge is a post supported roof over a roadway, over a conduit that connects a medium-sized, fenced in retention pond on the northwest side to a long retention pond, which runs behind the homes on the southeast side. Landscaped along both sides, the bridge borders homes and a concrete sidewalk along its southeast side.

The 41.0-foot long post supported roof is over an asphalt-paved roadway, with concrete gutters along each side. The retention pond overflow passes under the roadway, through a round galvanized-steel conduit. The sides, slanting inward toward the eaves, are covered with plywood paneling, with two large tent-shaped openings on each side. Large etched wood signs hang from the top of these openings on the northwest side, one inscribed *COVERED* and the other inscribed *BRIDGE*. A large banister-type railing runs down the outside of the sides at ground level. The portals are enclosed with plywood paneling, which is covered with a thin 1.5-inch wide wood lattice. A large, long, etched wood sign inscribed *COVERED BRIDGE* is mounted over the northeast entrance. Black asphalt shingles cover the roof, which is supported along each side by steel tubing posts, an 8-inch by 8-inch square post at the center and 8-inch by 8-inch square posts, flaring out to 8-inches by 16-inches at the top, at each end. The bridge is painted light blue on the exterior and on the interior, electric lighting illuminating the interior.

Directions: From Orlando, at the intersection of SR 436 (Semoran Boulevard) and SR 552 (Curry Ford Road), go east on SR 552 to the intersection with SR 551 (Goldenrod Road), continuing through 1.0 mile to Curry Ford Drive, at Curry Ford Woods Subdivision, on the right. Go 0.1 mile to Sagebrush Place, on the right; go 0.1 mile to the bridge. The bridge is visible from SR 552.

Old Horatio Avenue Covered Bridge

ORANGE COUNTY

Brief Statistics

Type: Non-authentic modern covered bridge. World Guide Covered Bridge Number: FL-48-B. Built in 1988 by City of Maitland. One-span bridge crossing Lake Nina Canal. Stringer bridge 20.0 feet long by 31.7 feet wide, with approximately 25.0-foot wide by 7.0-foot high portal openings. Alternate Name: None known. Located in Maitland on Old Horatio Avenue.

Old Horatio Avenue Covered Bridge was created when a cover was added, in 1988, over a two-lane concrete stringer bridge, built circa 1958 on Old Horatio Avenue. The 1988 cover was built by Max Mincey of the City of Maitland Public Works Department, and, following severe damage caused when a U-Haul truck struck the cover, rebuilt in 1995 by Max Mincey. The west-northwest/east-southeast aligned bridge was intentionally built with a low 7.0-foot clearance, to bar trucks from the neighborhood. The single-span bridge crosses Lake Nina Canal, which connects nearby Lake Minnehaha, to the north, with small Lake Nina. Residences run along the south side of Old Horatio Avenue up to the bridge, with several mature trees near the bridge and a small park on the north side extending to the new Horatio Avenue. Old Horatio Avenue Bridge is the shortest non-authentic covered motor vehicle bridge in Florida. Also, at 31.7 feet wide, the bridge is unusual in that it is wider than it is long.

The 20.0-foot long bridge is supported by a fourteen-ribbed concrete slab stringer, set on poured concrete abutments, with wing walls, with poured concrete guardrails inside the cover completing the circa 1958 bridge. The flooring consists of a concrete slab, covered with asphalt pavement, over the span and butted to poured cement extending out beyond the portals. The sides are open, with three 4-inch by 4-inch wood rails running down each side between the corner roof support posts. Weathered wood shakes cover the gables, which have metal *LOW CLEARANCE 6 FEET 10 INCHES NO TRUCKS* signs mounted over the entrances, replacing the previous 7-foot signs, in 2002, as the actual clearance in places is less than the measured 7.0 feet! The bottoms of the gables, and the enclosed roof rafter truss in the center of the bridge, wear numerous scars from encounters with motor vehicles, attesting to that fact. Vertical tongue and groove boards cover both sides of the interior roof truss, in the center of the bridge, and the inner side of the roof truss at the gables; these are the only roof trusses for the bridge. Wood shakes, fastened to 2-inch by 4-inch purlins, cover the low peaked roof, which is supported by four 8-inch by 8-inch timber posts in the corners of the bridge.

Directions: From Casselberry, at the intersection of SR 436 and US 17/US 92, go south on US 17/US 92 to Maitland and the intersection with Horatio Avenue. Go left 0.4 mile on Horatio Avenue to Oakleigh Drive, on the right, and take the first right onto Old Horatio Avenue; go 0.1 mile to the bridge.

Covered Bridge Community Covered Bridge

PALM BEACH COUNTY

Brief Statistics

Type: Non-authentic modern covered footbridge. World Guide Covered Bridge Number: FL-50-i. Built in 1973 by Hovnanian Developers. One-span footbridge crossing lakes interconnect. Stringer footbridge 30.8 feet long by 12.1 feet wide, with approximately 10.0-foot wide by 8.8-foot high portal openings. Alternate Name: Greenacres Bridge. Located west of Lake Worth in Covered Bridge Subdivision on Covered Bridge Boulevard.

Hovnanian Developers built Covered Bridge Community Bridge, in 1973, in Covered Bridge Subdivision, a footbridge crossing a wide and deep interconnect between North Lake on the north-northeast side and South Lake on the south-southwest side. The bridge has also been called Greenacres Bridge after the nearby community. The single-span bridge is very open, in a well-landscaped and maintained area, a schefflera tree at each portal. The clubhouse is near the east-southeast end, and the Covered Bridge Boulevard concrete bridge is near the north-northwest side. At 12.1 feet wide, Covered Bridge Community Bridge is the widest non-authentic covered footbridge in Florida.

The footbridge is supported by four large 8-inch by 24-inch timber stringers, set on two piers, constructed with three wood utility pole piles capped with a large timber crossbeam. The 30.8-foot long bridge has a 25.0-foot long ramp on the east-southeast end and a 24.9-foot long ramp on the west-northwest end, giving the bridge a total length of 80.7 feet. The ramps extend from the poured concrete abutments up to the piers and have sturdy wood handrails along each side. The flooring consists of transverse 2-inch by 6-inch planking that extends the length of the ramps. Gray weathered horizontal tongue and groove 2-inch by 6-inch planking covers the sides, including the alcoves, up to 4.5 feet, the sides open above to the eaves. Gray weathered vertical tongue and groove planking covers the gables, arch cut over the entrances. A multi-family birdhouse is mounted on each gable under the peak, the roof over the birdhouses extending out, offering protection from the elements. Weathered wood shingles on top of 2-inch by 6-inch planking cover the roof, which is supported along each side by six 8-inch by 8-inch timber posts. This bridge was built to last! An alcove extending out 1.1 feet is centered on each side. Electric lights illuminate the interior.

Directions: From Greenacres, at the intersection of SR 882 (Forest Hill Boulevard) and Jog Road, go west 0.7 mile on Forest Hill Boulevard to Pinehurst Drive, on the left; go 1.9 miles to Covered Bridge Boulevard, on the right; go 0.3 mile to the bridge, visible on the left, just beyond the clubhouse.

Ramblewood Village Covered Bridge

PASCO COUNTY

Brief Statistics

Type: Non-authentic modern covered bridge. World Guide Covered Bridge Number: FL-51-A. Built in 1990 by John Smith. Post supported roof over roadway crossing drainage ditch. Bridge 40.2 feet long by 26.3 feet wide, with approximately 24.7-foot wide by 12.0-foot high portal openings. Alternate Name: None known. Located in Zephyrhills in Ramblewood Village on Vinson Street.

John Smith, a local contractor, built Ramblewood Village Bridge in Ramblewood Village on Vinson Street, in 1990, construction actually starting in 1989. The two-lane bridge is a post supported roof over an asphalt-paved roadway, over a rectangular concrete culvert, allowing the passage of westward flowing drainage water. The north/south-aligned bridge has homes in the north corners, a relatively open wooded area in the south downstream corner, and a paved road paralleling the drainage ditch on the south upstream corner, with the subdivision clubhouse on the far side of the road. The banks of the drainage ditch are densely wooded.

The sides of the 40.2-foot long bridge have weathered vertical boarding with battens up to rail height and under the eaves, open for 5.8 feet between, exposing the roof support posts. The gables also have weathered vertical boarding with battens. The shiny tin roof is supported along each side by six 6-inch by 6-inch wood posts, the central panel and the end panels cross-braced. Four signs adorn the north end of the bridge: a large *Ramblewood/Village*, in red letters on a yellow sign, is centered on the gable; a small red-lettered yellow sign, centered over the entrance, admonishes *CAUTION BRIDGE CLEARANCE 11 FEET 6 INCHES*; a small private property sign on the portal, at the left end post; and a large red-lettered yellow sign, mounted on two posts at the right of the portal, welcome visitors with directions to the sales office.

Directions: From south of Zephyrhills, at the junction of US 301 and SR 39, go north 0.3 mile on US 301/SR 39 to Vinson Road, on the left; follow 0.2 mile to the bridge.

The Wilds Covered Bridge

PASCO COUNTY

Brief Statistics

Type: Non-authentic modern covered bridge. World Guide Covered Bridge Number: FL-51-B. Built in 1981 by unknown builder. Post supported roof over roadway crossing dry land. Bridge 50.7 feet long by 26.5 feet wide, with approximately 24.7-foot wide by 15.0-foot high portal openings. Alternate Name: None known. Located in New Port Richey off Main Street in The Wilds Condominiums.

The Wilds Bridge was built, in 1981, by the developer of The Wilds Condominiums, on the short entrance road. When originally built, the two-lane bridge crossed a stormwater runoff ditch, most likely over a conduit beneath the roadbed, but may have been a concrete slab stringer. In 1988, mounting complaints about a mosquito problem, caused by standing water near the bridge, moved the developer to fill in the drainage ditch. Thus, the bridge today

Bridges Restaurant Footbridge is between parking lots at Bridges Restaurant whose walls are adorned with pictures of covered bridges.

is a post supported roof over a concrete slab roadbed on dry ground. Surprisingly, the realism of a wood-floored bridge remains, as the concrete slab serving as the roadbed has transverse grooves spaced a typical wood-plank-width apart. When a vehicle runs through the bridge, the tires thumping across the grooves reverberate within the bridge, creating a loud clickety-clack sound like that of loose wood floor planks in a real covered bridge!

The 50.7-foot long bridge has gray weathered, worm-holed, vertical boarding on the sides and the portals, each side with a large, oval-topped, arched opening. Brown asphalt shingles cover the roof, which is supported along each side by similar gray weathered, worm-holed, 6-inch by 6-inch timber posts. A banister-style handrail extends along each side of the interior. Three fluorescent light fixtures, spaced down the center of the bridge, illuminate the interior, and an electric carriage lantern at each side of the south entrance, illuminates the portal.

The south/north-aligned bridge has landscape plantings along each side and at the south portal. Condominiums are at each side of the bridge; Main Street is at the north portal; and Devonshire Lane, within the Condominium Complex, is at the south portal. The Wilds Condominiums were formerly off Nebraska Avenue, but, effective January 1, 2002, Nebraska Avenue was renamed Main Street.

Directions: From New Port Richey, at the intersection of US 19 and Main Street, go east 1.4 miles on Main Street to The Wilds Condominiums. The bridge is on the right.

Bridges Restaurant Covered Bridge

PINELLAS COUNTY

Brief Statistics

Type: Non-authentic modern covered footbridge. World Guide Covered Bridge Number: FL-52-a. Built in 1985 by Tom Warren. One-span footbridge crossing drainage ditch. Stringer footbridge 24.9 feet long by 8.5 feet wide, with approximately 4.9-foot wide by 5.9-foot high portal openings. Alternate Name: None known. Located in Largo at 1999 Starkey Road.

Bridges Restaurant Bridge was built, in 1985, by Tom Warren, providing access over a drainage ditch to Bridges Restaurant. The bridge was dedicated on November 1,

1985, in memory of Lester Thompson. Cedar trees flank the north/south-aligned footbridge, with the restaurant on the north end and a parking lot on the south end. Paintings and photographs of covered bridges adorn the walls inside the restaurant.

Five wood stringers, resting on the banks of the shallow drainage ditch, support the 24.9-foot long footbridge. The flooring consists of transverse planking. Cedar shingles cover the lower part of the sides and the gables, the sides open above to the eaves, exposing the crossed roof supports. The portals have a significant outward slant. Dark weathered wood shingles cover the roof, which is supported along each side by six sets of crossed 2-inch by 8-inch plank posts, plus a 2-inch by 8-inch plank post at each portal, which establishes the outward slant of the portal. The 2-inch by 8-inch roof ridgepole, the longitudinal beam running along the peak, extends out about two feet on the south end, this extension ornately cut to embellish the bridge. The ornate extension on the north end has apparently been broken off. The single-span bridge is painted red on the exterior and on the interior.

Directions: From Largo, at the intersection of SR 688 (Ulmerton Road) and SR 693, go west 4.0 miles on SR 688 to Starkey Road, on the right; go 0.3 mile to Bridges Restaurant, on the right at No. 1999. The bridge is between parking lots.

Isla Key Covered Bridge

PINELLAS COUNTY

Brief Statistics

Type: Non-authentic modern covered bridge. World Guide Covered Bridge Number: FL-52-D. Built in 1997 by unknown builder. One-span bridge crossing man-made lake. Stringer bridge 36.7 feet long by 39.7 feet wide, with approximately 32.0-foot wide by 14.9-foot high portal openings. Alternate Name: None known. Located in St. Petersburg at Isla Key Condominiums on SR 682 (Pinellas Bayway).

Isla Key Bridge was built, in 1997, by an unknown builder, at the entrance to Isla Key Condominiums on Isla del Sol, spanning a man-made lake. The single-span bridge is picturesquely set amid many tall Royal Palm trees, its gleaming white painted stucco vividly contrasting against the brown-painted metal roof. The south/north-aligned bridge has Pinellas Bayway at the south end and the condominiums behind magnificently landscaped grounds at the north end, the man-made lake stretching out at both sides. At 39.7 feet wide, Isla Key Bridge is the widest non-authentic covered motor vehicle bridge in Florida and, for that matter, in the entire southeastern United States. The two-lane bridge is unusual in that it is wider than it is long.

A reinforced concrete slab stringer, set on poured concrete abutments, supports the 36.7-foot long stringer bridge. The floor consists of facsimile wood-plank bricks, placed over the span, with showy brick pavers at both portals. The sides are open, with a steel pipe railing along each side and a high white-painted stuccoed gable above to the roof peak, a colorful, ornate *ISLA KEY* sign centered on the gables. The stuccoed portals are painted white to the eaves and on the dormers above the entrances; a white-on-blue *ISLA KEY* plaque is centered on the dormer on the south portal. Four white-painted stuccoed corner posts support the brown-painted metal roof, with the peaks over the sides, rather than the ends. *See Isla Key Covered Bridge in the color photograph section: C-4.*

Directions: From St. Petersburg, at the junction of US 19 and SR 682, go west on SR 682 (Pinellas Bayway) to Isla Key Condominiums, on the right. The imposing bridge is at the entrance.

Tradewinds No. 1 Covered Bridge

PINELLAS COUNTY

Brief Statistics

Type: Non-authentic modern covered footbridge. World Guide Covered Bridge Number: FL-52-b. Built in 1980s by unknown builder. Two-span footbridge crossing man-made waterway. Stringer footbridge 22.5 feet long by 4.7 feet wide, with approximately 4.0-foot wide by 6.8-foot high portal openings. Alternate Name: None known. Located in St. Petersburg Beach at Tradewinds Island Grand Beach Resort at 5500 Gulf Boulevard.

Tradewinds No. 1 Bridge was built, in the early to mid 1980s, by an unknown builder, at Tradewinds Island Grand Beach Resort, over a man-made waterway. One of two covered footbridges located in the densely landscaped Hibiscus Courtyard, the 22.5-foot long Tradewinds No. 1 Bridge is supported by two 2-inch by 10-inch wood plank stringers, placed on edge and set on three piers. Each pier is constructed with two upright 4-inch by 4-inch wood posts that extend up to the roof, a 2-inch by 10-inch crossbeam bolted under the stringers to each side of the posts, and crossbeams on the center pier placed higher than the end piers, to give a rise to the bridge at the center. Crossbraces of 2-inch by 4-inch lumber, placed below the crossbeams, stabilize the piers. Five wood steps at the north-northwest portal and six wood steps at the south-southeast portal provide access to the two-span bridge. The flooring consists of transverse 2-inch by 4-inch planks, covered with anti-skid matting. The sides of the white painted bridge are open, a banister-style handrail running along each side and down the steps. A white colored fabric attached to the rafters comprises the roof, which is supported by six extended pier posts.

Directions: From St. Petersburg, at the junction of US 19 and SR 682 (Pinellas Bayway), go west on SR 682 to Gulf Boulevard (SR 699); go right to Tradewinds Island Grand Beach Resort at No. 5500, on the left. The bridge is in the courtyard, south of the tennis courts.

Tradewinds No. 2 Covered Bridge

PINELLAS COUNTY

Brief Statistics

Type: Non-authentic modern covered footbridge. World Guide Covered Bridge Number: FL-52-c. Built in 1989 by unknown builder. One-span footbridge crossing man-made waterway. Stringer footbridge 18.3 feet long by 4.6 feet wide, with approximately 3.5-foot wide by 6.3-foot high portal openings. Alternate Name: None known. Located in St. Petersburg Beach at Tradewinds Island Grand Beach Resort at 5500 Gulf Boulevard.

Tradewinds No. 2 Bridge was built, in 1989, by an unknown builder, at Tradewinds Island Grand Beach Resort, over a man-made waterway, one of two covered footbridges located in the densely landscaped Hibiscus Courtyard. The 4.6-foot wide bridge is the narrowest non-authentic covered footbridge in Florida.

The 18.3-foot long stringer footbridge is supported by two 2-inch by 10-inch wood plank stringers, placed on edge and set on two piers, each pier constructed with two upright 4-inch by 4-inch wood posts that extend up to the roof, a 2-inch by 10-inch crossbeam bolted under the stringers to each side of the posts, with 2-inch by 4-inch cross-braces below the crossbeams to stabilize the piers. The single-span bridge has six steps at each portal. Transverse planking forms the flat floor, which is covered with anti-skid matting. The sides are open, with 2-inch by 4-inch handrails along each side and cross-braces beneath the handrails between roof supports. Banister-style handrails extend down the steps. White-trimmed gray-green fabric attached to the rafters covers the hip-style roof, which is supported by four extended pier posts at the portals and six 2-inch by 4-inch wood posts, three evenly spaced along each side, between the portal posts. The west-northwest/east-southeast aligned bridge is painted white, with the top of the handrails painted gray-green.

Directions: From St. Petersburg, at the junction of US 19 and SR 682 (Pinellas Bayway), go west on SR 682 to Gulf Boulevard (SR 699); go right to Tradewinds Island Grand Beach Resort at No. 5500, on the left. The bridge is in the courtyard, south of the tennis courts.

Foxfire Covered Bridge

SARASOTA COUNTY

Brief Statistics

Type: Non-authentic modern covered footbridge. World Guide Covered Bridge Number: FL-58-a. Built in 1978 by unknown builder. One-span footbridge crossing pond. Stringer footbridge 24.2 feet long by 8.3 feet wide, with approximately 7.1-foot wide by 7.4-foot high portal openings. Alternate Name: None known. Located in Sarasota at Foxfire Golf Club at 7200 Proctor Road.

Foxfire Bridge was built, in 1978, by an unknown builder, at Foxfire Golf Club, on the front nine over a long pond, between the 3rd green and the 4th tee. The west/east-aligned footbridge is fully exposed, with the curving golfcart path passing through the bridge, split-leaf philodendrons (selloums) near each portal and scattered mature Long-Needle Pine trees nearby.

A cement slab stringer, set on poured concrete abutments, supports the 24.2-foot long bridge. The cement slab also forms the flooring. Vertical lapped boarding covers the sides up to rail height, open above to the eaves, exposing the cross-braced roof support posts. Vertical lapped boarding also covers the gables. Old black asphalt shingles cover the roof, which is supported along each side by five 4-inch by 4-inch wood posts, cross-braced with 2-inch by 4-inch lumber. The single-span bridge is painted white on the exterior and on the interior.

Directions: From I-75, east of Sarasota, go south to the SR 72 exit; go east on SR 72 (Clark Road) 1.7 miles to Proctor Road, on the left; go 0.9 mile to Foxfire Golf Club at No. 7200, on the left. The bridge is between the 3rd green and the 4th tee.

Sherwood Forest Covered Bridge

SARASOTA COUNTY

Brief Statistics

Type: Non-authentic modern covered bridge. World Guide Covered Bridge Number: FL-58-C. Built in 1984 by unknown builder. Post supported roof over roadway crossing unnamed creek. Bridge 48.1 feet long by 24.0 feet wide, with approximately 21.5-foot wide by 16.5-foot high portal openings. Alternate Name: None known. Located in Sarasota at Sherwood Forest Subdivision on Sherwood Forest Boulevard.

Sherwood Forest Bridge was built, in 1984, by an unknown builder, on Sherwood Forest Boulevard in Sherwood Forest Subdivision. The two-lane bridge is a post supported roof over an asphalt roadway that is over a square concrete culvert. The north-northeast/south-southwest aligned bridge has homes in all four corners, with a heavily wooded, small, unnamed creek passing beneath on its way to nearby Philippe Creek, which flows into Roberts Bay and the Gulf of Mexico.

The 48.1-foot long brown painted bridge has open sides and vertical boarding in the gables. Asphalt shingles cover the roof, which is supported along each side by four 12-inch by 14-inch timber posts, with a smaller wood post and two braces, resembling a kingpost truss, in the three openings between the larger posts. Two white-trimmed brown-painted wood signs are mounted on the gables above each entrance: a long, narrow, white-lettered *SHERWOOD FOREST*, with a deer at each end, sign above a small white-numbered *1984* sign. Electric carriage lanterns are mounted on both portals, at both sides of the entrances, and a large carriage lantern, hanging in the center of the bridge, illuminates the interior.

Directions: From Sarasota, go south on US 41 to the major intersection at Bahia Vista Road; go left 3.2 miles to Sherwood Forest Subdivision, on the left, turning in on Sherwood Forest Boulevard and continuing 0.3 mile to the bridge.

Lake Lotus Covered Bridge

Seminole County

Brief Statistics

Type: Non-authentic modern covered bridge. World Guide Covered Bridge Number: FL-59-A. Built in 1972 by unknown builder. Post supported roof over roadway crossing unnamed creek. Bridge 56.4 feet long by 32.4 feet wide, with approximately 28.5-foot wide by 14.3-foot high portal openings. Alternate Name: None known. Located in Altamonte Springs on Country Creek Parkway in Country Creek Development.

Lake Lotus Bridge was built, in 1972, by an unknown builder, on Country Creek Parkway in Country Creek Development. The two-lane bridge is a post supported roof over an asphalt-paved roadway that is over a massive galvanized steel pipe, which allows the waters of a small, wet-weather, unnamed creek to pass by into nearby Lake Lotus. The galvanized steel pipe is encased in poured concrete, restraining the roadbed at both sides of the southwest/northeast-aligned bridge. Thick vine-choked woods following the creek spread out from both sides of the bridge. The roadway has a short landscaped median near the southwest portal.

The 56.4-foot long post supported roof rests along each side on poured concrete bases about two feet high. The siding boxes in the roof support posts at the portals, with five open panels along each side, exposing four vertical posts evenly spaced between the portals with five X braces between the posts and between the posts and boxed in end posts. The sides and portals are covered with facsimile vertical board paneling, the gables with a slight arch cut over the entrances. Brown asphalt shingles cover the steeply peaked roof, which overhangs the sides, offering some weather protection for the two open exterior walkways. The walkways have handrails along the outer sides. The bridge is painted dark brown on the exterior and on the interior. Electric lights illuminate the interior.

Directions: From Altamonte Springs, at the intersection of SR 434 and SR 436, go south on SR 434 to the first right, West Town Parkway, following to Bunnell Road, bearing left to Country Creek Development, on the left, and taking the entrance road, Country Creek Parkway, to the bridge.

Georgia

Prior to 1900, Georgia had some two hundred fifty covered bridges spread across the state, providing travelers shelter during storms, locals places for community meetings and dances, highwaymen hideaways where they may lie in wait, or lovers privacy where they may steal a kiss. These monuments to the bridge builder's skill dwindled in number, until ninety were left in 1954, twenty-eight remained in 1968 and only sixteen survive today.

Of the forty-seven covered bridges in Georgia, sixteen are authentic and thirty-one are non-authentic, sixteen of the forty-seven being historic. These forty-seven covered bridges were constructed, between the 1840s and 2000, in thirty-four of the one hundred fifty-nine counties in Georgia.

By far, the most commonly constructed truss in Georgia was the Town lattice truss, still supporting eleven of the remaining historic bridges and used in one of the modern bridges. The state also has three queenpost truss bridges and one kingpost truss bridge.

In 1997-1998, Georgia contracted major repair work for ten of their historic covered bridges, plus one bridge was rebuilt, following its destruction by a flood in 1994. Five historic covered bridges are in private hands. Georgia's well maintained covered bridges are destined to be around for the enjoyment of future generations, five of them still resounding to the daily clickety-clack of passing motorists.

Lula Covered Bridge

Banks County

Brief Statistics

Type: Authentic historic covered bridge. World Guide Covered Bridge Number: GA-06-06 #2. Built in 1976 by Harry Holland. One-span bridge crossing small creek. Modified kingpost truss bridge 34.0 feet long by 14.9 feet wide, with approximately 10.9-foot wide by 8.8-foot high portal openings. Alternate Names: Blind Susie Bridge, Garrison Bridge, Hyder Bridge. Located southeast of Lula off Antioch Road.

Rebuilt in 1976 by Harry Holland, Lula Bridge, at a length of 34.0 feet, is the shortest historic covered bridge remaining in Georgia. This privately owned bridge also has the distinction of being the only covered bridge in Georgia with a kingpost truss. Located in Banks County, the county seat of which is Homer, Lula Bridge was originally built, in 1915, by an unknown bridge builder. It was relocated to its present site from over the same creek about one mile away. Named after the nearby City of Lula, the bridge, also known as Hyder Bridge and Garrison Bridge, has been referred to as Blind Susie Bridge, presumably after a colorful Banks County woman who sat on her porch selling jars of moonshine hidden under her skirt.

Lula Bridge's single span crosses a small creek on mortared natural stone abutments. The flooring is transverse planking. There are no wheel treads nor approaches, as the bridge is not used for traffic. The sides and portals are covered with weathered wood shingles; the roof is covered with old mossy, weathered wood shingles. The truss is a two-panel kingpost over a half height M support and has three steel tension rods, one in the center and one outside each leg of the M support. The bridge is situated a short distance off Antioch Road, in the dense wooded margin to a former agricultural field. It is not readily visible during summer foliage, especially when the area becomes overgrown. The portals align on a south-southwest/north-northeast orientation. There are concrete picnic tables on the north-northeast side of the small creek, a feeder stream to nearby Reservoir 59, through which Grove Creek flows on its way to the Hudson River at Erastus.

Lula Bridge is the last authentic historic covered bridge remaining in Banks County, the county having at least eighteen in 1954!

Directions: From Lula, at the junction of SR 51 and SR 52, go east 3.0 miles on SR 51 to Antioch Road (CR 109), on the right; go down Antioch Road 1.1 miles to the bridge, on the right, at the dip in the road. There is no parking available at the bridge. Private property: request permission to visit.

This bridge maker's mark is located on the truss member inside the left downstream portal of the Euharlee Bridge.

Euharlee Covered Bridge

Bartow County

Brief Statistics

Type: Authentic historic covered bridge. World Guide Covered Bridge Number: GA-08-01. Built in 1886 by Washington W. King. One-span bridge crossing Euharlee Creek. Town lattice truss bridge 137.8 feet long by 16.3 feet wide, with approximately 13.5-foot wide by 9.0-foot high portal openings. Alternate Names: Euharlee Creek Bridge, Lowry (or Lowery) Bridge. Located in downtown Euharlee in a park off Covered Bridge Road.

Euharlee Bridge was built over Euharlee Creek, in 1886, by Washington W. King, upstream from a mill owned and operated by Daniel Lowry (or Lowery), who may have assisted in the construction of the bridge, following the collapse of the previous bridge. According to the *North Georgia Journal*, the previous bridge collapsed, causing the death of a local man, Mr. Nelson, a mule and a horse. His two young sons emerged from the disaster unscathed. This led to the construction of the much sturdier bridge by Washington W. King. Euharlee Bridge is one of four covered bridges remaining in Georgia built by Washington W. King, son of Horace King, freed slave of covered bridge fame. Many lower chord members and truss members still have legible white numbers at the lower chord, indicating that the bridge had been pre-assembled at the factory, to assure a proper fit when reassembled at the bridge site. The 78 numbers on the upstream members had an X added to distinguish them from the downstream members. This is the only historic bridge in Georgia with a bridge maker's mark. As this mark has an X at the left on the upstream side and no X on the downstream side, as the numbered truss members have, the author deduces this mark to represent the factory. The bridge maker's marks are on the truss members at the upper chord, at both sides, just inside the east or left downstream portal. Euharlee Bridge is sometimes referred to as Euharlee Creek Bridge, after the creek below, or Lowry (or Lowery) Bridge, after the former owner of the nearby mill ruins. Under the ISTEA (Intermodal Surface Transportation Efficiency Act of 1991) Bridge Restoration Project in Georgia, a contract was awarded in November 1997 for $162,998 to make extensive repairs to the Euharlee Bridge, the repairs being completed in 1998.

The present bridge, owned by Bartow County, the county seat of which is Cartersville, was constructed on raised piers, to eliminate the risk of washing away during floods. The higher bridge necessitated long approach ramps, 30.9 feet at the left downstream (east) portal and 89.4 feet at the right downstream (west) portal; the covered bridge itself being 137.8 feet long, the second longest single-span bridge in Georgia. Euharlee Bridge rests on two mortared cut-stone piers, with the east approach ramp resting on a poured concrete abutment and one bridge pier supported by an additional, centrally located poured-concrete pier, and, due to its extensive length, the west approach ramp resting on a poured concrete abutment and the other bridge pier supported by two additional, centrally located poured-concrete piers. As with eleven other covered bridges in Georgia, this bridge has a Town lattice truss, with chin braces inside each portal. The flooring consists of transverse planking, with wheel treads, extending the length of the approach ramps. Dark weathered vertical board siding with battens and a corrugated tin roof complete the bridge. Euharlee Bridge is equipped with interior lights, exterior portal lights, a fire sprinkler system and a surveillance system. A bridge informational sign is near the right downstream portal.

Closed to motor traffic, Euharlee Bridge carries the bypassed roadway through the downtown park, wherein it stands. The park, enhanced with a museum about the bridge and the historic City of Euharlee, also contains the old Lowry Mill ruins, downstream of the bridge. Beneath the bridge, a cement stairway descends each bank to Euharlee Creek, the creek flowing into the nearby Etowah River and eventually emptying into the Coosa River, at Rome. The upstream side of the bridge is heavily wooded, obscuring the bridge during foliage season. Euharlee Bridge is the only authentic historic covered bridge in Bartow County.

Directions: From I-75, exit 288, take SR 113 west 6.0 miles to Covered Bridge Road, on the right, continuing 3.2 miles to the bridge in the park, on the left.

Shiloh Walking Trail Covered Bridge

CARROLL COUNTY

Brief Statistics

Type: Authentic modern covered bridge. World Guide Covered Bridge Number: GA-22-01. Built in 1993 by Dennis Crews and church members. One-span bridge crossing Sams Creek. Town lattice truss bridge 42.4 feet long by 13.8 feet wide, with approximately 11.0-foot wide by 9.7-foot wide portal openings. Alternate Name: None known. Located in Burwell at Shiloh United Methodist Church.

Behind the Shiloh United Methodist Church, in Smith Bottoms, former farmland that, in 1949, was converted into a lake, church members, in 1991, undertook the draining of the lake to create graveled walking trails. These meander by newly planted apple, willow, oak and other in-memoriam trees, cross Sams Creek on an arched bridge, and pass through a covered bridge, reminiscent of the past. Built in 1993 by church members, under the leadership of Dennis Crews, Shiloh Walking Trail Bridge was constructed in the style of a 19th century Town lattice truss bridge, thereby producing a modern authentic covered bridge. Part of the bridge is made from 100-year-old white oak and poplar timbers from the old Davenport Mill Road Bridge. The 42.4-foot long single-span bridge rests on a mortared natural-stone abutment on the north-northwest end, a poured concrete abutment on the south-southeast end and two mortared natural-stone piers in between. This bridge truly was built to carry the heaviest of loads, albeit only intended for foot traffic! The interior of the bridge has two chin braces at each end, unusual in that they attach to the floor planking, transverse planked flooring, and a bench along each side. The exterior has natural vertical boarding with battens, on the sides and portals, and a silvery corrugated tin roof. The roof and gables extend beyond the flooring to provide shelter for the entrances. A unique feature of the Shiloh Walking Trail Bridge is the antique barn ornaments on the roof peak above each gable, called roof pigeons, which were obtained from the Proctor barn in nearby Bowdon.

An antique barn ornament called a roof pigeon is on the roof peak above each portal of the Shiloh Walking Trail Bridge.

Shiloh Walking Trail Bridge spans Sams Creek, which joins nearby Turkey Creek, then Big Indian Creek and Little Tallapoosa River, before emptying its waters into the Tallapoosa River at Lake Wedowee in Alabama. Adjacent to a road on the downstream side, the Carroll

County bridge is relatively open all around, with a low wooded hill on the upstream south-southeast end.

Directions: From south of Tallapoosa, at the I-20 and SR 100 intersection, go south on SR 100 through Kansas to Smithfield Road, on the left; follow to the center of Burwell (Smithfield Road becomes Burwell Road). The bridge is behind Shiloh United Methodist Church, on the right. This is a private bridge.

Concord Covered Bridge

Cobb County

Brief Statistics

Type: Authentic historic covered bridge. World Guide Covered Bridge Number: GA-33-02. Built in 1872 by Robert Daniell and Martin L. Ruff. Two-span bridge crossing Nickajack Creek. Modified queenpost truss bridge 132.1 feet long by 16.0 feet wide, with approximately 14.8-foot wide by 9.5-foot high portal openings. Alternate Names: Nickajack Creek Bridge, Ruff's Mill Bridge. Located southwest of Smyrna on Concord Road.

Built in 1872 by Robert Daniell and Martin L. Ruff, the current Concord Bridge replaced a bridge built circa 1842 that was destroyed in 1864, during the Civil War. Concord Bridge was built using a dry natural-stone center pier that may have been from the previous bridge. This pier appears to have been widened for the current bridge. In the 1950s, a concrete pier was added on each side of the center pier for additional support and gives the bridge the appearance of having four spans. At the same time, eight longitudinal steel I-beams were added under the floor beams to strengthen the bridge. On the east-northeast end of the bridge a timber support was added between the abutment and the added concrete pier, probably predating the added pier. The 132.1-foot long Concord Bridge is the longest two-span covered bridge remaining in Georgia. Named after Concord Road that passes through it, Concord Bridge is also known as Nickajack Creek Bridge, after the creek flowing beneath, and Ruff's Mill Bridge, after the upstream mill ruins, at one time owned by Martin L. Ruff. The mill was the scene of a Civil War battle.

Concord Bridge was one of the ten covered bridges in Georgia that had restoration work performed under the ISTEA (Intermodal Surface Transportation Efficiency Act of 1991) Bridge Restoration Project in Georgia. These repairs were completed in 1999 at a cost of $151,188.

The exterior of the bridge has dark weathered, vertical boards with battens, a wood-shingled roof and three small wood-shingle-roofed window openings on each side. There is a steel height restrictor at each portal, the portals aligning on a west-southwest/east-northeast orientation. Each span of the two-span bridge has a modified queenpost truss, protected by steel guardrails running the length of the bridge. This bridge is the oldest of the three queenpost truss bridges remaining in Georgia, the other two being Coheelee Creek Bridge and Stovall Mill Bridge. The flooring consists of three-plank-wide wheel treads on transverse planking. The roof and gables extend beyond the flooring affording protection from the elements for the entrances. Engraved wood signs mounted above each entrance read *CONCORD/BRIDGE/CA 1850*; however, the construction date of 1872 is more generally accepted. A bridge informational sign is near the right downstream portal.

Concord Bridge is open to traffic, carrying the highest traffic rate of Georgia's covered bridges, making it a dangerous bridge to visit. Furthermore, Concord Road is narrow and winding at the bridge site, affording no parking at or near the bridge. The creek is heavily wooded, with moderate exposure of the bridge at close range. Bedrock-bottomed Nickajack Creek, flowing south-southeast at the bridge site, drains into the Chattahoochee River at Oakdale.

The bridge was listed on the National Register of Historic Places on November 24, 1980. Concord Bridge is the only authentic historic covered bridge remaining in Cobb County. Marietta is the county seat of Cobb County.

Directions: From I-285, northwest of Atlanta, take exit 10 onto SR 280 North; go 4.0 miles to Concord Road, and go left 2.4 miles to the bridge. The bridge is southwest of Smyrna.

Covered Bridge Subdivision Covered Bridge

Cobb County

Brief Statistics

Type: Non-authentic modern covered footbridge. World Guide Covered Bridge Number: GA-33-c. Built in 1979 by unknown builder. One-span footbridge crossing wet weather stream. Stringer footbridge 22.7 feet long by 8.2 feet wide, with approximately 7.0-foot wide by 7.3-foot high portal openings. Alternate Name: None known. Located southeast of Marietta in Covered Bridge Subdivision.

Covered Bridge Subdivision Bridge, built in 1979 by an unknown builder, crosses a wet weather stream in a low-lying area that is prone to frequent flooding, as evidenced by the sediment and decay on the bridge timbers. The single-span bridge aligns on a west/east orientation at the rear of a cleared park area, with scattered mature trees left standing. A swampy area with dense woods extends beyond the 22.7-foot long bridge, which is adjacent to the Covered Bridge Subdivision clubhouse and tennis courts.

The Cobb County stringer bridge has three 8-inch by 8-inch wood timber stringers resting on concrete barriers, set on the ground, with transverse planking spanning the stringers. The sides are covered with natural horizontal boarding; the ends are covered with natural vertical boarding, with a duplex bird nest box mounted in each gable; the roof is covered with brown asphalt shingles. Six 4-inch by 4-inch wood posts along each side support the roof. There are transverse-planked upward-sloping approach ramps at each end.

Directions: From the junction of I-75 and I-575, north of Marietta, take I-75 south to exit 261 SR at 280 (Delk Road); go east 0.9 mile to Powers Ferry Road; go left 0.8 mile to Chadds Ford Road (unmarked) at Covered Bridge Subdivision, on the left. Follow Chadds Ford Road to the clubhouse, on the left. The bridge is in the adjoining park.

Woodbridge Covered Bridge

COLUMBIA COUNTY

Brief Statistics

Type: Non-authentic modern covered bridge. World Guide Covered Bridge Number: GA-36-A #2. Built in 1994 by Columbia County. Post supported roof over roadway crossing Mt. Ema Branch. Bridge 60.2 feet long by 36.8 feet wide, with approximately 35.2-foot wide by 13.6-foot high portal openings. Alternate Name: None known. Located in Evans on Covered Bridge Road in Woodbridge Subdivision.

The previous covered bridge built at this site, in 1975, by the development of Woodbridge Subdivision, was washed away by the 1990 hurricane. The present Woodbridge Bridge was built, in 1994, by Columbia County, the cover being paid for by the residents of Woodbridge Subdivision. The southeast/northwest-aligned bridge consists of a post supported roof above an asphalt roadway over dual concrete culverts, with the roof supported by six wood posts along each side. The exterior of the two-lane bridge has natural vertical boarding with battens on the sides and portals, topped by a brownish asphalt-shingled roof. There are five opposing large cathedral-style window openings on each side. Poured concrete sidewalks extend along both sides of the interior for the length of the 60.2-foot long bridge.

Woodbridge Bridge is over Mt. Ema Branch, which flows northward through Silver Lake, just downstream of the bridge, into the nearby Savannah River. The Columbia County bridge offers a great view of man-made Silver Lake downstream, is visible from State Road 104 upstream, has residences on the southeast end and has Woodbridge Drive beyond some woods on the northwest end.

Directions: From Evans, at the junction of SR 383 and SR 104, go west 1.8 miles on SR 104 to the second Woodbridge Subdivision entrance (Woodbridge Drive), on the right; take the first right on Covered Bridge Road to the bridge.

Burt Farm No. 1 Covered Bridge

DAWSON COUNTY

Brief Statistics

Type: Non-authentic modern covered bridge. World Guide Covered Bridge Number: GA-42-A. Built in 1988 by unknown builder. One-span bridge crossing Amicalola Falls Creek. Stringer bridge 32.2 feet long by 12.2 feet wide, with approximately 11.0-foot wide by 10.0-foot high portal openings. Alternate Name: None known. Located in Amicalola at 4801 SR 52.

Located in Amicalola, adjacent to the parking lot at Burt Pumpkin Farm, Burt Farm No. 1 Bridge was built in 1988 over Amicalola Falls Creek. Amicalola Falls Creek, originating on the flank of Amicalola Mountain above Amicalola Falls State Park, flows through nearby Lake Laurel into Little Amicalola Creek, which empties into Amicalola Creek before joining the Etowah River, southwest of Dawsonville.

Aligned southwest/northeast, Burt Farm No. 1 Bridge is a single-span stringer bridge supported by nine 4-inch by 12-inch timbers laid on edge upon poured concrete abutments. The flooring consists of transverse planking, only over the length of the narrow span, and poured concrete flooring inside each portal. The exterior has dark weathered, vertical boarding with battens on the sides and portals, and the roof, supported by 2-inch by 4-inch studding along each side, has gray weathered cedar shingles. There are two opposing vertical rectangular window openings on each side. A *FOOT/TRAFFIC/ONLY* sign is mounted on the right downstream portal.

Burt Farm No. 1 Bridge is in an open area, with a large parking lot on the southwest end, State Road 52 on the upstream side, scattered trees on the northeast end and picnic tables beside Amicalola Falls Creek on the downstream side. The 32.2-foot long bridge is closed to motor traffic.

Directions: From the junction of SR 52 and SR 183, go 1.0 mile east on SR 52 to Burt Pumpkin Farm (No. 4801), on the right. The bridge is at the far end of the large parking lot.

The 150.9-foot long Stone Mountain Park Bridge flaunts its Town lattice truss that was built in 1891.

Burt Farm No. 2 Covered Bridge

Dawson County

Brief Statistics

Type: Non-authentic modern covered bridge. World Guide Covered Bridge Number: GA-42-B. Built in 1995 by unknown builder. Post supported roof over roadway crossing Amicalola Falls Creek. Bridge 43.2 feet long by 20.4 feet wide, with approximately 13.9-foot wide by 14.5-foot high portal openings. Alternate Name: None known. Located in Amicalola at 4801 SR 52.

Located in Amicalola, one-half mile behind Burt Pumpkin Farm, in a heavily wooded area, Burt Farm No. 2 Bridge was built in 1995 over Amicalola Falls Creek. Amicalola Falls Creek, originating on the flank of Amicalola Mountain above Amicalola Falls State Park, flows through nearby Lake Laurel into Little Amicalola Creek, which empties into Amicalola Creek, before joining the Etowah River, southwest of Dawsonville.

Aligned west-northwest/east-southeast, Burt Farm No. 2 Bridge is a post supported roof above a poured concrete roadway over four galvanized steel culverts, which allow the passage of the waters of Amicalola Falls Creek. The Dawson County bridge is enclosed on the sides and portals by natural vertical boarding with battens, and the roof is red-painted tin. There are four opposing vertical rectangular window openings on each side, aligned above the four culverts.

Dense woods in a low-lying swampy area surround the 43.2-foot long bridge. There is a high hill on the right downstream or east-southeast end and a small open field on the west-northwest end.

Directions: From the junction of SR 52 and SR 183, go 1.0 mile east on SR 52 to Burt Pumpkin Farm (No. 4801), on the right. The bridge is 0.5 mile behind the Farm Stand, on the paved road branching to the left. Private property: ask permission to visit.

Stone Mountain Park Covered Bridge

DeKalb County

Brief Statistics

Type: Authentic historic covered bridge. World Guide Covered Bridge Number: GA-44-01 (Formerly

GA-29-01). Built in 1891 by Washington W. King. Three-span bridge crossing Stone Mountain Lake. Town lattice truss bridge 150.9 feet long by 20.3 feet wide, with approximately 17.5-foot wide by 12.8-foot high portal openings. Alternate Names: College Avenue Bridge, Effie's Bridge. Located east of Stone Mountain in Stone Mountain Park.

Stone Mountain Park Bridge was originally built in Athens over the North Fork of the Oconee River, between College Avenue and Hobson Avenue, in 1891, by Washington W. King, the contract being awarded by Clarke County on March 26, 1891, for $2,470. While at this location, the bridge was called College Avenue Bridge, after the nearby avenue, or Effie's Bridge, after an Athens bordello owner. On February 18, 1964, Stone Mountain Park purchased the bridge for $1.00 and, in March 1965, moved it 60 miles to its present location, at a cost of $18,000, where it was rebuilt to a shortened 150.9-foot length. Stone Mountain Park Bridge is one of four covered bridges built in Georgia by Washington W. King, son of Horace King of covered bridge fame. Stone Mountain Park Bridge has the distinction of being the widest single-lane historic covered bridge remaining in Georgia, measuring 20.3 feet wide.

The three-span bridge rests upon mortared cut-stone abutments and two mortared cut-stone piers. The Town lattice truss, common to eleven other Georgia covered bridges, is fastened with treenails that protrude through the truss members, some four or more inches, and has chin braces inside each portal, the portals aligning on a west-northwest/east-southeast orientation. Stone Mountain Park Bridge has transverse planked flooring, with three-plank-wide wheel treads, and an interior walkway, five planks wide, on the north side, set apart by a handrail. The exterior has dark weathered, vertical boarding with battens, on the portals and the lower part of the sides; the upper part is open, exposing the Town lattice truss. An old weathered cedar-shingled roof completes the structure. A bridge informational sign is at the west-northwest portal.

The only authentic historic covered bridge in DeKalb County, Stone Mountain Park Bridge is open to park visitor motor traffic, providing access to the picnic area on Indian Island in Stone Mountain Lake. Only four other historic covered bridges in Georgia remain open to motor traffic. This bridge, next to the old grist mill off Robert E. Lee Boulevard, has a picturesque setting spanning the lake channel. There is parking available at both ends of the bridge.

Directions: From Decatur, the county seat of DeKalb County, take I-285 north to exit 30b; go 7.5 miles east on US 78 to Stone Mountain Park, on the right. The bridge location is shown in the brochure, available at the gate. Admission.

Stone Mountain Park Golf Club Covered Bridge

DEKALB COUNTY

Brief Statistics

Type: Non-authentic modern covered footbridge. World Guide Covered Bridge Number: GA-44-b. Built in 1993 by unknown builder. One-span footbridge crossing cove of Stone Mountain Lake. Stringer footbridge 60.2 feet long by 11.0 feet wide, with approximately 9.0-foot wide by 8.8-foot high portal openings. Alternate Name: None known. Located east of Stone Mountain in Stone Mountain Park.

The longest single-span covered footbridge in Georgia at 60.2 feet, Stone Mountain Park Golf Club Bridge was completed by an unknown builder on January 13, 1993. The DeKalb County stringer bridge is supported by two steel I-beam stringers, set on abutments constructed from poured concrete, faced with mortared rough-cut stone blocks. The transverse-planked floor is protected from the elements by a gray weathered cedar-shingled roof, supported by sixteen posts along each side, evenly spaced with crossed braces between, a railing above the floor and natural vertical boarding with battens below the floor. Natural vertical boarding with battens also enclose the gables above the portal entrances.

Stone Mountain Park Golf Club Bridge, aligned west-southwest/east-northeast, is situated within the golf course confines, along the edge of a narrow wooded area, spanning the placid waters of a cove of Stone Mountain Lake. The exposed south-southeast or lake side and west-southwest portal offer a splendid view in the late afternoon sun.

Directions: From Decatur, take I-285 north to exit 30b; go 7.5 miles east on US 78 to Stone Mountain Park, on the right. The bridge is located on the golf course, on the right, off Stonewall Jackson Drive, the first left after the admissions gate. Admission.

Coheelee Creek Covered Bridge

EARLY COUNTY

Brief Statistics

Type: Authentic historic covered bridge. World Guide Covered Bridge Number: GA-49-02. Built in 1891 by J.W. Baughman. Two-span bridge crossing Coheelee Creek. Modified queenpost/kingpost combination truss bridge 120.1 feet long by 12.6 feet wide, with approximately 11.1-foot wide by 9.2-foot high portal openings. Alternate Names: Hilton Bridge, McDonald's Ford Bridge. Located north-northeast of Hilton off Old River Road.

The 120.1-foot long Coheelee Creek Bridge crosses tumbling Coheelee Creek in two spans. *See Coheelee Creek Covered Bridge in the color photograph section: C-6.*

The southernmost-remaining historic covered bridge in the United States, Coheelee Creek Bridge was built, in 1891, at a total cost of $490.41, where Old River Road crossed Coheelee Creek at McDonald's Ford. Its construction was authorized by the Early County Board of Commissioners on July 7, 1891. Named after the creek it spans, Coheelee Creek Bridge was the first bridge erected at this site. J.W. Baughman was supervisor of the bridge construction, completed in November 1891. Bypassed, diverting Old River Road (CR 80), the bridge is closed to motor traffic. Coheelee Creek Bridge was one of the ten covered bridges in Georgia that had restoration work performed under the ISTEA (Intermodal Surface Transportation Efficiency Act of 1991) Bridge Restoration Project in Georgia. These repairs, totaling $128,007.00, were completed in 1999. The bridge was also known as McDonald's Ford Bridge, derived from the location where the bridge crosses the creek, and Hilton Bridge, after the nearby community.

The second longest two-span covered bridge remaining in Georgia at 120.1 feet, Coheelee Creek Bridge rests on three rock-filled poured-concrete piers and poured concrete abutments, the abutments added July 12, 1958. Each span has a queenpost/kingpost combination truss, modified by the addition of three steel tension rods, a diagonal tension rod on each side of a central vertical tension rod. Only three queenpost truss bridges remain in Georgia; the other two are Concord Bridge and Stovall Mill Bridge. The exterior of the Early County owned bridge has natural vertical boarding with battens and opposing rectangular window openings in each span. Flooring consists of transverse planking with six-plank-wide wheel treads. A wood-shingled roof replaced the tin roof during 1999 repairs. Alignment of portals is on a north-northwest/south-southeast orientation. Coheelee Creek Bridge has four bridge informational signs, two near the north-northwest or left downstream portal, one mounted on the left downstream gable, and one at State Road 62 and Damascus Hilton Road.

Westward flowing Coheelee Creek, on its way to the nearby Chattahoochee River, tumbles over a small waterfall just below the bridge. When the creek waters are high, the waterfall adds to the beauty of the bridge setting. Coheelee Creek is heavily wooded at the bridge site adjoining Fannie Askew Williams Park, a county maintained picnic park to the north. Ample parking is available at the park.

Watson Mill Bridge passes through the near side of the 1880 Watson Mill, mirrored here in Watson Pond.

The only authentic historic covered bridge remaining in Early County since the loss of Blakely Bridge in 1971, Coheelee Creek Bridge was listed on the National Register of Historic Places on May 13, 1976. *See Coheelee Creek Covered Bridge in the color photograph section: C-6.*

Directions: From Blakely, the county seat of Early County, take SR 62 southwest 9.0 miles to Martin Road, on the right (loops back to SR 62 at Damascus Hilton Road); go right 0.3 mile to Old River Road, a dirt road on the right; go 1.6 miles to the bridge, on the right.

Watson Mill Covered Bridge

EMANUEL COUNTY

Brief Statistics

Type: Non-authentic historic covered bridge. World Guide Covered Bridge Number: GA-53-M1. Built in 1880 by Alexander Hendricks and James M. Parrish. Multiple-span bridge crossing Fifteenmile Creek. Post and beam bridge 101.3 feet long totally within mill building, with approximately 9.1-foot wide by 7.3-foot high portal openings. Alternate Name: Parrish Mill Bridge. Located in George L. Smith II State Park on former Old Statesboro Road.

Watson Mill was planned early in 1879 by Alexander Hendricks, when he purchased 200 acres of land to build a mill. By late spring 1880, Alexander Hendricks had a partner, his nephew, James M. Parrish, and together they constructed the dam and mill, completing both by the end of 1880. Construction of the dam was considered an engineering feat of the day, as no heavy machinery was available, thus requiring the hiring of local African Americans and Irish immigrants to haul the enormous amount of dirt and stone with shovels and wheelbarrows. The mill was built atop the dam at a point of the greatest water depth, to provide the waterpower necessary to operate the mill. A road traversing the dam passed through the mill building, providing access to the United States Post Office, Bay Branch, located on the east side of the mill, thereby creating a unique covered bridge within a combination grist mill, saw mill and cotton gin building. The covered bridge has doors at both portals, allowing the mill to be secured after hours. The site was also known as Parrish Mill and Parrish Mill Bridge, named after James M.

Parrish, who owned and operated the mill until his death around 1927.

The grist mill was acquired in 1945 by Hubert Watson, hence was called Watson Mill and the bridge, Watson Mill Bridge. The mill was closed in 1973, and, one year later, the site was purchased by the State of Georgia and turned into George L. Smith II State Park. The covered bridge remained open to motor traffic until 1984.

Watson Mill Bridge passes through the upstream side of the 101.3-foot long mill building, with the portals aligning in a west/east orientation. The bridge and mill are supported by posts and the concrete sides of the spillway. The flooring, set on joists, is transverse planking, with two-plank-wide wheel treads transiting the bridge. The mill exterior has natural horizontal lapped siding on the sides and ends, with shiny tin on the roof. The portals have double swinging doors. There are four windows opening onto Watson Pond, on the upstream side of the bridge. Informational signs are mounted inside the mill and on the west or left downstream end of the mill.

Watson Mill Bridge is within Watson Mill, located in George L. Smith II State Park in Emanuel County, the county seat of which is Swainsboro. Owned by the Georgia Department of Natural Resources, this bridge is not to be confused with Watson Mill Bridge in Watson Mill Bridge State Park, in Madison and Oglethorpe Counties.

The only historic covered bridge in Emanuel County, Watson Mill Bridge is picturesquely situated on top of the spillway of the dam on Fifteenmile Creek that forms 412-acre Watson Pond. Fifteenmile Creek flows into the Canoochee River, their combined waters entering the Ogeechee River at Burroughs, just before reaching the Atlantic Ocean.

Directions: From Twin City, at the intersection of US 80/SR 26 and SR 23, go 4.0 miles south on SR 23 to George L. Smith II State Park Road (CR 29); go left 2.0 miles to the park. Admission.

Burnt Covered Bridge

Forsyth County

Brief Statistics

Type: Authentic historic covered bridge. World Guide Covered Bridge Number: GA-58-02. Built in 1895 by unknown builder. One-span bridge crossing wet weather stream. Town lattice truss bridge 34.9 feet long by 11.3 feet wide, with approximately 9.4-foot wide by 9.2-foot high portal openings. Alternate Name: Mashburn Estate Bridge. Located north of Cumming at 515 Pilgrim Mill Road.

The name "Burnt" originated from an earlier bridge, spanning Settendown Creek, which was burned in 1864 to discourage Sherman's Raiders (forces under Union General William Tecumseh Sherman that were spreading destruction in the south) from crossing the high banks of the creek. Built in 1895, at an original length of 75 feet, by an unknown builder, Burnt Bridge spanned Settendown Creek on Burnt Bridge Road, in Forsyth County, north of Poole's Mill Bridge, in Heardville. Following Burnt Bridge's collapse into the creek, Dr. Jim Mashburn, around 1970, removed the timbers and reconstructed a shortened 34.9-foot long bridge on his property on Pilgrim Mill Road.

The 11.3-foot wide Burnt Bridge is the narrowest historic covered bridge remaining in Georgia and, for that matter, in the entire Southeastern United States. Owing to its present location, the bridge is also known as the Mashburn Estate Bridge.

The sides and gables of the bridge have gray weathered vertical boarding; the upper 28 inches of the sides are open, exposing the Town lattice truss fastened together with treenails. Burnt Bridge is one of the twelve Town lattice truss bridges remaining in Georgia. The northeast/southwest aligned single-span bridge rests on solid cement blocks with no abutments. Transverse planking, without wheel treads, constitutes the floor. The low peaked roof is covered with light brown painted tin.

Burnt Bridge is currently located over a usually dry stream on the private property of the Mashburn Estate, in relatively open old-growth woods near some relocated old cabins, creating a turn-of-the-century scene. The bridge is closed to motor traffic. Forsyth County has only two authentic historic covered bridges, Burnt Bridge and Poole's Mill Bridge.

Directions: From Cumming, the county seat of Forsyth County, at the junction of SR 9 and SR 20, go 0.1 mile east on SR 9, then left on Pilgrim Mill Road. The bridge is on the private property of the Mashburn Estate, at 515 Pilgrim Mill Road. Private property: request permission to visit.

Poole's Mill Covered Bridge

Forsyth County

Brief Statistics

Type: Authentic historic covered bridge. World Guide Covered Bridge Number: GA-58-01. Built in 1906 by Dr. Marcus L. Poole. One-span bridge crossing Settendown Creek. Town lattice truss bridge 93.8 feet long by 14.5 feet wide, with approximately 11.9-foot wide by 11.7-foot high portal openings. Alternate Names: Pool's Mill Bridge, Settendown Creek Bridge. Located north of Heardville in Poole's Mill Bridge Park.

A Town lattice truss member of Poole's Mill Bridge displaying the empty misdrilled holes displaced to the right of the treenails.

Following the loss of a previous bridge in 1900 by a flash spring flood, the construction of Poole's Mill Bridge was contracted to a millwright named John Wofford. When he attempted to assemble the bridge, he realized he had drilled all the treenail holes in the wrong places and, dejected, abandoned the project. Dr. Marcus L. Poole undertook the completion of the construction by drilling new holes in the Town lattice truss members and chord members, finishing in 1906. Inside the bridge, one can see the empty misdrilled holes adjacent to the treenail-filled new holes in the truss members, as well as unusable chord and truss members in the wheel treads, those with side-by-side holes intended for the chord and those with holes offset diagonally intended for the truss. Poole's (or Pool's) Mill Bridge derived its name from the nearby mill, originally constructed circa 1820 and eventually owned by Dr. Marcus L. Poole. The 93.8-foot long bridge has also been called Settendown Creek Bridge, after the creek flowing beneath. In 1998, extensive repairs costing $283,403.00 were made under the ISTEA (Intermodal Surface Transportation Efficiency Act of 1991) Bridge Restoration Project in Georgia.

Poole's Mill Bridge rests on dry natural stone abutments, sporadic mortar being added in later repairs, and a central concrete pier, added during the 1998 repairs. The single-span bridge is one of the twelve covered bridges in Georgia with a Town lattice truss. There are chin braces inside each portal, the portals aligning on a north/south orientation. The flooring consists of transverse planking, with multi-plank-wide wheel treads made from unusable misdrilled timbers. The exterior has a new wood-shingled roof and new natural vertical boarding with battens on the sides, the upper two feet of the sides open, exposing the lattice truss. The portals are clad with old weathered vertical boarding that wraps around into the interior. The roof and gables overhang the entrances, providing shelter from the weather.

Poole's Mill Bridge, owned by Forsyth County, the county seat of which is Cumming, is in a relatively exposed setting in Poole's Mill Bridge Park, offering a paved parking area, a pavilion and restrooms. Now closed to motor traffic, the bridge originally carried Poole's Mill Road, prior to its rerouting, across Settendown Creek, which cascades over bedrock, just downstream from the bridge on its way to the nearby Etowah River.

One of two authentic historic covered bridges remaining in Forsyth County, Poole's Mill Bridge, was listed on the National Register of Historic Places on April 1, 1975.

Directions: From SR 20, in Ducktown, go north on Heardsville Road (turns into Heardsville Circle) to Poole's Mill Road, on the right. The bridge is in Poole's Mill Bridge Park, at the left corner of this junction.

Cromers Mill Covered Bridge

FRANKLIN COUNTY

Brief Statistics

Type: Authentic historic covered bridge. World Guide Covered Bridge Number: GA-59-01. Built in 1905 by James M. "Pink" Hunt. One-span bridge crossing Nails Creek. Town lattice truss bridge 111.2 feet long by 16.4 feet wide, with approximately 10.5-foot wide by 9.8-foot high portal openings. Alternate Name: Nails Creek Bridge. Located south-southeast of Sandy Cross on abandoned old SR 106.

Settling near Nails Creek in 1845, the Cromer family owned and operated a woolen mill nearby, prior to the Civil War. The area developed into the former community called Cromers Mill, the name carried on in Cromers Mill Bridge. The Franklin County owned bridge was built

Red factory assembly numbers appear on the truss members at the lower chord of Cromers Mill Bridge, such as this "20 X."

at its present site in 1905 by James M. "Pink" Hunt, who, with Washington W. King, also built Big Clouds Creek Bridge. Will Cromer, a descendant of the original settlers, built the cut stone abutments. The 111.2-foot long bridge has also been called Nails Creek Bridge, after the creek flowing beneath. State Road 106 originally transited the bridge; however, when State Road 106 was relocated nearby, the bridge was abandoned, falling into disrepair. Under the ISTEA (Intermodal Surface Transportation Efficiency Act of 1991) Bridge Restoration Project in Georgia, a contract was awarded in September 1998 for $147,815 to make extensive repairs to Cromers Mill Bridge, the work being completed one year later.

Cromers Mill Bridge rests on mortared cut-stone abutments, in a single span, over swift-flowing, rock-strewn Nails Creek. One of twelve Town lattice truss bridges in Georgia, Cromers Mill Bridge's truss, fastened together with treenails, has timber chin braces inside each portal, with four sets of galvanized steel chin braces that were added, equally spaced between the timber chin braces, during the 1999 restoration. A practice of assembling the truss and chords at the factory and pre-numbering the members assured a proper fit when reassembled at the bridge site. Most of the truss members of Cromers Mill Bridge have red numbers visible at the lower chord, each side numbered 8 through 77, with the numbers on the downstream side preceded with an X to distinguish them from the upstream side. What happened to members numbered 1 through 7? Was the bridge shortened? In 1955, the recorded length of the bridge was 129 feet, as compared to the present length of 111.2 feet.

The flooring consists of transverse planking with wheel treads. Wood approaches extend from each portal, the portals aligning in a southwest/northeast orientation. The exterior has mostly new natural vertical boarding with battens and gleaming corrugated tin on the low peaked roof. In 2000, a bridge informational sign was erected near the left downstream or southwest portal.

Cromers Mill Bridge crosses heavily wooded Nails Creek, the creek being exposed near the bridge due to the recent restoration work. The rocky creek flows southeast, into the nearby Hudson River, before entering the Broad River, southwest of Franklin Springs. The bridge is on an abandoned roadway, not open to traffic, with picnic benches near the southwest portal.

Cromers Mill Bridge was listed on the National Register of Historic Places on August 17, 1976, the only authentic historic covered bridge remaining in Franklin County since the loss of Kesler Mill Bridge, in 1979.

Directions: From Carnesville, the county seat of Franklin County, take SR 106 south 8.0 miles to Cromer's Mill Bridge Road, on the left. The bridge is immediately on the right, on an abandoned road.

Ye Olde Gap Covered Bridge

Fulton County

Brief Statistics

Type: Non-authentic modern covered footbridge. World Guide Covered Bridge Number: GA-60-b. Built in 1977 by Fulton County. Three-span footbridge crossing Tanyard Creek. Stringer footbridge 77.2 feet long by 6.5 feet wide, with approximately 3.4-foot wide by 6.2-foot high portal openings. Alternate Name: None known. Located in Atlanta at Yorke Downs Apartment Homes.

Located at Yorke Downs Apartment Homes in Atlanta, Ye Olde Gap Bridge is the enclosed sidewalk on the upstream side of a three-span stringer bridge, carrying motor traffic across cement-lined Tanyard Creek, which flows into nearby Peachtree Creek before entering Nancy Creek and the Chattahoochee River, about 3.5 miles west of the bridge. The 77.2-foot long covered footbridge was constructed, in 1977, upon the existing stringer bridge, built in the 1950s by Fulton County. Aligned east-southeast/west-northwest, the asphalt-paved stringer bridge rests on poured concrete abutments and two sets of four, square, poured concrete pillars. The cement-floored covered sidewalk has a beige-painted stuccoed exterior, with brown trim and reddish-brown asphalt shingles on the roof, which extends about 4.75 feet beyond the portals, providing shelter for the entrances. There are six opposing square window openings on each side, two electric lanterns mounted on each portal, and scrolled veneer fascia boards on the roof overhang. The stuccoed interior has crossed boards forming an X between the wood framed window openings and four fluorescent lights mounted on the ceiling.

Ye Olde Gap Bridge is fully exposed, with access drives running parallel to the creek and the main entry roads crossing the Fulton County bridge, with apartment buildings along each side before and after the bridge. *See Ye Olde Gap Covered Bridge in the color photograph section: C-8.*

Directions: From I-75, north of I-85, take US 41 north to Collier Road; go right to US 19; go right to 26th Street, the second right, and follow to the end, at Yorke Downs Apartment Homes. The bridge is straight ahead.

Blackberry Mountain Covered Bridge

GILMER COUNTY

Brief Statistics

Type: Non-authentic modern covered bridge. World Guide Covered Bridge Number: GA-61-B. Built in 1983 by Shadowood Construction. Two-span bridge crossing Cartecay River. Stringer bridge 96.9 feet long by 19.9 feet wide, with approximately 11.1-foot wide by 10.9-foot high portal openings. Alternate Name: None known. Located south of Cartecay on Blackberry Mountain Drive.

Blackberry Mountain Bridge, a private bridge built in 1983 by Shadowood Construction of Chattanooga, Tennessee, serves the communities of Blackberry Mountain and Rivers Edge. The 96.9-foot long bridge carries Blackberry Mountain Drive across the Cartecay River and is owned by Blackberry Mountain Management, Inc. Ownership is scheduled to be turned over, jointly, to the Blackberry Mountain Association and the Rivers Edge Association.

Located in Gilmer County, Blackberry Mountain Bridge consists of two spans, each span comprised of three longitudinal steel I-beams, joining the concrete abutments with the central concrete pier. The flooring is constructed from planks placed transversely, overlaid with three-plank-wide wheel treads. Internal electric lights were added at a later date.

The sides are open, each side having an external, uncovered, four-foot wide walkway, with a wood handrail and seven buttress braces placed at alternating post locations between the roof and the outside of the walkway flooring. Aligned on a north-northwest/south-southeast orientation, the portals have vertical boarding. The left downstream portal has a sign over the entrance, which was obtained from an old covered bridge, that reads *WALK YOUR HORSES OR PAY TWO DOLLAR FINE.* Above this sign is another sign stating *BLACKBERRY/ MOUNTAIN.* A green-painted tin roof completes the olive-green painted bridge.

Blackberry Mountain Bridge has a very picturesque setting, spanning a clear, wide, rock-bottomed mountain river, banked by numerous trees, its tumbling waters joining the Ellijay River at Ellijay, whose confluence gives birth to the Coosawattee River, which flows into Carters Lake.

Directions: From East Ellijay, take SR 52 east 3.2 miles from the SR 5 overpass to Mulkey Road, on the right; go to the end of the road. The bridge is just ahead, beyond the control gate. Request permission to visit at the sales office, just to the left of the control gate.

Rainbow Farm Covered Bridge

GILMER COUNTY

Brief Statistics

Type: Non-authentic modern covered bridge. World Guide Covered Bridge Number: GA-61-C. Built in 1988 by David Cantrell. One-span bridge crossing inlet to Rainbow Lake. Stringer bridge 17.6 feet long by 14.7 feet wide, with approximately 11.9-foot wide by 8.6-foot high portal openings. Alternate Name: None known. Located southeast of Cartecay off SR 52 at Rainbow Farm.

David Cantrell built Rainbow Farm Bridge, in 1988, over the inlet to man-made Rainbow Lake, a trout-fishing pond. Wood plank stringers set on the ground support the privately owned stringer bridge. The flooring consists of transverse planking with three-plank-wide wheel treads. The sides are open, with double rails along each side for the length of the 17.6-foot long bridge, the upper rail painted red. Four 2-inch by 6-inch wood posts along each side support the shiny, tin covered roof.

The Gilmer County single-span bridge, aligned north-northwest/south-southeast, is well exposed, with the trout pond on the downstream side and SR 52 on the upstream side. Two trees protrude through the tin roof overhang on the upstream side, an unusual feature.

Directions: From the Dawson/Gilmer County line, go 0.9 mile north on SR 52. The bridge is on the left, at Rainbow Farm. Private property: request permission to visit.

Harbor Club Covered Bridge

GREENE COUNTY

Brief Statistics

Type: Non-authentic modern covered bridge. World Guide Covered Bridge Number: GA-66-A. Built in 1997 by unknown builder. One-span bridge crossing wet weather wash. Stringer bridge 32.3 feet long by 36.6 feet wide, with approximately 26.1-foot wide by 15.2-foot high

The unique Harbor Club Bridge features a gable window, a cupola with a weathervane atop, a gambrel roof, external walkways and bright colors.

portal openings. Alternate Name: None known. Located south of Greensboro in Harbor Club Subdivision.

The vibrantly colored Harbor Club Bridge is the landmark of the Harbor Club Subdivision, clearly visible from State Road 44 from its vantage, high up the entrance road. The two-lane bridge is unusual in that it is wider than it is long, owing to the covered walkways at both sides; however, Haralson Mill Bridge exceeds the width of this bridge by 6 inches. A rounded-top multi-paned window in the gable, a gambrel roof, and a horse weathervane-topped cupola are the distinctive features of this stringer bridge, built in 1997.

The 32.3-foot long single-span bridge, crossing a wet weather drainage wash, has open sides with external covered walkways, also open-sided. The light yellow painted structure, with deep red trim and a green roof, held up by seven posts along each side, is over a concrete slab supported by steel I-beam stringers, set on poured concrete abutments. Horizontal lapped siding and asphalt shingles on the wraparound gambrel roof, which is surmounted by an octagonal, windowed cupola, complete the structure. Below the attic of the bridge is a ceiling, with four recessed fluorescent lights.

Harbor Club Bridge has office buildings on the west end of the upstream side and pine woods on the east end, separated by the grassed-in wash. There is an open horse pasture on the downstream or north side. The Greene County bridge accommodates oversized vehicles.

Directions: From Greensboro, take SR 44 south 3.3 miles past I-20 to Harbor Club Subdivision, on the left. The bridge is visible up the entrance road.

Hub's Creek Covered Bridge

Hall County

Brief Statistics

Type: Non-authentic modern covered bridge. World Guide Covered Bridge Number: GA-69-A. Built in 1977 by unknown builder. Post supported roof over roadway crossing Hub's Creek. Bridge 60.0 feet long by 28.0 feet wide, with approximately 24.5-foot wide by 17.3-foot high

Built in 1870, Callaway Gardens Bridge sits in storage awaiting its return to public display. Photograph courtesy Callaway Gardens.

portal openings. Alternate Name: None known. Located north of Gainesville on Woodlane Road.

Named after the creek flowing below, Hub's Creek Bridge is a two-lane open sided bridge that has the sides slanting outward, up to the roof, giving the portals a width of 28.0 feet at the ground and 33.5 feet at the eaves. Built in 1977, the Hall County bridge was severely damaged by a 1999 flood that dislodged the posts supporting the roof. Shored up by cables and transporting timbers underneath, the bridge cover was temporarily relocated, by two tow-motors, up the road a sufficient distance to allow the rebuilding of the culvert. Upon completion, the bridge was returned, the decayed lower portion of the posts removed, and set on concrete barriers, raising the clearance height of the cover the prescribed distance. A new cedar shingle roof was installed at this time.

Hub's Creek Bridge is actually a post supported roof over a roadway that crosses Hub's Creek. Eleven posts support the 60.0-foot long bridge above the asphalt roadway, which goes over a concrete culvert with concrete wingwalls, through which flows the narrow creek. The sides are open, but dark natural vertical boarding covers the portals that align west/east. A wood sign with *HUB'S CREEK* carved into it is mounted above the right downstream entrance.

The bridge site is wooded on the south or downstream side and unobstructed on the north side, homes being at a distance. The waters of Hub's Creek flow into the nearby Chattahoochee River and Lake Sidney Lanier.

Directions: From Gainesville, take SR 60 (Thompson Bridge Road) north across Lake Sidney Lanier and past the junction with SR 283; go left on Southers Road 0.2 mile after the junction; go 0.4 mile to Woodlane Road, on the right (there is a small bridge replica at this junction). The bridge is 0.3 mile down Woodlane Road.

Callaway Gardens Covered Bridge

HARRIS COUNTY

Brief Statistics

Type: Authentic historic covered bridge. World Guide Covered Bridge Number: GA-72-01 (formerly GA-141-02). Built in 1870 by Horace King and sons. One-span bridge in storage. Town lattice truss bridge 57.7 feet long by 16.7 feet wide, with approximately 13.6-foot wide by

10.5-foot high portal openings. Alternate Names: Cofield Bridge, Harmony Road Bridge, Howell's Bridge, Neeley Bridge, Wehadkee Creek Bridge. Located in Pine Mountain at Callaway Gardens.

Originally known as Wehadkee Creek Bridge, this bridge was built in 1870, by Horace King and his sons, over Wehadkee Creek, west of Abbottsford in Troup County, near the Alabama border. As construction of the West Point Dam would have inundated the bridge with the waters of West Point Lake, Callaway Gardens acquired the 173-foot long bridge in 1965, had it moved onto its property at Pine Mountain, and shortened to 57.7 feet, to accommodate its new site. The bridge was open to the public until 1984. Callaway Gardens Bridge is currently in storage, fully erected on dry land in an open field, pending the raising of funds to relocate and return the bridge to public display.

The single-span bridge, while at its original location, was also called Neeley Bridge, Cofield Bridge, Howell's Bridge and Harmony Road Bridge (probably the name of the road crossing the bridge).

Callaway Gardens Bridge has a Town lattice truss, fastened together with treenails, with chin braces inside each portal, one of twelve Town lattice truss bridges remaining in Georgia. The bridge is enclosed with gray weathered, vertical boarding with battens, on the sides and gables, and shiny tin on the roof.

Directions: The bridge is in storage at Callaway Gardens in Pine Mountain, restricted from public view.

Author's Note: Measurements for this bridge were provided by Callaway Gardens.

Old Oglesby Covered Bridge

Henry County/Rockdale County

Brief Statistics

Type: Non-authentic modern covered bridge. World Guide Covered Bridge Number: GA-75-A/GA-122-B. Built in 1997 by Richard Turner. Four-span bridge crossing Cottonwood Creek. Stringer bridge 91.2 feet long by 12.6 feet wide, with approximately 12.6-foot wide by 11.2-foot high portal openings. Alternate Name: None known. Located northeast of Kelleytown off Oglesby Bridge Road.

Originally an uncovered, plank-floored four-span stringer bridge, exceeding 50 years in age, Old Oglesby Bridge carried Oglesby Bridge Road across Cottonwood Creek, the county line between Henry County and Rockdale County. When Oglesby Bridge Road was rerouted, the bridge site become private property. The present four-span covered bridge was built in 1997 by Richard Turner, a local contractor, upon the original six steel I-beam stringers, resting on the original poured concrete abutments and three poured concrete piers. This construction required the replacement of the old rotted flooring and the adding of the cover.

The sturdy cover, supported by posts, has natural vertical boarding with battens, on the exterior sides and portals, and the same natural vertical boarding without battens on the entire interior. The sides have eight 9.5-foot wide by 3.2-foot high opposing window openings on each side. The steeply peaked roof is covered with shiny tin.

Aligned west-northwest/east-southeast, Old Oglesby Bridge spans Cottonwood Creek, with a dam-created waterfall adjacent to the bridge on the upstream side and a ledge-rock creek bed beneath and on the downstream side, the site enhanced by the falling waters that then swiftly cascade downstream into the nearby South River. There are tall, mature trees on both sides of the 91.2-foot long bridge, the old original roadway being used as a driveway onto the property.

Directions: From Klondike, take SR 212 south 4.5 miles past the SR 138 intersection to Oglesby Bridge Road, on the right; go 2.7 miles to the terminus at Union Church Road. The bridge is on private property, on the left. Request permission to visit.

Forest Hill Covered Bridge

Houston County

Brief Statistics

Type: Non-authentic modern covered footbridge. World Guide Covered Bridge Number: GA-76-a. Built circa 1974 by Elton Hardy. One-span footbridge crossing tributary to Big Indian Creek. Stringer footbridge 25.0 feet long by 10.4 feet wide, with approximately 8.8-foot wide by 7.1-foot high portal openings. Alternate Name: None known. Located in Perry in Hardy Park.

Located in Perry, at Hardy Park, Forest Hill Bridge, named after Forest Hill Drive adjacent to the park, was built around 1974 by Elton Hardy, owner of the park. The single-span footbridge, although large enough to accommodate motor vehicles, was apparently intended solely for foot traffic, as no motor traffic throughway exists to or from the bridge and the 1-inch by 6-inch floor boarding lacks the strength to support vehicles. Aligned north-northwest/south-southeast, the 25.0-foot long stringer bridge has three massive 12-inch by 12-inch wood timber stringers, set on bed timbers originally resting on the ground, but now shored up with cement blocks and stones in an effort to combat ground erosion, which already has caused the north-northwest end to sink at one corner.

The massive wood timber stringers were obtained from the old Baptist Church, torn down just prior to the construction of the bridge. The flooring is comprised of transverse boards; the sides and portals are covered by green-painted 2-inch by 12-inch vertical planking with battens, the gable planking scalloped over the entrances; and the hip style roof is covered by green-painted asphalt shingles. There is a rectangular window opening on each side of the Houston County bridge.

Forest Hill Bridge is in a small park, with US 341 on the left downstream end and scattered mature trees in the park on the right downstream end. Also in the park, giving the appearance of a small chapel, is the steeple from the old Baptist Church, which supplied the timbers for the stringers. The footbridge spans a tributary to nearby Big Indian Creek, which flows eastward, forming the Houston County/Pulaski County boundary just before entering the Ocmulgee River.

Directions: From Perry, at the intersection of US 41 and US 341 (Main Street), go 0.8 mile south on US 341 to Forest Hill Drive, on the right. The bridge is on the right, in privately owned Hardy Park.

Phillips Family Covered Bridge

LINCOLN COUNTY

Brief Statistics

Type: Non-authentic modern covered bridge. World Guide Covered Bridge Number: GA-90-A. Built in 1994 by Jerry Phillips. One-span bridge crossing wet weather stream. Stringer bridge 41.3 feet long by 12.2 feet wide, with approximately 10.0-foot wide by 8.0-foot high portal openings. Alternate Name: None known. Located west of Lincolnton on Covered Bridge Road.

Jerry Phillips built the Phillips Family Bridge in 1994 and, to make the bridge unique, the Phillips family decorates the bridge every Christmas with lights and animated displays in each of the ten windows and two central openings, which have been boxed in for the occasion. (The Phillips family extends invitations to all to visit their bridge at Christmastime and to share hot chocolate and good conversation.) The single-span stringer bridge crosses a small wet weather stream, which flows into a man-made pond, just below the bridge.

Phillips Family Bridge, aligned south-southwest/north-northeast, is supported by three utility-pole stringers, resting on the stream banks and the ground of a little peninsula, near the center of the stream. The flooring is transverse 2-inch by 6-inch wood planking placed directly on the stringers, which are located on the interior side of the walls; consequently, the flooring is buckled due to the lack of support under the walls. The sides and portals are enclosed with natural horizontal lapped siding and the roof is covered with brown asphalt shingles. The roof is supported by 2-inch by 4-inch studding along each side. Nine 2-inch by 4-inch wood buttress braces along each side stabilize the bridge. There are five opposing vertical rectangular window openings on each side, with opposing 4.2-foot by 6.8-foot central openings. An electric lamp is mounted on the right downstream or north-northwest gable.

The Lincoln County covered bridge is in front of the residence of the Phillips family, well exposed, with several trees along both sides of the stream. A man-made pond is on the downstream side. Due to the length of Covered Bridge Road, the private driveway of the Phillips family, the 41.3-foot long bridge is not visible from the main road.

Directions: From Lincolnton, at the junction of SR 47 and US 378 (Washington Street), go toward downtown Lincolnton on US 378 to Metasville Road, on the left; go 1.6 miles to Aycock-Norman Road, on the right; go 0.3 mile to Covered Bridge Road (private driveway), on the right; follow to the bridge. Private property: request permission to visit.

Wimpy Covered Bridge

LUMPKIN COUNTY

Brief Statistics

Type: Non-authentic modern covered bridge. World Guide Covered Bridge Number: GA-93-A. Built in 1990 by Al Wimpy. One-span bridge crossing Cane Creek. Stringer bridge 37.4 feet long by 16.8 feet wide, with approximately 12.7-foot wide by 13.5-foot high portal openings. Alternate Name: Cross Creek Farm Bridge. Located north of Dahlonega on Covered Bridge Lane at Cross Creek Farm.

A stringer bridge featuring a decorative half-height queenpost truss, Wimpy Bridge has an interesting interior, festooned with antique tools, farm implements and horse tack. Three steel I-beam stringers, resting on poured concrete abutments, support the west/east aligned single-span bridge. Built in 1990 by Al Wimpy, the 37.4-foot long bridge has a slightly diagonal planked floor and weathered vertical boarding with battens, on the exterior sides and portals. Gray weathered cedar shingles cover the steeply peaked roof. The roof and gables extend beyond the flooring, forming a weather shelter for the entrances. Wimpy Bridge, named after its owner, Al Wimpy, is also known as Cross Creek Farm Bridge, named after its location in Cross Creek Farm.

Wimpy Bridge spans wooded Cane Creek, which

Wimpy Bridge is a stringer bridge sporting a decorative half-height queenpost truss and has antique farm implements adorning the walls.

flows southerly into the Chestatee River, south of Dahlonega, a feeder to Lake Sidney Lanier. The Lumpkin County bridge is located in an open area, with a small pond nearby.

Directions: From the circle in Dahlonega Center, go 1.6 miles north on US Business 19/SR Business 60 to Oak Grove Road, on the left; go 0.6 mile to a road (Cross Creek Trail) and mailbox, on the left, marked Cross Creek Farm; go 0.2 mile to Covered Bridge Lane, on the left; follow to the bridge. Private property: request permission to visit.

Watson Mill Covered Bridge

Madison County/Oglethorpe County

Brief Statistics

Type: Authentic historic covered bridge. World Guide Covered Bridge Number: GA-97-01/GA-109-02. Built in 1885 by Washington W. King. Three-span bridge crossing South Fork of the Broad River. Town lattice truss bridge 229.0 feet long by 17.9 feet wide, with approximately 16.0-foot wide by 11.0-foot high portal openings. Alternate Names: Broad River Bridge, Carlton Bridge. Located southeast of Comer in Watson Mill Bridge State Park on Covered Bridge Road.

Constructed in 1885 by Washington W. King, at a cost just over $3,000, Watson Mill Bridge is one of four covered bridges surviving in Georgia that Mr. King built. Washington W. King was the son of Horace King, a famous covered bridge builder in the south. Watson Mill Bridge derived its name from the grist mill, several hundred feet downstream from the bridge, which was owned and operated by Gabriel Watson. The bridge was also known as Carlton Bridge, after the nearby town, and Broad River Bridge, after the river flowing beneath. The present dam and raceway wall below the bridge were constructed around 1905, as part of the old hydro-electric plant, which supplied the electric power to Jefferson Mills, a textile mill in Crawford, about ten miles away.

Watson Mill Bridge has the distinction of being the longest covered bridge remaining in Georgia, the bridge measuring 229.0 feet. Its three spans join Madison County with Oglethorpe County, across the South Fork of the Broad

Built in 1885, the 229.0-foot long Watson Mill Bridge spans the South Fork of the Broad River just above the dam for the old hydroelectric plant. *See Watson Mill Covered Bridge in the color photograph section: C-8.*

River. This river forms the Madison County/Oglethorpe County southern boundary, flowing eastward to the Broad River, which forms the Madison County/Elbert County eastern boundary, thence continuing eastward to the Savannah River, which is the Georgia/South Carolina border.

Watson Mill Bridge was one of ten covered bridges in Georgia that had restoration work performed under the ISTEA (Intermodal Surface Transportation Efficiency Act of 1991) Bridge Restoration Project in Georgia. A contract was awarded in April 1997 for $165,880, with repairs completed later that year.

Owned by the Georgia Department of Natural Resources, Watson Mill Bridge rests on mortared natural stone abutments, later repairs made with cut stones, and four piers, the south-southwest center pier of concrete construction, the other three piers made from natural stone mortared together. The bridge is supported by a Town lattice truss (the most frequently used truss in Georgia being common to eleven other bridges), fastened together with treenails, with chin braces inside each portal. The flooring is made of transverse planking with four-plank-wide wheel treads. The exterior has natural vertical boarding with battens, on the sides and portals; the upstream side has a wide, weathered wood-shingle-roofed central window opening and a narrow, rusted tin-roofed window opening to the left; the downstream side has only the wide, weathered wood-shingle-roofed central window opening. A gray weathered, wood-shingled roof completes the exterior. *WATSON MILL BRIDGE/EST. 1885* wood signs, white letters on brown background, are mounted above each entrance. A bridge informational sign is near the left downstream or south-southwest portal.

Watson Mill Bridge, one of five historic covered bridges in Georgia still open to motor traffic, spans a wide river, which tumbles over the old dam and cascades down the ledge rock riverbed, creating a photogenic scene. Aligned on a south-southwest/north-northeast orientation, the bridge is fully exposed from all directions.

Watson Mill Bridge was listed on the National Register of Historic Places on September 5, 1991. The bridge is the last authentic historic covered bridge remaining in Madison County, but is one of two remaining in Oglethorpe County, the other being Big Clouds Creek Bridge. *See Watson Mill Covered Bridge in the color photograph section: C-8.*

Directions: From Comer, proceed south on SR 22 for 3.0 miles after the junction with SR 72, to Watson Mill

Bridge Road, on the left (no sign—the road is across from Collier Church Road); go 3.2 miles to Watson Mill Bridge State Park. Admission.

Red Oak Creek Covered Bridge

Meriwether County

Brief Statistics

Type: Authentic historic covered bridge. World Guide Covered Bridge Number: GA-99-02. Built in 1840s by Horace King. One-span bridge crossing Red Oak Creek. Town lattice truss bridge 127.5 feet long by 17.0 feet wide, with approximately 14.0-foot wide by 11.1-foot high portal openings. Alternate Names: Big Red Oak Creek Bridge, Gay Bridge, Imlac Bridge. Located northeast of Woodbury on Huel Brown Road/Covered Bridge Road.

Named after the creek flowing below, Red Oak Creek Bridge was built in the 1840s, making it the oldest covered bridge in Georgia. The Meriwether County-owned bridge was built by Horace King, a former slave, legally freed, who continued to work for his former owner, John Godwin, building about 50 bridges in Georgia and Alabama. Under the ISTEA (Intermodal Surface Transportation Efficiency Act of 1991) Bridge Restoration Project in Georgia, a contract for $176,253.00 was awarded in November 1998 for extensive repairs to Red Oak Creek Bridge, the work being completed in March 1999. The bridge is also known as Gay Bridge and Imlac Bridge, after nearby communities.

Red Oak Creek Bridge is set on poured concrete piers to form a single span over Red Oak Creek, a major tributary to the nearby Flint River. The east approach ramp is extremely long, measuring 252.7 feet! The covered bridge length of 127.5 feet, together with an 11.5-foot west approach ramp, gives the bridge and ramps a total length of 391.7 feet. Red Oak Creek Bridge is one of twelve covered bridges in Georgia with a Town lattice truss, this one fastened together with treenails. There are chin braces inside each portal, the portals aligning on a west/east orientation. The flooring consists of transverse planking within the bridge and extending the entire length of the ramps, with five-plank-wide wheel treads within the covered portion of the bridge. The exterior has a new wood-shingled roof, replacing the old tin roof in 1999, and mostly new natural vertical boarding with battens, on the sides and portals. The roof and gables extend beyond the entrances, providing shelter for the portals from the weather.

Red Oak Creek Bridge is one of only three remaining historic covered bridges in Georgia still open to public motor traffic; the other two are Concord Bridge and Elder's Mill Bridge. The bridge is located in an expansive wooded area, which, at high water, becomes inundated near the bridge. During the flood of July 7, 1994, the water rose several feet inside the bridge, as indicated by the high water mark metal sign, nailed to a truss member inside the west end, on the downstream side. At the time, the bridge must have appeared to be floating in the middle of a lake, and it is related that someone actually paddled a canoe through the bridge! The area is cleared on the downstream side of the bridge, permitting an unobstructed view and allowing access to normally placid-flowing Red Oak Creek. There is no off road parking at the bridge. The county seat of Meriwether County is Greenville.

On May 7, 1973, Red Oak Creek Bridge became the first covered bridge in Georgia to be placed on the National Register of Historic Places. Red Oak Creek Bridge is the last authentic historic covered bridge remaining in Meriwether County, since the loss of White Oak Creek Bridge in 1985.

Directions: From Woodbury, at the junction of SR 74 and SR 85, go 2.8 miles north to Covered Bridge Road, on the right; go 1.5 miles to the bridge (on a dirt road).

Elder's Mill Covered Bridge

Oconee County

Brief Statistics

Type: Authentic historic covered bridge. World Guide Covered Bridge Number: GA-108-01. Built in 1897 by Nathaniel Richardson. One-span bridge crossing Rose Creek. Town lattice truss bridge 99.0 feet long by 13.1 feet wide, with approximately 11.3-foot wide by 13.1-foot high portal openings. Alternate Names: Elder's Bridge, Rose Creek Bridge. Located south-southeast of Watkinsville on Elder's Mill Road.

In 1897, Nathaniel Richardson built a covered bridge over Cass Creek on the Watkinsville-Athens Road, just north of Watkinsville. In 1924, this bridge was dismantled and taken by wagon to be reconstructed, by John Chandler, at its present site on Elder's Mill Road, whence it became known as Elder's Mill Bridge. This is the only extant bridge built by Nathaniel Richardson. This Oconee County-owned bridge is one of three remaining historic covered bridges in Georgia still open to public motor traffic; the other bridges are Concord Bridge in Cobb County and Red Oak Creek Bridge in Meriwether County. Stone Mountain Park Bridge in DeKalb County and Watson Mill Bridge in Madison/Oglethorpe Counties are open only to park visitor motor traffic. Elder's Mill Bridge has been called Elder's Bridge and also Rose Creek Bridge, after the creek cascading beneath. Elder's Mill

Bridge was one of ten covered bridges in Georgia that had restoration work performed under the ISTEA (Intermodal Surface Transportation Efficiency Act of 1991) Bridge Restoration Project in Georgia. These repairs, costing $162,030.00, were completed in 1999.

The single-span Elder's Mill Bridge rests on two concrete piers located at each end, just inside the abutments, upon which rests the approach ramps. The Town lattice truss has the truss members fastened together with treenails and has chin braces, protected by vehicle barriers, inside each portal. Truss members still bear numbers assigned in the 1924 relocation, which assisted in reconstruction of the bridge. The flooring consists of transverse planking, with 3- or 4-plank-wide wheel treads, that extends to the end of the approach ramps. The exterior of Elder's Mill Bridge has natural vertical boarding with battens, on the sides and portals, and shiny tin covers the low peaked roof. Mounted on the left side of the right downstream portal is a bridge informational sign. The bridge portals align on a southwest/northeast orientation.

Rose Creek has a boulder-strewn streambed on each side of the bridge and heavily wooded banks. Rose Creek flows into the Oconee River, above Lake Oconee. A local garden club maintains a flower and shrubbery planted area at the northeast end of the bridge, an attractant for butterflies and local fauna.

The only authentic historic covered bridge remaining in Oconee County, Elder's Mill Bridge was listed on the National Register of Historic Places on May 5, 1994. *See Elder's Mill Covered Bridge in the color photograph section: C-7.*

Directions: From Watkinsville, the county seat of Oconee County, at the junction of US 129/US 441 and SR 15, go 5.7 miles south on SR 15 to Elder's Mill Road, on the right; go 0.7 mile to the bridge.

Northwest Woods Covered Bridge

OCONEE COUNTY

Brief Statistics

Type: Non-authentic modern covered bridge. World Guide Covered Bridge Number: GA-108-A. Built in 1975 by unknown builder. Post supported roof over roadway crossing small creek. Bridge 50.0 feet long by 25.5 feet wide, with approximately 23.2-foot wide by 13.9-foot high portal openings. Alternate Name: None known. Located west of Watkinsville on Robin Hood Road.

Built by an unknown builder in 1975, Northwest Woods Bridge is the oldest non-authentic covered bridge in Georgia. Northwest Woods Bridge is a post supported roof over Robin Hood Road, an asphalt roadway over three galvanized conduits, which allow the passage of a small creek. The two-lane Oconee County bridge features five vertical rectangular window openings on each side. The 50.0-foot long bridge cover is supported by six posts along each side; the sides and portals are enclosed by gray weathered, vertical boarding with battens; and the roof is covered with dark weathered cedar shingles. In 1994, the footings for the sides were replaced.

The east/west-aligned bridge is in Northwest Woods Subdivision, densely wooded on the downstream or south side and lightly wooded on the north side. A small unnamed creek flows northerly beneath the roadway into nearby Barber Creek, which enters McNutt Creek near Athens, before joining the Middle Oconee River.

Directions: From Watkinsville, at the junction of US 129/US 441 and SR 53, go west on SR 53 to Robin Hood Road, on the right (at Briarwood Church); go 1.3 miles to the bridge.

Big Clouds Creek Covered Bridge

OGLETHORPE COUNTY

Brief Statistics

Type: Authentic historic covered bridge. World Guide Covered Bridge Number: GA-109-01. Built in 1905 by Washington W. King and James M. "Pink" Hunt. One-span bridge crossing Big Clouds Creek. Town lattice truss bridge 161.7 feet long by 16.3 feet wide, with approximately 13.1-foot long by 11.3-foot high portal openings. Alternate Names: Howard's (or Howard) Bridge, Smithonia Bridge. Located southeast of Smithonia on abandoned roadway.

At a length of 161.7 feet, this Oglethorpe County-owned bridge is the longest single-span covered bridge remaining in Georgia and the longest authentic historic single-span covered bridge in the southeastern United States. Named after the creek that flows beneath, Big Clouds Creek Bridge was built in 1905, at its present site, by Washington W. King and James M. "Pink" Hunt. Washington W. King, son of Horace King, a master covered bridge builder of his time, also built the following covered bridges in Georgia: Euharlee Bridge, Stone Mountain Park Bridge and Watson Mill Bridge (Madison/Oglethorpe Counties). James M. "Pink" Hunt, in addition to this bridge, built Cromers Mill Bridge. Hand carved into the truss member, inside the left downstream portal, on the upstream side near the chin brace, are the initials A.D. and 1905, attesting to the age of the bridge. Also known as Howard's or Howard Bridge, after the Howard family, who settled near the creek in the late 18th century, and Smithonia Bridge, after the nearby community, Big Clouds Creek Bridge was one of ten covered bridges in Georgia that had restoration work performed under the

The initials "A.D." and "1905" are hand carved into a truss member attesting to the age of Big Clouds Creek Bridge.

ISTEA (Intermodal Surface Transportation Efficiency Act of 1991) Bridge Restoration Project in Georgia. These repairs were completed, at a cost of $125,840.00, in April 1998.

Big Clouds Creek Bridge has dry natural-stone abutments with two mortared cut-stone piers. Two timber supports, located between the piers, were added at a later date. These supports are built on a small island, temporarily dividing the northeast flowing creek that joins the South Fork of the Broad River, just east of Watson Mill Bridge State Park. The Town lattice truss, common to eleven other Georgia covered bridges, is fastened together with treenails and has chin braces inside each portal. A practice of assembling the truss and chords at the factory and pre-numbering the members assured a proper fit when reassembled at the bridge site. Most of the truss members of Big Clouds Creek Bridge have red numbers visible at the lower chord, each side numbered 1 through 92, with the numbers on the downstream side followed by an X to distinguish them from the upstream side. The southeast/northwest-aligned portals have plank approach ramps of 17.1 feet and 19.8 feet, respectively. Triple-plank-wide wheel treads atop transverse planking forms the flooring, which extends to the ends of the approach ramps. The sides and portals have natural vertical boarding with battens, many boards replaced in 1998. The shiny corrugated tin roof replaced the old metal roof, during the 1998 restoration.

Big Clouds Creek Bridge is located in an isolated woodland area of Oglethorpe County, on an abandoned roadway crossing heavily wooded Big Clouds Creek. The bridge, bypassed in 1986, is closed to motor traffic. An informational sign was erected at the northwest end in 2000.

Big Clouds Creek Bridge was listed on the National Register of Historic Places on July 1, 1975. The bridge is one of the two authentic historic covered bridges remaining in Oglethorpe County, the other being Watson Mill Bridge.

Directions: From Smithonia, go south on Smithonia Road (CR 2164) to Chandler Silver Road, on the left, then 0.1 mile to the bridge, on the right. From Lexington, the county seat of Oglethorpe County, at the junction of US 78 west and SR 22 north, go north 7.4 miles on SR 22 to Smithonia Road (dirt), on the left; go 2.3 miles to Chandler Silver Road, on the right, then go 0.1 mile to the bridge, on the right.

Big Canoe Covered Bridge

Pickens County

Brief Statistics

Type: Non-authentic modern covered bridge. World Guide Covered Bridge Number: GA-112-A. Built in 1976 by McDavit & Street Construction. One-span bridge crossing Blackwell Creek. Stringer bridge 32.9 feet long by 25.0 feet wide, with approximately 21.7-foot wide by 12.6-foot high portal openings. Alternate Name: None known. Located north of Holcomb off Steve Tate Highway in Big Canoe Development.

McDavit & Street Construction built Big Canoe Bridge, in 1976, in Big Canoe Development. The single-span bridge crosses the waters of Blackwell Creek, which join Long Swamp Creek near Tate, before entering the Etowah River, southeast of Ball Ground.

The 32.9-foot long stringer bridge is supported by five steel I-beam stringers, which are anchored on abutments constructed from seven upright 12-inch by 12-inch timbers, set in individual concrete bases and capped by a 12-inch by 12-inch cross timber. These upright timbers restrain 9-inch by 9-inch timbers, placed horizontally to form a retaining wall, which extends around the sides, forming wing walls. The flooring consists of reinforced concrete slabs, placed on the stringers and covered with asphalt. Vertical boarding with battens covers the sides, to rail height, and above a central opening running the length of the bridge; the opening is filled in with a widely spaced 1-inch by 6-inch board lattice; and the vertical boarding with battens encloses the underside of the eaves. The gables are also covered with vertical boarding with battens. Reddish-brown wood shingles cover the low peaked roof, which is supported by three 11-inch by 11-inch timber posts along each side. The two-lane bridge is painted brown on the outside and also on the entire interior.

A wide, rock- and boulder-strewn, sandy-bottomed

creek runs under Big Canoe Bridge, which is situated in an old growth, deciduous forest, which has been thinned out a comfortable distance around the bridge. Asphalt-paved Wilderness Parkway makes a U-turn as it passes through the west-southwest/east-northeast aligned bridge. The ruins of marble abutments, from an earlier bridge erected at this site, lie beneath the present bridge, marble being plentiful, as this part of Georgia abounds with marble quarries.

Directions: From Dawsonville, go west on SR 53 to Holcomb to Steve Tate Highway, on the right; go 1.7 miles to the Big Canoe Development main entrance, on the left. Follow Wilderness Parkway, the entrance road, to the security gate and request permission to visit the bridge, which is 0.3 mile beyond the gate.

Rockmart Covered Bridge

POLK COUNTY

Brief Statistics

Type: Non-authentic modern covered footbridge. World Guide Covered Bridge Number: GA-115-a. Built in 1979 by City of Rockmart. Three-span footbridge crossing Euharlee Creek. Stringer footbridge 87.3 feet long by 5.7 feet wide, with approximately 5.1-foot wide by 7.0-foot high portal openings. Alternate Name: None known. Located in Rockmart in downtown park.

At 87.3 feet, Rockmart Bridge is the longest non-authentic covered footbridge in Georgia. The City of Rockmart built Rockmart Bridge over Euharlee Creek in 1979. The three-span stringer bridge, aligned southeast/northwest, has steel I-beam stringers supported by four piers. Starting from the northwest end, there is a double utility pole pier, set in a large concrete block; then, in midstream, there is a double steel I-beam pier, set on a single cement-filled pipe, followed by two more piers, of the same construction as the first pier. The northwest end has a longitudinal-planked ramp set on steel I-beams; the southeast end has a concrete ramp. The bridge floor has longitudinal planking, the sides are open, and the roof is gray asphalt shingled. Twenty-three 4-inch by 4-inch weathered wood posts along each side support the low peaked roof. A double-rail wood railing extends the length of the bridge on both sides, with double-rail welded steel pipe handrails along the right downstream or northwest ramp. There are three sets of buttress-type wood wing braces on each side, each set anchored to a common steel I-beam.

The Polk County bridge is in an open area of a downtown city park, with pavilions, picnic tables and a gazebo on the northwest end and an open grassy area and parking lot on the southeast end. Crossing placidly flowing Euharlee Creek, upstream and downstream of Rockmart Bridge, are main downtown streets. Euharlee Creek flows into the Etowah River, thence entering the Coosa River at Rome.

Directions: From the junction of US 278 and SR 113 in Rockmart, take US 278 west, turning left on US Business 278, toward downtown. At the third traffic light, US Business 278 goes right (Elm Street); follow across railroad tracks. The bridge is on the left, in a small park.

Big Creek Flying Ranch Covered Bridge

RABUN COUNTY

Brief Statistics

Type: Non-authentic modern covered bridge. World Guide Covered Bridge Number: GA-119-H. Built in 1998 by John Daniel Benson. One-span bridge crossing Stekoa Creek. Stringer bridge 49.7 feet long by 14.3 feet wide, with approximately 11.0-foot wide by 9.0-foot high portal openings. Alternate Name: Benson Bridge. Located southeast of Tiger off Rickman Airfield Road.

John Daniel Benson, with the aid of friends, built Big Creek Flying Ranch Bridge, in 1998, at the entrance to his private property. Deriving the name Big Creek from a nickname of Stekoa Creek, the single-span bridge crosses Stekoa Creek, whose waters flow into the Chatooga River, the eastern boundary between Rabun County, Georgia, and Oconee County, South Carolina. The rustic bridge has a covered alcove on each side; the upstream side has a double wide swing seat; and a cupola sits atop the gray weathered cedar-shingled roof, all features shared by other privately owned covered bridges in Rabun County. Also known as Benson Bridge, after the builder, this bridge has a lovely duck weathervane atop the cupola, mounted on the peak of the roof at the center of the bridge. A *BIG CREEK FLYING RANCH* sign, gold letters on a green background, hangs over the entrance, on the right downstream portal.

Aligned northwest/southeast, the 49.7-foot long stringer bridge has transverse planking laid on four large, green painted, steel I-beam stringers, resting on poured concrete abutments. Bark-covered log beams, set on four bark-covered log posts along each side, with bark-covered log railings between the posts, support the cover. A rail fence with four boards butts up to the bridge at both portals. An electric security gate at the entrance portal maintains the privacy of the residence. A low shrub landscaping around the bridge, together with a long, board-rail fenced, winding driveway across the open front yard, creates a pretty bridge site, indeed.

Directions: From Clayton, take US 441 south to

Big Creek Flying Ranch Bridge displays the rustic log construction, side alcoves and cupola features shared by several other Rabun County bridges.

Rickman Airfield Road, the second left after the milepost 9 marker, on the left side of the highway. Go 1.0 mile to the bridge, on the right. Private property: request permission to visit.

Blanchard Covered Bridge

Rabun County

Brief Statistics

Type: Non-authentic modern covered bridge. World Guide Covered Bridge Number: GA-119-J. Built in 1999 by Madison McCracken. One-span bridge crossing Stekoa Creek. Stringer bridge 40.1 feet long by 15.5 feet wide, with approximately 13.7-foot wide by 9.9-foot high portal openings. Alternate Name: None known. Located southeast of Tiger off Rickman Airfield Road.

Blanchard Bridge was built, in 1999, by Madison McCracken, who later built a covered bridge on his own property. Built in the same style as nearby Big Creek Flying Ranch Bridge and McCracken Bridge, the rustic stringer bridge is constructed from natural logs, with a cupola centered on the peak of the roof and covered alcoves on the sides. Aligned south/north, the single-span bridge crosses Stekoa Creek, swiftly flowing on its way to the Chatooga River, which forms the eastern boundary between Rabun County, Georgia, and Oconee County, South Carolina.

Blanchard Bridge has transverse planking laid on four large, rusty, steel I-beam stringers, resting on poured concrete abutments. Bark-covered log beams, set on four barkless log posts along each side, with log railings between the posts, support the cover. The portal gables have natural vertical boarding with battens, and the roof has gray weathered cedar shingles. Rustic log gates at both portals secure the Rabun County bridge, presumably from the pastured cattle. Situated on the private property of J. Blanchard, the 40.1-foot long bridge has a wooded hill on the north end, open pasture on the south end, a few trees along the creek and wooden fencing along the entry roadway.

Directions: From Clayton, take US 441 south to Rickman Airfield Road, the second left after the milepost 9 marker, on the left side of the highway. Go 1.9 miles to the bridge, on the left. Private property: request permission to visit.

Covered Bridge Farm Covered Bridge

RABUN COUNTY

Brief Statistics

Type: Non-authentic modern covered bridge. World Guide Covered Bridge Number: GA-119-G. Built in 1990 by Gordon E. Clark. One-span bridge crossing Scott Creek. Stringer bridge 25.0 feet long by 13.0 feet wide, with approximately 11.4-foot wide by 10.6-foot high portal openings. Alternate Name: J. G. Clark Bridge. Located west of Clayton at Covered Bridge Farm.

Covered Bridge Farm Bridge, also known as J. G. Clark Bridge, was built by Gordon E. Clark, to a length of 25.0 feet, in 1990, over Scott Creek at Covered Bridge Farm, west of Clayton. Covered Bridge Farm Bridge was modeled after Stovall Mill Bridge, in White County. A poured concrete slab, resting on mortared stone abutments, supports the single-span bridge. Gray weathered, vertical boarding with battens covers the sides and portals. The sides are open at the floor, for the breadth of the creek banks. Seven 4-inch by 4-inch wood posts along each side hold aloft the dark weathered cedar-shingled roof that, with the gables, extends beyond the end posts, to shelter the entrances from the weather. An electric carriage lantern is mounted under the peak on the gable, above the construction date, 1990, burned into a wood sign over the entrance of the right downstream portal. A nearby flagpole holds aloft a United States flag.

Aligned south-southeast/north-northwest, the Rabun County bridge is fully exposed, with open pasture bisected by Scott Creek on the downstream side and a second driveway running parallel to the bridge driveway, on the upstream side. Scott Creek joins nearby Stekoa Creek, on its way to the Chatooga River, which forms the eastern boundary between Rabun County, Georgia, and Oconee County, South Carolina.

Directions: From Clayton, at US 441, take US 76 west 3.4 miles to Covered Bridge Farm, on the left. Private property: request permission to visit.

Handley Covered Bridge

RABUN COUNTY

Brief Statistics

Type: Non-authentic modern covered bridge. World Guide Covered Bridge Number: GA-119-K. Built in 1999 by Bobby Handley. One-span bridge crossing Tiger Creek. Stringer bridge 44.5 feet long by 13.0 feet wide, with approximately 12.2-foot wide by 10.9-foot high portal openings. Alternate Name: None known. Located south of Tiger off Old 441 South.

Built by Bobby Handley in 1999, Handley Bridge follows the style found in other Rabun County covered bridges. This 44.5-foot long stringer bridge has the covered alcoves on the sides and a cupola on the roof, but the similarity ends there. Supported by four large, red painted, steel I-beam stringers on poured concrete abutments, the bridge has a transverse planked floor and a steeply-peaked aluminum roof, held aloft by six wood posts on each side, all metal painted red and all wood gleaming natural color. Each alcove is equipped with a hanging, double-seat swing and two candle-lit lanterns. The portal gables have facsimile tongue and groove paneling, while the sides are open, with red steel banister-style railings between the wood posts and around the alcoves. The bridge is secured on the highway end, with a red gate, mimicking the railings.

Tiger Creek passes under the single-span Handley Bridge, on its way to the Tallulah River, at Lakemont. Aligned west/east, the picturesque bridge has dense woods on three sides, with US 441 South on the west end.

Directions: From Clayton, go south on US 441 to Tiger Road, the first right after milepost 9 marker, on the left side of the highway. Go 1.2 miles to Old 441 South; go left 1.6 miles to the bridge, on the left. Private property: request permission to visit.

McCracken Covered Bridge

RABUN COUNTY

Brief Statistics

Type: Non-authentic modern covered bridge. World Guide Covered Bridge Number: GA-119-I. Built in 2000 by Madison McCracken. One-span bridge crossing Stekoa Creek. Stringer bridge 50.9 feet long by 15.2 feet wide, with approximately 13.9-foot wide by 10.8-foot high portal openings. Alternate Name: None known. Located southeast of Tiger off Rickman Airfield Road.

McCracken Bridge is the longest, non-authentic single-span covered bridge in Georgia, at 50.9 feet. Madison McCracken built the stringer bridge around 1985, adding the cover in 2000. Built in the same style as nearby Big Creek Flying Ranch Bridge and Blanchard Bridge, this rustic bridge is constructed from bark-covered logs, with a cupola centered on the roof peak; covered alcoves on the sides are absent. Wooded Stekoa Creek races under the single-span bridge, on its way to the Chatooga River, the eastern boundary between Rabun County, Georgia, and Oconee County, South Carolina.

Aligned south/north, McCracken Bridge has a cement slab floor, set on four rusty steel I-beam stringers resting upon poured concrete abutments. Bark-covered log beams, set on seven bark-covered log posts along each side, log railings running between the posts, support the cover. Natural vertical boarding covers the portal gables, and tin with rusty patches covers the roof. Surrounded by pasture, the Rabun County bridge is on the private property of Madison McCracken.

Directions: From Clayton, take US 441 south to Rickman Airfield Road, the second left after the milepost 9 marker, on the left side of the highway. Go 1.6 miles to the bridge, on the left. Private property: request permission to visit.

Sky Valley Resort Covered Bridge

Rabun County

Brief Statistics

Type: Non-authentic modern covered footbridge. World Guide Covered Bridge Number: GA-119-l (lower case L). Built c. 1975 by Frank Rickman. One-span footbridge crossing tributary to Mud Creek. Stringer footbridge 21.7 feet long by 7.0 feet wide, with approximately 4.0-foot wide by 8.0-foot high portal openings. Alternate Name: None known. Located in Sky Valley at Sky Valley Resort.

Frank Rickman, of Clayton, Georgia, built Sky Valley Resort Bridge, circa 1975, near the Lodge at Sky Valley Resort. Sky Valley Resort Bridge is the shortest non-authentic covered footbridge in Georgia, at 21.7 feet long.

The stringer bridge is supported by two heavy steel I-beam stringers, welded in five sections to form an archway. The stringers are anchored on large poured concrete block abutments. There are nine hand-hewn half-log steps, at each end of the bridge, which ascend to the transverse 2-inch by 8-inch plank flooring. Wood shakes cover the sides—inside, outside and on the ends—to rail height, the rail capped with two side-by-side 2-inch by 12-inch planks. The outside shakes extend down the sides of the log steps and are placed radially to form a continuous arch, from the bottom of the steps on one side of the bridge to the bottom of the steps on the other side of the bridge, the bottom row scalloped. The gables are covered with wood shakes, boxed on the inside to form a large birdhouse with nine compartments, which is highly decorated on the outside. The bottom row of gable shakes is scalloped. The roof is also covered with wood shakes and extends halfway out over the log steps. Three bark-covered log posts along each side support the roof. Log handrails extend down both sides of the log steps, ending at a decorative log post. Several timbers used in the construction of the bridge are hand hewn. All logs and timbers in the bridge have weathered to a pleasant gray color, and the shakes have aged to a dark color. Moss and lichen abundantly cover the log steps and railings.

A narrow, sand- and rock-bottomed tributary to Mud Creek passes beneath the single-span bridge, on its way to nearby Mud Creek, which enters the Little Tennessee River. The golf course is at the north-northwest or left downstream end of the bridge; a golf path, roadways and the Sky Valley Chapel are at the south-southeast end. Overgrown landscape plantings crowd the upstream side of the bridge; the downstream side is unobstructed. The oldest graffiti carving found on the bridge is dated 6/11/84.

Directions: From Scaly, NC, go south on SR 106 to Old Mud Creek Road to Sky Valley, on the left; follow 0.9 mile to the Sky Valley Resort gated entrance, on the right (Old Mud Creek Road becomes Bald Mountain Road at the Georgia State line). Follow Sky Valley Way, the entrance road, going to the left down to the Lodge and Golf Pro Shop. The bridge is across the parking lot.

Sky Valley Stables Covered Bridge

Rabun County

Brief Statistics

Type: Non-authentic modern covered bridge. World Guide Covered Bridge Number: GA-119-E. Built in 1980 by Frank Rickman. One-span bridge crossing Mud Creek. Stringer bridge 29.9 feet long by 21.9 feet wide, with approximately 11.4-foot wide by 13.6-foot high portal openings. Alternate Name: None known. Located in Sky Valley northeast of Dillard just off SR 246.

Sky Valley Stables Bridge, built in 1980 by Frank Rickman of Clayton, Georgia, owes its name to a former, nearby riding stable. The single-span stringer bridge is constructed in a rustic style from bark-covered logs, over the top of Estatoah Falls on Mud Creek; a magnificent site, albeit the beauty is lost to dense foliage that hides a view of the bridge with the waterfall. The unique feature of the south-southeast/north-northwest aligned bridge is the separate cedar-shingled roofs over each side, attached under the central roof to its supporting posts.

Sky Valley Stables Bridge has a reinforced concrete slab floor, set on poured concrete abutments, with bark-covered log posts and beams supporting the dark weathered cedar-shingled roofs, the sides and portal ends all open. The fasciae on the sides of both roofs are covered with short lengths of weathered vertical boards, rounded on the downward ends, giving a scalloped appearance. The main roof extends several feet beyond the slab floor, providing weather protection for the entrances. There are two hand-hewn log benches inside the bridge, along the downstream side, wherefrom a seated person may see straight down the waterfall.

The forsaken Rabun County bridge, becoming obscured by encroaching foliage from the surrounding woods, no longer carries any traffic. Although adjacent to State Road 246, separated by the guard railing, the barely noticeable 29.9-foot long bridge spans Mud Creek, which flows parallel to State Road 246 and empties into the Little Tennessee River nearby.

Directions: From the junction of US 441 and SR 246, north of Dillard, take SR 246 for 3.1 miles up the mountain. The bridge is on the right, about 100 yards beyond the scenic overlook.

Thomas Ramey Covered Bridge

RABUN COUNTY

Brief Statistics

Type: Non-authentic modern covered bridge. World Guide Covered Bridge Number: GA-119-D. Built in 1978 by Thomas Ramey. One-span bridge crossing small creek. Stringer bridge 22.6 feet long by 10.6 feet wide, with approximately 9.2-foot wide by 6.7-foot high portal openings. Alternate Name: None known. Located in Clayton in Leaning Chimney Development.

Built by Thomas Ramey in 1978, Thomas Ramey Bridge is a stringer bridge with a concrete slab floor, set on log stringers resting on poured concrete and large stone abutments. The Rabun County single-span bridge crosses a small creek feeding a man-made pond. The south/north-aligned bridge has gray weathered, vertical boarding on the sides and portals, with four log posts on each side supporting the dark weathered cedar-shingled roof. The roof and gables extend beyond the end posts, to provide shelter from the elements for the entrances.

Thomas Ramey Bridge is in a park area in Leaning Chimney Development and, ironically, the bridge leans significantly toward the upstream side. The 22.6-foot long bridge is fully exposed, with the pond on the downstream side and homes very close to the bridge, at the south end.

Directions: From Clayton, at US 76 West and South Main Street (Old US 441), go south 0.8 mile to Leaning Chimney Road, on the left. The bridge is in the park, on the right. Private property: request permission to visit.

Haralson Mill Covered Bridge

ROCKDALE COUNTY

Brief Statistics

Type: Non-authentic modern covered bridge. World Guide Covered Bridge Number: GA-122-A. Built in 1997 by APAC, Inc. Three-span bridge crossing Mill Rock Creek. Stringer bridge 151.5 feet long by 37.1 feet wide, with approximately 34.3-foot wide by 15.1-foot high portal openings. Alternate Names: Rockdale Bridge, Rockdale County Bridge. Located northeast of Zingara on Haralson Mill Road.

Haralson Mill Bridge was built in 1997, at a cost of $880,000, by Rockdale County, contracted to APAC, Inc., and Cummings Grading. The construction of a bridge at this site was necessitated due to the planned raising of the water level of Big Haynes Reservoir, that at certain times of the year would have inundated the roadway. This Rockdale County-owned bridge replaced a historic ford on Haralson Mill Road, immediately north of the Haralson Mill historic district, which consists of the Haralson Mill House, a general store, the old mill site and a blacksmith shop. This is a two-lane covered bridge, constructed to accommodate multi-axle vehicles, the portals having a minimum clearance of 15.1 feet! In addition, Haralson Mill Bridge, named after the nearby historic district, is the longest non-authentic covered bridge in Georgia, at 151.5 feet, and, for that matter, the longest in the southeastern United States. The stringer bridge is also known as Rockdale Bridge or Rockdale County Bridge.

Haralson Mill Bridge is comprised of three 50-foot spans, resting on spill-through concrete abutments and two solid concrete piers, all encased in cut granite veneer. The bridge is a stringer type, with a decorative lattice imitating the Town lattice truss design of 19th century covered bridges (patented by Ithiel Town). A unique feature of the Rockdale County bridge is the flooring, which is made up of 12-foot by 12-foot sections of stress-laminated two-inch thick wood planks, placed on edge, positioned transversely to the flow of traffic. Inside the bridge, at each side, are pedestrian walkways, separated from the two travel lanes by double planks, placed on edge and serving as tire bumpers. The bridge interior is equipped with a sprinkler system, smoke detectors and three overhead rafter lights, illuminating the underside of the roof. The sides have tongue and groove natural vertical boarding above and below the open decorative lattice. Tongue and groove natural vertical boarding is also on the portals that align on a south/north orientation. A dark weathered cedar-shingled roof completes the bridge.

Haralson Mill Bridge, located in Rockdale County, the county seat of which is Conyers, spans Mill Rock Creek, one of two tributaries feeding Big Haynes Reservoir, just downstream from the bridge. The area around the bridge is open, access to the spring-fed creek is easy and there is a trail at the northeast corner of the bridge. Parking is available at the north end of the bridge. *See Haralson Mill Covered Bridge in the color photograph section: C-5.*

Directions: From Loganville, at the intersection of US 78 and SR 20, take SR 20 south 7.0 miles to Bethel Road; go left to Haralson Mill Road; go left to the bridge.

Henderson Falls Park Covered Bridge

Stephens County

Brief Statistics

Type: Non-authentic modern covered footbridge. World Guide Covered Bridge Number: GA-127-a. Built in 1976 by City of Toccoa. One-span footbridge crossing Henderson Creek. Stringer footbridge 41.6 feet long by 12.8 feet wide, with approximately 11.7-foot wide by 8.0-foot high portal openings. Alternate Name: None known. Located in Toccoa in Henderson Falls Park.

The oldest non-authentic covered footbridge in Georgia, Henderson Falls Park Bridge was built in 1976 by the City of Toccoa, in city-owned Henderson Falls Park. Located about 250 feet downstream from Henderson Falls, this 41.6-foot long single-span bridge is the only covered bridge of the several footbridges spanning Henderson Creek within the park.

The brown painted bridge is supported by three steel I-beam stringers, resting on poured concrete abutments and lateral steel I-beams on two sets of double concrete posts, located near the abutments. The transverse planked flooring is placed on the stringers. Vertical facsimile tongue and groove paneling is used on the closed areas of the sides and portals. Brown asphalt shingles cover the roof, which has two small cupolas mounted on the peak, equidistant from the ends. Electric lighting is installed inside the bridge.

The Stephens County bridge, aligning east/west, is located in an open area of the park that has walkways and pavilions on the east side of the creek and woods on the steeply sloped west side of the creek. Henderson Creek drains into nearby Toccoa Creek, on its way to the Tugaloo River, the border between Stephens County, Georgia, and Oconee County, South Carolina. The park has tennis courts and a nature educational center.

Directions: Approaching Toccoa from the southeast on Alternate SR 17, go right on Pond Street 0.2 mile to a Y, bear left on Henderson Falls Park Road; go 0.5 mile to Henderson Falls Park, on the right. The bridge is in the park, toward the back.

Butts Mill Farm No. 1 Covered Bridge

Troup County

Brief Statistics

Type: Non-authentic modern covered bridge. World Guide Covered Bridge Number: GA-141-A. Built in 1996 by Neil and Trish Liechty. Three-span bridge crossing Turkey Creek. Stringer bridge 94.4 feet long by 14.2 feet wide, with 7.8-foot wide by 7.9-foot high portal opening. Alternate Name: None known. Located in Pine Mountain at 2280 Butts Mill Road.

Built in 1996 by Neil and Trish Liechty, Butts Mill Farm No. 1 Bridge is occupying the site of an earlier covered bridge, dating back to circa 1870, that was washed away during a flood around 1910. Its successor, an uncovered concrete bridge, was also washed away by a flood, in 1945. Still visible, scattered downstream, are broken slabs of concrete from this bridge. Rebuilt again in 1945, the bridge remained in use until abandoned in 1975, when Butts Mill Road was rerouted. Adjacent, downstream, to the bridge site is Butts Mill, a grist mill dating back to 1843. During the 1945 flood, the grist mill was damaged and abandoned until acquired by the present owner, in 1994. Today, the rebuilt bridge, closed to traffic, and the renovated grist mill are part of a serene and beautiful farm setting called Butts Mill Farm, which provides family and company outing activities for all ages and two covered bridges!

Butts Mill Farm No. 1 Bridge, extending 94.4 feet long, was built upon the abutments, piers, stringers and floor planking from the 1945 bridge. These triple-span stringers rest upon poured concrete abutments and two poured concrete piers, plus two double steel-angle pillar supports added, one near each abutment, during the 1945 construction or shortly thereafter. There are five steel angle-beam stringers at the right downstream span and five steel I-beam stringers at the middle and left downstream spans. Longitudinal planking was added, in 1996, on top of the earlier transverse-planked floor. The exterior has natural vertical boarding with battens on the sides, and natural horizontal lapped siding on the right downstream or west-southwest portal. The east-northeast end is totally boarded up, to accommodate weddings. Shiny tin covers the roof. The right downstream end has the roof overhanging the portal, coming to a point at the peak, to provide shelter from the weather for the entrance.

There are four opposing window openings along each side, with pink blossoming allamanda plants, in window boxes placed outside each window opening, and hanging fern baskets in each opening. The interior is boxed in with vertical boarding, three lighted ceiling fans hang from the rafters, evenly spaced down the center of the bridge, and pew benches are placed down the sides, all for the accommodation of wedding parties, as it has become a popular site for such gatherings.

Crossing Turkey Creek, which flows into Flat Shoals Creek near Smiths Mill, eventually emptying into the Chattahoochee River at the Alabama border, Butts Mill Farm No. 1 Bridge is situated amid the open surroundings of Butts Mill Farm, where farm animals mill about, horses await riders, children bounce on trampolines, fish scurry over bedrock creek bottoms under the watchful eye of the kingfishers, and a little train pulls cars loaded with smiling children. An astonishing place, well worth the visit.

Directions: From Pine Mountain, take US 27 north about 2 miles to North Butts Mill Road, on the left; go 1.2 miles to the second stop sign, and turn right. Butts Mill Farm is on the right, at 2280 Butts Mill Road. Admission.

Butts Mill Farm No. 2 Covered Bridge

TROUP COUNTY

Brief Statistics

Type: Non-authentic modern covered footbridge. World Guide Covered Bridge Number: GA-141-b. Built in 1998 by Neil and Trish Liechty. Three-span footbridge crossing unnamed creek. Stringer footbridge 48.0 feet long by 6.4 feet wide, with approximately 5.4-foot wide by 7.1-foot high portal openings. Alternate Name: None known. Located in Pine Mountain at 2280 Butts Mill Road.

Butts Mill Farm No. 2 Bridge was built, in 1998, by Neil and Trish Liechty, a short distance upstream from Butts Mill Farm No. 1 Bridge. The stringer footbridge spans an unnamed creek just above its confluence with Turkey Creek, which then passes beneath Butts Mill Farm No. 1 Bridge on its way to Flat Shoals Creek, near Smiths Mill. Like the larger downstream bridge, this triple-span bridge also accommodates weddings, being situated in an enchanting barnyard setting designed for family outings.

Butts Mill Farm No. 2 Bridge is supported by three 4-inch by 6-inch wood timber stringers, set on railroad ties resting on the ground and two piers, comprised of 4-inch by 6-inch wood posts with cross braces. Transverse planking forms the floor. Nine posts along each side support the red-painted, steeply peaked, corrugated fiberglass roof. The roof fasciae is trimmed in white. The sides have natural vertical boarding up to railing height, the open area above being festooned with hanging ferns. A stag weathervane is mounted atop the roof.

The 48.0-foot long east/west aligned bridge reposes in an open area under the shade of mature trees, surrounded by the family-oriented activities of Butts Mill Farm.

Directions: From Pine Mountain, take US 27 north about 2 miles to North Butts Mill Road, on the left; go 0.2 miles to the second stop sign, and turn right. Butts Mill Farm is on the right, at 2280 Butts Mill Road. Admission.

Auchumpkee Creek Covered Bridge

UPSON COUNTY

Brief Statistics

Type: Authentic historic covered bridge. World Guide Covered Bridge Number: GA-145-02 #2. Built in 1997 by Arnold M. Graton Associates. One-span bridge crossing Auchumpkee Creek. Town lattice truss bridge 121.7 feet long by 16.3 feet wide, with approximately 14.6-foot wide by 12.6-foot high portal openings. Alternate Names: Hootenville Bridge, Zorn's Mill Bridge. Located southeast of Thomaston on Old Allen Road.

Named after the creek it spans, Auchumpkee Creek Bridge was originally built in 1892 by Dr. James Wiley Herring and Warren Jackson Alford, of the firm Herring and Alford, at a cost of $1,199. The former small community of Hootenville thrived in the area of the bridge, in its early days, hence the bridge was also known as Hootenville Bridge. Another name for the old bridge was Zorn's Mill Bridge, after the former mill located nearby. The Upson Preservation Commission restored the bridge in 1985. The original bridge was destroyed by floodwaters from tropical storm "Alberto" in July 1994 and was totally rebuilt in 1997 by Arnold M. Graton Associates of Ashland, NH, at a cost of $209,000.

Owned by Upson County, the single-span bridge is set on dry natural-stone piers on opposite banks of Auchumpkee Creek, with the ends joined to the abutments by handrail lined ramps, held up by braced, double wood post supports at each end. A Town lattice truss, fastened together with treenails that protrude as much as four inches from the truss members, supports the bridge. Transverse planking, with no wheel treads, forms the flooring, which extends to the ends of long, sloping approach ramps. Weathering wood shingles on the roof and natural vertical boarding with battens on the sides and portals, cover the bridge exterior. The portals align west-southwest/east-northeast.

Washed away by the 1994 flood, the 1997 rebuilt Auchumpkee Creek Bridge serenely reposes under the hot summer sun. *See Auchumpkee Creek Covered Bridge in the color photograph section: C-5.*

Auchumpkee Creek Bridge is located in a large, open, county-maintained area, with picnic tables at the upstream side of the bridge. Bypassed by a new bridge on the downstream side, Auchumpkee Creek Bridge is no longer open to motor traffic. Ample parking is provided at both ends of the bridge. Auchumpkee Creek flows lazily under the 121.7-foot long bridge, in a south-southeast direction, eventually joining Ulcohatchee Creek, not far from its confluence with the Flint River.

The original bridge was listed on the National Register of Historic Places on April 1, 1975. The present bridge is the last authentic covered bridge remaining in Upson County. *See Auchumpkee Creek Covered Bridge in the color photograph section: C-5.*

Directions: From Thomaston, the county seat of Upson County, take US 19 south 12.0 miles to Allen Road, on the left; follow Allen Road to the bridge. From the junction of US 80 and US 19, south of Thomaston, take US 19 north 2.0 miles to Allen Road, on the right, then go 0.7 mile to the bridge (Allen Road loops back to US 19).

Steve Kennedy Covered Bridge

WEBSTER COUNTY

Brief Statistics

Type: Non-authentic modern covered bridge. World Guide Covered Bridge Number: GA-152-A. Built in 1984 by Steve Kennedy. One-span bridge crossing wash. Stringer bridge 15.9 feet long by 12.2 feet wide, with approximately 10.5-foot wide by 7.3-foot high portal openings. Alternate Name: None known. Located in Preston off US 280.

The shortest non-authentic covered bridge in Georgia, at 15.9 feet, Steve Kennedy Bridge was built in 1984 by Steve Kennedy, on the driveway to his home across a wash in the front yard. Six timber stringers resting on earth abutments, restrained by boards, and one pier, consisting of six posts, support the single-span, red-painted, stringer bridge. The flooring consists of transverse 2-inch by 8-inch wood planks, the sides and ends are enclosed with facsimile tongue and groove paneling, and the roof is covered with old tin.

The southwest/northeast-aligned bridge, situated in the front yard, has scattered trees between the bridge and US 280, dense woods on the northwest side, and the Kennedy residence at the northeast end.

Directions: From Preston, at the intersection of SR 41 and US 280, go 1.1 miles east on US 280. The bridge is on the left. Private property: request permission to visit.

Stovall Mill Covered Bridge

White County

Brief Statistics

Type: Authentic historic covered bridge. World Guide Covered Bridge Number: GA-154-03. Built in 1895 by Will Pardue. One-span bridge crossing Chickamauga Creek. Queenpost truss bridge 37.8 feet long by 11.8 feet wide, with approximately 10.4-foot wide by 11.1-foot high portal openings. Alternate Names: Chickamauga Bridge, Chickamauga Creek Bridge, Helen Bridge, Nacoochee Bridge, Sautee Bridge. Located north-northwest of Sautee off SR 255 on old Rabun Road.

Built in 1895, at the site where an earlier covered bridge was lost in a flood in the early 1890s, Stovall Mill Bridge is the only covered bridge constructed by Will Pardue. The bridge carried old Rabun Road, connecting Rabun County with Gainesville, across Chickamauga Creek, in a single span. Stovall Mill Bridge is the third shortest, at 37.8 feet, and second narrowest, at 11.8 feet (Burnt Bridge measures 11.3 feet wide), of the remaining historic covered bridges in Georgia. Stovall Mill Bridge gained fame from its appearance in the 1951 Henry King directed movie *I'd Climb the Highest Mountain*, starring Susan Hayward and William Lundigan, which was filmed on location in Georgia.

Stovall Mill Bridge obtained its name from the nearby grist mill, saw mill and shingle mill, all owned and operated by Fred Stovall, Sr. The mills have been lost to the ages; however, the upstream waterfalls are the remains of a dam, which impounded the water to power the turbines. The bridge is also known as Chickamauga Bridge or Chickamauga Creek Bridge, after the creek flowing beneath; Helen Bridge, after nearby Helen, a tourist destination; Nacoochee Bridge, after the nearby community, west of Sautee; and Sautee Bridge, after historic Sautee, down the road.

Stovall Mill Bridge was one of ten covered bridges in Georgia that had restoration work performed under the ISTEA (Intermodal Surface Transportation Efficiency Act of 1991) Bridge Restoration Project in Georgia. A contract was let in November 1997 for $89,107; however, termite damage proved to be greater than assessed, requiring additional funding prior to completion, in 1998.

Set on mortared natural-stone abutments, Stovall Mill Bridge is supported by the only half-height queenpost truss of the three queenpost truss bridges remaining in Georgia; Coheelee Creek Bridge and Concord Bridge each have a full height queenpost truss. This three-panel queenpost truss has a small second queenpost truss in the middle panel, under the horizontal member of the larger truss. The flooring consists of transverse planking, without wheel treads, set on timber floor joists that are *hand-hewn*, unusual for a bridge of such a young age. Perhaps they were salvaged from the previous bridge which was washed away in the flood. Vertical boarding with battens, mostly new boards replaced during the 1997 restoration, covers the sides and portals; old weathered wood shingles cover the roof. The bridge, aligned north-northwest/south-southeast, has the roof and gables extending beyond the entrances, to provide protection from inclement weather.

The only authentic historic covered bridge remaining in White County, Stovall Mill Bridge, owned by the White County Historical Society, located in Cleveland, the county seat, is situated in a small roadside picnic area, very visible, except for the downstream side, which is heavily foliated. Closed to motor traffic, the bridge spans Chickamauga Creek, a tributary to nearby Sautee Creek, eventually flowing into the headwaters of the Chattahoochee River, near Sautee. Ample parking is available at the bridge.

Directions: From Sautee, at the junction of SR 17 and SR 255, go north on SR 255 for 2.7 miles, to the bridge, on the right.

Kentucky

Around 1900, Kentucky had approximately four hundred covered bridges dotting the landscape. This number diminished to two hundred, by 1924, and one hundred twenty-five, by 1937. Only twenty of these historic structures remained in 1964. In 1997, Walcott Bridge was dismantled and placed in storage; in 1998, Switzer Bridge was rebuilt; in 1999, Oldtown Bridge was rebuilt; and, in 2000, Colville Bridge was rebuilt, leaving only nine historic covered bridges in Kentucky. Hopefully, Walcott Bridge will be re-erected. Floods and arson have taken their toll on Kentucky's links to a bygone era.

Of the twenty-eight covered bridges in Kentucky, twelve are authentic and sixteen are non-authentic stringer type bridges; nine of the twenty-eight are historic, with all nine being authentic. These twenty-eight bridges were constructed between circa 1820 and 2002 in eighteen of the one hundred twenty counties in Kentucky. The most predominant truss remaining in Kentucky is the kingpost, with seven examples. The other five authentic covered bridges utilize five different trusses: Howe, queenpost, Smith, Town lattice and Wheeler.

Kentucky's nine historic covered bridges, built between 1820 and 1874, are primarily located in the northeastern part of the state, in Robertson, Mason, Fleming, Lewis and Greenup Counties.

Colville Covered Bridge

Bourbon County

Brief Statistics

Type: Authentic historic covered bridge. World Guide Covered Bridge Number: KY-09-03 #2. Built in 2000 by unknown builder. One-span bridge crossing Hinkston Creek. Multiple kingpost truss bridge 122.8 feet long by 18.0 feet wide, with approximately 13.0-foot wide by 11.1-foot high portal openings. Alternate Name: Colville Pike Bridge. Located south of Colville on Colville Road.

Deriving its name from the nearby community of Colville, Colville Bridge has also been called Colville Pike Bridge, after a former name for Colville Road, which now crosses through the bridge. Originally built in 1877 by Jacob N. Bower, the bridge had major restoration work by the son of Jacob N. Bower, Louis Stockton Bower, Sr., in 1913, followed by more restoration work by his son, Louis Stockton "Stock" Bower, Jr., in 1937 and 1976. Interestingly, Jacob N. Bower also built Johnson Creek Bridge, in Robertson County, that was likewise restored by his son. During the 1937 restoration, Louis Stockton Bower, Jr., added 1¼-inch diameter iron rods to the four panels at each end of the twelve-panel multiple kingpost truss. An iron rod, acting as a counterbrace, was added to each panel passing diagonally between the center of the double timber kingpost diagonal braces, forming an X, effectively converting the multiple kingpost truss to a Childs truss. A Childs truss typically utilized a kingpost truss in the center panel. This same conversion to a Childs truss occurs in Cabin Creek Bridge, in Lewis County.

On March 1, 1997, a record flood severely damaged Colville Bridge, causing its closure to motor traffic. The 122.8-foot long bridge was completely rebuilt by an unknown builder, with new materials, retaining the Childs truss conversion, and the bridge reopening to motor traffic in 2000. The new bridge displays a noticeable camber. Colville Bridge is one of only four historic covered bridges in Kentucky still open to motor traffic, the other three being Dover Bridge, Goddard Bridge and privately owned Valley Pike Bridge.

One of seven multiple kingpost truss bridges remaining in Kentucky, Colville Bridge's truss is unusual, in that all of the truss timbers are doubled, with the end posts

Flagg Spring Golf Course No. 1 Bridge has a twin bridge at the end of the golfcart path leading off to the right.

quadrupled. This same unusual truss is employed in Cabin Creek Bridge, Hillsboro Bridge, Oldtown Bridge and Ringo's Mill Bridge. The bridge rests on poured concrete that surrounds and reinforces the original limestone abutments. The flooring consists of transverse planking with six-plank-wide wheel treads. The sides have natural vertical boarding, open under the eaves. The portals have white painted vertical boarding, with green painted battens. The cedar-shingled roof and the gables extend approximately two feet beyond the entrances, forming a weather panel for the portals.

Crossing Hinkston Creek in a single span, Colville Bridge is in a heavily wooded area, the foliage trimmed back on the upstream side at the south portal. Deep, muddy Hinkston Creek joins Stoner Creek, approximately 2.5 miles downstream at Ruddles Mills, to form the South Fork Licking River, which flows into the Licking River, at Falmouth. Colville Road passes through the bridge, making a sharp left turn after exiting the south portal.

Listed on the National Register of Historic Places on December 30, 1974, Colville Bridge is the only historic covered bridge remaining in Bourbon County. Paris is the county seat of Bourbon County.

Directions: From Cynthiana, at the junction of US 27 and SR 36/SR 32, go east 5.6 miles on SR 36/SR 32 to Colville Road (next road after Colville Lane), on the right; go 2.8 miles to the bridge.

Flagg Spring Golf Course No. 1 Covered Bridge

CAMPBELL COUNTY

Brief Statistics

Type: Non-authentic modern covered footbridge. World Guide Covered Bridge Number: KY-19-b. Built in 1999 by unknown builder. One-span footbridge crossing waterway connecting ponds. Stringer footbridge 39.9 feet long by 8.0 feet wide, with approximately 7.0-foot wide by 7.5-foot high portal openings. Alternate Name: None known. Located in Flagg Spring on Flagg Spring Golf Course.

Flagg Spring Golf Course has two identical covered golfcart bridges, Flagg Spring Golf Course No. 1 Bridge, at No. 14 tee, and Flagg Spring Golf Course No. 2 Bridge, between No. 2 tee and No. 9 tee. An unknown builder built Flagg Spring Golf Course No. 1 Bridge, in 1999.

The 39.9-foot long footbridge is supported by two steel I-beam stringers, set on poured concrete abutments. Transverse 2-inch by 6-inch planks form the floor. The sides and gables are enclosed with natural, vertical, facsimile tongue and groove paneling, the sides up to railing height and open above. The red-painted aluminum roof is supported by four 6-inch by 6-inch wood posts along each side.

Aligned north-northwest/south-southeast, the single-span bridge carries a winding asphalt paved golfcart path across a short interconnecting waterway, between two ponds, the larger pond on the west-southwest side. The bridge appears starkly alone, sitting on the open fairways, with no foliage or trees in proximity.

Directions: From Mentor, go west on SR 735 to Flagg Spring Golf Course, on the right, just before the intersection with SR 9. The bridge is at No. 14 tee.

Flagg Spring Golf Course No. 2 Covered Bridge

Campbell County

Brief Statistics

Type: Non-authentic modern covered footbridge. World Guide Covered Bridge Number: KY-19-c. Built in 1999 by unknown builder. One-span footbridge crossing tributary to Flagg Spring Creek. Stringer footbridge 40.0 feet long by 9.3 feet wide, with approximately 8.3-foot wide by 7.4-foot high portal openings. Alternate Name: None known. Located in Flagg Spring on Flagg Spring Golf Course.

Flagg Spring Golf Course has two identical covered golfcart bridges, Flagg Spring Golf Course No. 2 Bridge, between No. 2 tee and No. 9 tee, and Flagg Spring Golf Course No. 1 Bridge, at No. 14 tee. Flagg Spring Golf Course No. 2 Bridge was built, in 1999, by an unknown builder. The 40.0-foot long bridge is the longest footbridge in Kentucky.

The stringer footbridge is supported by two steel I-beam stringers, set on poured concrete abutments. Transverse 2-inch by 6-inch planks form the floor. The sides and gables are enclosed with natural, vertical, facsimile tongue and groove paneling, the sides up to railing height and open above. The red-painted aluminum roof is supported by four 6-inch by 6-inch wood posts along each side.

Aligned north-northeast/south-southwest, the single-span bridge crosses a tributary to Flagg Spring Creek, which flows into the Ohio River, at Mentor. A few trees line the tributary at each side of the bridge.

Directions: From Mentor, go west on SR 735 to Flagg Spring Golf Course, on the right, just before the intersection with SR 9. The bridge is between the No. 2 and No. 9 tees.

Paul Kidd Covered Bridge

Campbell County

Brief Statistics

Type: Non-authentic modern covered bridge. World Guide Covered Bridge Number: KY-19-A. Built in 1970 by Paul Kidd. One-span bridge crossing Pooles Creek No. 2. Stringer bridge 18.3 feet long by 11.3 feet wide, with approximately 10.3-foot wide by 10.8-foot high portal openings. Alternate Name: None known. Located in Cold Spring at 234 Pooles Creek Road No. 2.

In October 1970, Paul Kidd completed the cover to Paul Kidd Bridge, over the entrance to the concrete driveway onto his property. The stringer portion of the bridge was built at an earlier date. Paul Kidd Bridge is the oldest non-authentic motor vehicle covered bridge remaining in Kentucky. At 18.3 feet long, Paul Kidd Bridge also has the distinction of being the shortest motor vehicle covered bridge in Kentucky.

Resting on poured concrete abutments, the stringer bridge is supported by three steel I-beam stringers. A reinforced concrete slab forms the floor. Weathered vertical boarding with battens covers the sides and the gables, the bottom of the gable boards scalloped to enhance the bridge's appearance. Three opposing, small, square window openings adorn each side. Moss-covered wood shingles cover the roof, which is supported along each side by four 4-inch by 4-inch wood posts. A three-rail steel pipe railing passes through each side of the bridge, flaring out several feet at each portal. A chain-link-fence gate may be used to secure the right downstream or southeast portal. An oxen yoke adorns the left downstream or northwest gable. The interior of the bridge was originally painted red.

The single-span bridge crosses narrow Pooles Creek No. 2, which joins Pooles Creek No. 1, before flowing into the nearby Licking River, which then flows into the Ohio River, at Covington. The tree-flanked bridge and foliated creek banks separate Pooles Creek Road No. 2 from the open front yard of the residence.

Directions: From Alexandria, go north on US 27

to the intersection with SR 9, continuing 1.2 miles on SR 27 to Murnan Road, on the left. Go 0.4 mile to Pooles Creek Road No. 2, on the right; go 1.0 mile to No. 234, on the left. Private property: request permission to visit.

Willis Covered Bridge

CARTER COUNTY

Brief Statistics

Type: Non-authentic modern covered footbridge. World Guide Covered Bridge Number: KY-22-a. Built in 2002 by Ben Rice. One-span footbridge crossing small unnamed creek. Stringer footbridge 26.1 feet long by 8.2 feet wide, with approximately 7.2-foot wide by 7.6-foot high portal openings. Alternate Name: None known. Located in Grayson at Ponderosa Farms.

At the time of the author's visit, in 2001, Ben Rice was constructing Willis Bridge, completion scheduled in 2002. The bridge is named Willis Bridge to honor the maiden name of the owner's wife. The stringer footbridge is supported by four steel I-beam stringers, set on two piers, each consisting of four upright 6-inch by 6-inch wood posts set in concrete footings. The ramps, yet to be built, will have mortared, natural-flat-stone combination retaining/guard walls running from ground level up to the elevated bridge flooring, at each end. The flooring of the 26.1-foot long bridge consists of transverse boards. Vertical boarding covers the sides, up to railing height and under the eaves, leaving the middle open for the length of the bridge. The boarding is cut along the bottom of the sides to form an arch. The gables also are covered with vertical boarding. Green-painted aluminum covers the roof, which has an elevated secondary roof at the center, invoking memories of an old time train caboose, the whole supported by four 4-inch by 4-inch wood posts along each side.

The single-span bridge crosses a small, unnamed creek, with mortared, flat-stone-lined banks and creek bed, that flows into the Right Prong, before entering the nearby Little Sandy River. The south/north-aligned bridge is fully exposed, with a mature tree near the downstream side; however, ongoing landscaping is altering the site.

Willis Bridge is the shortest non-authentic covered footbridge in Kentucky; the state has no authentic covered footbridges.

Directions: From Grayson, go north on SR 1 to the junction with SR 7, at Pactolus; continue 1.7 miles north on SR 1 to Ponderosa Farms, on the left. The bridge is visible from the highway. Private property: request permission to visit.

Goddard Covered Bridge

FLEMING COUNTY

Brief Statistics

Type: Authentic historic covered bridge. World Guide Covered Bridge Number: KY-35-06. Built circa 1820 by unknown builder. One-span bridge crossing Sand Lick Creek. Town lattice truss bridge 62.8 feet long by 14.8 feet wide, with approximately 12.2-foot wide by 11.9-foot high portal openings. Alternate Name: White Bridge. Located in Goddard on Maddox Road.

Originally built, circa 1820, by an unknown builder, about one mile upstream, Goddard Bridge was relocated to its present site in 1932, utilizing the abutments of an earlier bridge that it replaced. The historic bridge is the oldest covered bridge remaining in the southeastern United States. The aging bridge was restored in 1968 by Fleming County, under the supervision of Louis Stockton "Stock" Bower, Jr., of Flemingsburg, Kentucky. Named after its location in the community of Goddard, which derived its name from Joseph Goddard, an early settler, Goddard Bridge has also been called White Bridge, due to an earlier practice of keeping the bridge whitewashed. The 14.8-foot wide bridge shares with Switzer Bridge, in Franklin County, the distinction of being the narrowest authentic covered bridge remaining in Kentucky. Goddard Bridge joins Colville Bridge, Dover Bridge and privately owned Valley Pike Bridge in being the only historic covered bridges in Kentucky still carrying motor traffic. Under the National Historic Covered Bridge Preservation Program, $464,640 was approved on September 26, 2001, by the Federal Highway Administration, for restoration work on Goddard Bridge.

Goddard Bridge is the only example of a Town lattice truss bridge in Kentucky. Pegged together with two-inch diameter treenails, the truss has three one-inch diameter vertical iron rods, located in the center of the truss at the small diamond window openings. Most likely these rods were added at a later date. The upper chord and truss members have been pre-numbered, 1 through 19 on the upstream side, and 20 through 38 on the downstream side. This numbering was most likely done during the 1932 move, to facilitate re-assembly. Such numbering was performed at the factory, to assure a proper fit at the bridge site. In these instances, the numbering was repeated on both sides, with one side's numbering distinguished with an X. An example of this factory numbering may be seen on Euharlee Bridge, in Bartow County, Georgia.

The 62.8-foot long bridge rests on a dry, large creek-stone abutment—the mortar added at a later date—on the left downstream or west end, and a poured concrete pier, added in 1932, on the east end. Two steel I-beam supports, constructed from two I-beams in the form of an inverted

Goddard United Methodist Church is framed by the portal of Goddard Bridge (circa 1820), the oldest covered bridge in the southeastern United States.

V with a cross-brace, set in a concrete base, are located under each lower chord, just inside the west abutment and the east pier, added for additional support, probably in 1968. There is a 30.3-foot long approach ramp, constructed in 1932, at the east end, spanning from the concrete pier to the other original dry, large creek stone abutment, supported at the center by two mortared creek-stone pillars, with additional support from upright steel I-beams at each side. The flooring is made from transverse planking with four-plank-wide wheel treads that extends to the end of the ramp. Weathered vertical boarding covers the sides and gables; a bridge informational sign is mounted on the west gable. The tin covered roof extends approximately four feet beyond the portal entrances, affording extra shelter from the elements.

The single-span bridge crosses overgrown Sand Lick Creek, at an open grassy area, the creek joining Big Run, at Plummers Mill, to form Fox Creek, which flows into the Licking River, approximately two miles west of Hillsboro Bridge. State Road 32 is at the west end of the bridge; white painted Goddard United Methodist Church is at the east end, producing magnificent photo opportunities viewed through the bridge. A bridge information sign is located at State Road 32, at the west portal of the bridge.

Listed on the National Register of Historic Places on August 22, 1975, Goddard Bridge is one of three historic covered bridges remaining in Fleming County. Attesting to the age of the bridge, hand carved on a truss member inside the bridge is the date 1835.

Directions: From Flemingsburg, the county seat, take SR 32 south toward Goddard, continuing 1.1 miles after the junction with SR 156, on the right, to the bridge, on the left at Maddox Road.

Hillsboro Covered Bridge

FLEMING COUNTY

Brief Statistics

Type: Authentic historic covered bridge. World Guide Covered Bridge Number: KY-35-05. Built circa 1867 by unknown builder. One-span bridge crossing Fox Creek. Multiple kingpost truss bridge 94.1 feet long by 18.2 feet wide, with approximately 14.1-foot wide by 12.3-foot high portal openings. Alternate Name: Grange City Bridge. Located south of Hillsboro on old SR 111.

Hillsboro Bridge has sometimes been called Grange City Bridge, both names derived from nearby communities. Hillsboro Bridge was built, circa 1867, by an unknown builder, most likely the same builder of Ringo's Mill Bridge, a few miles upstream, as the size and construction details are almost identical. Hillsboro Bridge has survived six major floods since 1935 (and, most certainly, several before 1935). These floods and the high water levels of each are recorded inside the bridge, on timbers on the downstream side; in descending order, from the highest water level, in feet, above the floor planking, they are: March 1, 1997—7 feet 3 inches; January 25, 1937—5 feet 9 inches; February 2, 1950—3 feet 10 inches; 1962 (no month or day recorded)—3 feet 7 inches; 1948 (no month or day recorded)—2 feet 11 inches; and March 14, 1935—1 foot 2 inches. Amazingly, the bridge is still there!

The 94.1-foot long bridge is supported by an eight-panel, multiple kingpost truss, constructed with double yellow-pine truss members, securely attached to large, cut, red sandstone-block abutments, which were stucco-coated at a later date. The popular kingpost truss is common to seven covered bridges in Kentucky, five of these bridges having unusual double truss members: this bridge, Cabin Creek Bridge, Colville Bridge, Oldtown Bridge and Ringo's Mill Bridge. The flooring of Hillsboro Bridge is transverse oak planking with five-plank-wide wheel treads. Vertical boarding covers the natural, gray weathered sides and the white-painted portals, replacing the corrugated sheet metal siding in 1983; tin covers the roof. There is a framed metal sign, over the left downstream entrance, proclaiming, *GRANGE CITY/COVERED BRIDGE/86 FT. SPAN—BUILT 1865-1870.*

The high water mark of 5 feet 9 inches above the floor was recorded inside Hillsboro Bridge for the January 25, 1937, flood.

The single-span bridge crosses Fox Creek, which joins the Licking River, approximately two miles west of the bridge. The concrete bypass bridge is close by, on the upstream side; heavy woods and thickets choke the downstream side. The north/south-aligned bridge was bypassed and closed to motor traffic in 1968. A bridge informational sign is at State Road 111, near the left downstream or north portal.

Hillsboro Bridge, one of three historic covered bridges remaining in Fleming County, was listed on the National Register of Historic Places on March 26, 1976. Flemingsburg is the county seat of Fleming County.

Directions: From Hillsboro, go south on SR 111 to Ringo Mills-Grange City Road, on the left. The bridge is across the highway, on the right.

Ringo's Mill Covered Bridge

FLEMING COUNTY

Brief Statistics

Type: Authentic historic covered bridge. World Guide Covered Bridge Number: KY-35-04. Built in 1867 by unknown builder. One-span bridge crossing Fox Creek. Multiple kingpost truss bridge 87.5 feet long by 18.3 feet wide, with approximately 14.5-foot wide by 12.8-foot high portal openings. Alternate Name: None known. Located in Ringos Mills on old SR 158.

Ringo's Mill Bridge derived its name from a grist mill, operating on Fox Creek in the mid-nineteenth century, which was served by the bridge. The bridge was built in 1867, most likely by the same builder of Hillsboro Bridge, a few miles downstream, as the bridges are nearly identical in size and construction. Ringo's Mill Bridge underwent a major restoration, begun in 1983, by Louis Stockton "Stock" Bower, Jr., who was forced to stop the effort due to poor health. The restoration was resumed and completed in November 1984, by L. A. Thompson, of Flemingsburg, Kentucky.

The 87.5-foot long bridge is supported by an eight-panel multiple kingpost truss, constructed with double yellow-pine truss timbers resting on large, cut, red sandstone-block abutments, which were later stucco-coated and capped with poured concrete under each lower chord, during the 1984 restoration. The popular kingpost truss is common to seven covered bridges in Kentucky, five of these bridges having unusual double truss members: Ringo's Mill Bridge, Cabin Creek Bridge, Colville Bridge, Hillsboro Bridge and Oldtown Bridge. Transverse oak planking constitutes the floor of Ringo's Mill Bridge. Replacing the original 1867 very wide poplar boards, new vertical boarding was installed, on the sides and portals, during the 1984 restoration, which is now a weathered gray color. The roof is covered with tin. A metal sign is mounted on the right downstream gable stating, *RINGOS MILL/COVERED BRIDGE/BUILT IN 1867/RESTORED IN 1984/BY: L. A. THOMPSON.*

Spanning Fox Creek, which joins the Licking River, approximately four miles west of the bridge, the single-span bridge was bypassed by a new concrete bridge, in 1969, and closed to motor traffic. The bypass bridge runs diagonally to the historic west-northwest/east-southeast aligned bridge, on the downstream side, coming very close to the covered bridge at the west-northwest or left downstream end. The upstream side of the bridge faces wooded creek banks. There is a bridge informational sign at State Road 258, near the bridge.

Listed on the National Register of Historic Places on March 26, 1976, Ringo's Mill Bridge is one of three historic covered bridges remaining in Fleming County, the county possessing eighteen in the 1950s. Flemingsburg is the county seat of Fleming County.

Directions: From Hillsboro, go east on SR 158 to Ringos Mills. The bridge is on the left, on a bypassed section of SR 158.

Switzer Covered Bridge

Franklin County

Brief Statistics

Type: Authentic modern covered bridge. World Guide Covered Bridge Number: KY-37-01 #2. Built in 1998 by Intech Contracting, Inc. One-span bridge crossing North Elkhorn Creek. Howe truss bridge 124.1 feet long by 14.8 feet wide, with approximately 10.7-foot wide by 11.7-foot high portal openings. Alternate Name: None known. Located in Switzer on old Jones Lane (SR 1262).

Switzer Bridge was originally built, in 1855, by George Hockensmith; restored, in 1906, by Louis Stockton Bower, Sr.; bypassed, in 1954, by a concrete bridge downstream; and restored again, in 1990, at which time the tin roof gave way to wood shingles. Unfortunately, heavy rains brought about the record flood of March 1, 1997, which washed Switzer Bridge from its abutments and rammed it against the concrete bypass bridge, a short distance downstream, where it came to rest and was prevented from being swept further downstream, to its ultimate destruction. When the flood subsided, remnants of Switzer Bridge were salvaged. In 1998, Intech Contracting, Inc., of Lexington, Kentucky, rebuilt the bridge, replicating the previous bridge. Some materials salvaged from the old bridge were utilized and some were replaced, e.g., the vertical iron tension rods in the Howe truss were replaced with stronger steel rods.

The new Switzer Bridge is the only Howe truss covered bridge remaining in Kentucky. Each panel of the fourteen-panel Howe truss consists of a double timber brace sloping toward the center of the bridge, with a single timber counterbrace sloping the other way to form an X, with double vertical steel tension rods between the Xs. Switzer Bridge, at 14.8 feet wide, shares with Goddard Bridge, in Fleming County, the distinction of being one of the two narrowest authentic covered bridges in Kentucky.

The 124.1-foot long bridge is supported by a Howe truss, set on rough-cut limestone block abutments; the unmortared abutments were rebuilt with mortar in June 1999. The moderately cambered bridge has a longitudinally planked floor, vertical boarding on the sides, vertical boarding with battens on the gables—the bottom of the boards V cut, to make a sawtooth edge—and cedar shingles on the roof. The portals slope outward to form a weather shelter for the portal.

Closed to motor traffic, the single-span bridge crosses North Elkhorn Creek, which joins South Elkhorn Creek, east of Frankfort, to form Elkhorn Creek, thence flowing north into the Kentucky River. The northwest/southeast-aligned bridge is open at both portals, the bypassed section of State Road 1262 affording parking at either end. The wide muddy creek has wooded banks upstream of the bridge and is open downstream to the concrete bypass bridge, which is a comfortable distance away. There is a cataract, just upstream of the bypass bridge. An engraved stone monument and a plaque, set on a low cement base at the left downstream or northwest portal, are illuminated at night by an overhead light.

The only authentic covered bridge remaining in Franklin County, Switzer Bridge was listed on the National Register of Historic Places on September 6, 1974, the first covered bridge in Kentucky so listed.

Directions: From Frankfort, the county seat of Franklin County and the state capital, take US 460 east to SR 1689 (Switzer Road); go left to SR 1262 (Jones Lane); go right 1.0 mile to the bridge, on the left. Parking is available at either end of the bridge.

Bennett's Mill Covered Bridge

GREENUP COUNTY

Brief Statistics

Type: Authentic historic covered bridge. World Guide Covered Bridge Number: KY-45-01. Built in 1855 by B. F. Bennett and Pramley Bennett. One-span bridge crossing Tygarts Creek. Wheeler truss bridge 159.2 feet long by 20.9 feet wide, with approximately 16.7-foot wide by 13.4-foot high portal openings. Alternate Name: None known. Located southwest of Grays Branch on Tygarts Creek Road.

The brothers B. F. Bennett and Pramley Bennett operated Bennetts Mill on Tygarts Creek. To accommodate customers on the other side of the creek, they built Bennett's Mill Bridge, in 1855. The bridge they built is the longest single-span authentic historic covered bridge remaining in Kentucky. Bennett's Mill Bridge, at 20.9 feet wide, shares the distinction of being the widest authentic historic covered bridge in Kentucky with Cabin Creek Bridge, in Lewis County. The record flood of March 1, 1997, washed out one end of the historic bridge, bringing about its closure to motor traffic.

The 159.2-foot long bridge is supported by the only Wheeler truss in the southeastern United States. The sixteen-panel truss has a secondary chord (third chord) running down the middle of the truss. The truss is secured, by iron tie rods, to dry, large sandstone-block abutments, which have a small poured concrete cap under both ends of each lower chord. The sandstone blocks for the abutments were salvaged from the abandoned Globe Furnace, built near the bridge site, in 1830. The unusual flooring is constructed with oak planking, laid diagonally, in blocks that change direction every 16 to 18 feet. Seven-plank-wide wheel treads run the length of the floor. Weathered vertical poplar boarding covers the sides and the portals, the boarding on the portals with battens, and the bottom of the boards over the entrance V cut, to make a sawtooth edge. The roof is covered with rusted corrugated tin, which extends approximately five feet beyond the ends of the bridge, to shelter the portals.

The northeast/southwest-aligned bridge spans Tygarts Creek, which flows into the Ohio River, at South Shore. The wide creek has heavily wooded banks, the trees overhanging the creek near the bridge, impeding a clear view during the foliage seasons. There is a residence near the left downstream or northeast end.

One of two authentic historic covered bridges remaining in Greenup County, Bennett's Mill Bridge was listed on the National Register of Historic Places on March 26, 1976. The county seat of Greenup County is Greenup. *See Bennett's Mill Covered Bridge in the color photograph section: C-9.*

Directions: From Greenup, take US 23 north to the intersection with SR 10; go left 3.3 miles on SR 10 to SR 7; go right 0.9 mile on SR 7 to the bridge, on the right. Parking is available at the bridge.

Oldtown Covered Bridge

GREENUP COUNTY

Brief Statistics

Type: Authentic historic covered bridge. World Guide Covered Bridge Number: KY-45-02 #2. Built in 1999 by Intech Contracting, Inc. Two-span bridge crossing Little Sandy River. Multiple kingpost truss bridge 199.6 feet long by 18.8 feet wide, with approximately 15.0-foot wide by 14.3-foot high portal openings. Alternate Name: None known. Located in Oldtown on Frazer Road.

An unknown builder built the original Oldtown Bridge, in 1880, at a cost of $4,000. The historic bridge was bypassed by a new concrete bridge on the upstream side, in 1987, and closed to motor traffic. Intech Contracting, Inc., of Lexington, Kentucky, rebuilt Oldtown Bridge, from 1998 to 1999, with new materials, salvaging a few truss members and about 75 feet of decking. The two-span bridge is one of seven multiple kingpost truss covered bridges remaining in Kentucky, this unique bridge having a different length truss in each span. The 41-foot short span on the right downstream or northwest end consists of four panels of single vertical posts and braces; the 145-foot long span on the other end consists of fourteen panels of double posts and braces, with three sets of double vertical iron rods added at a later date.

The 199.6-foot long bridge is set on dry, large sandstone-block abutments that have dry, natural sandstone wing walls. The flooring is constructed with diagonal planking that has six-plank-wide wheel treads. Natural vertical boarding with battens covers the sides and gables, the sides wrapped around inside the portals. Red painted aluminum covers the roof, the previous structure having oak shakes. The roof and the gables extend beyond each portal, sheltering the entrances from inclement weather. A bridge informational plaque is at the northwest portal.

Century House Bridge is on the circular driveway at the left side of the residence in the background.

The heavily wooded Little Sandy River flows under the southeast/northwest-aligned bridge, before entering the Ohio River, at Greenup, the county seat. The new bypass bridge is a comfortable distance upstream. The northwest end overlooks open agricultural fields, while the southeast end has open sloping terrain.

One of two historic covered bridges remaining in Greenup County, Oldtown Bridge was listed on the National Register of Historic Places on March 26, 1976.

Directions: From Greenup, take US 23 south to SR 1; go right, through Oldtown, to Frazer Road, on the left, following 0.2 mile to the bridge, on the left.

Century House Covered Bridge

HARDIN COUNTY

Brief Statistics

Type: Non-authentic modern covered bridge. World Guide Covered Bridge Number: KY-47-A. Built in 1974 by unknown builder. One-span bridge crossing wet weather stream. Stringer bridge 18.8 feet long by 13.0 feet wide, with approximately 9.9-foot wide by 8.1-foot high portal openings. Alternate Name: None known. Located in Elizabethtown at 548 Bates Road.

Century House Bridge was built, in 1974, by an unknown builder, on the circular asphalt paved driveway to a residence, spanning a wet weather feeder stream to Freeman Lake. Four wood utility pole stringers, set on poured concrete abutments, support the 18.8-foot long stringer bridge. A 6-inch by 6-inch crossbeam, propped up by two double 4-inch by 4-inch wood posts with a concrete pillar between, all set on a concrete footing located near each abutment, lends additional support to the stringers. Transverse planking makes up the floor. The sides and portals, inside and outside, are covered with light yellow, facsimile-board plastic panels, mounted vertically. Opposing each other along both sides is a large central window opening, flanked by two smaller window openings; fern baskets hang in the four smaller window openings, adding color to the bridge. A *WELCOME* sign hangs from the right downstream or east-northeast entrance. Red-brown asphalt shingles, contrasting with the light yellow sides, cover the roof, which is supported by six 4-inch by 4-inch wood posts along each side. The roof eaves and window frames are painted white. Electric lanterns are mounted on each portal, at each side of the entrance, and an electric lantern hangs from the inside center of the bridge.

Aligned west-northwest/east-southeast, the single-span bridge, surrounded by mature trees, crosses the normally dry, bedrock streambed of an unnamed stream in the open side yard.

Directions: From Elizabethtown, at the intersection of US 31W (not Bypass US 31W) and SR 251, go 3.6 miles north on SR 251 (North Miles Street) to Forest Springs Subdivision, on the left, following the entrance road (Bates Road) 0.6 mile to No. 548, on the right. Private property: request permission to visit.

Cloverdown Farm Covered Bridge

JEFFERSON COUNTY

Brief Statistics

Type: Non-authentic modern covered bridge. World Guide Covered Bridge Number: KY-56-A. Built in 1986 by unknown builder. Two-span bridge crossing Long Run. Stringer bridge 40.2 feet long by 15.1 feet wide, with approximately 12.0-foot wide by 12.0-foot high portal openings. Alternate Name: None known. Located north of Boston at 1005 Long Run Road.

Built in 1986 by an unknown builder, Cloverdown Farm Bridge crosses Long Run in two spans, on the driveway to property marked *Lindale Place*.

The 40.2-foot long stringer bridge has six steel I-beam stringers, set on poured concrete abutments and one poured concrete pier. A concrete slab forms the floor. Vertical facsimile tongue and groove paneling covers the sides and portals of the gray painted bridge. Black asphalt shingles cover the roof. A garage-type overhead door secures the southwest or roadside portal. An electric lamp is mounted on the southwest portal, just below the peak.

A long fence-lined unpaved driveway leads to the southwest/northeast-aligned bridge. The bridge spans Long Run, which flows into Floyds Fork, near Fisherville, and then into the Salt River, near Shepherdsville, in Bullitt County. Cloverdown Farm Bridge is very exposed at both ends, but the heavily wooded banks of the run conceal the sides.

Directions: From Eastwood, go east on US 60 to Boston; go left on Long Run Road 1.3 miles to No. 1005, on the right. The bridge is visible from the roadway. Private property: request permission to visit.

Cabin Creek Covered Bridge

LEWIS COUNTY

Brief Statistics

Type: Authentic historic covered bridge. World Guide Covered Bridge Number: KY-68-03. Built in 1873 by unknown builder. One-span bridge crossing Cabin Creek. Multiple kingpost truss with Burr arch bridge 115.4 feet long by 20.9 feet wide, with approximately 15.9-foot wide by 11.1-foot high portal openings. Alternate Names: Hughes Farm Bridge, Jones Farm Bridge, Mackey Bridge, Mackey-Hughes Bridge, Rectorville Bridge. Located east of Plumville on old Cabin Creek Road.

Cabin Creek Bridge has had several names over the years; the origins of some are obscure. It is assumed that Hughes Farm Bridge and Jones Farm Bridge relate to adjoining farmsteads, and Mackey Bridge could also have been named after an adjacent property owner, whereas Rectorville Bridge derived its name from the nearby community of Rectorville, in Mason County. Built in 1873 by an unknown builder, Cabin Creek Bridge had major restoration work done, in 1914, by Louis Stockton Bower, Sr. At that time, iron rod counterbraces and the Burr arch were added. The 1¼-inch diagonal iron rods, acting as counterbraces, were added to the six panels at each side of the center kingpost, by passing a rod diagonally between the center of each of the double timber kingpost diagonal braces, forming an X, thus converting the multiple kingpost truss to a Childs truss. This same Childs truss conversion also occurred in Colville Bridge, in Bourbon County. (A Childs truss is a series of X panels formed from iron rod counterbraces and timber braces, utilizing a kingpost or an open panel in the center.) An example of a Childs truss with a kingpost center panel may be seen in the Harshman Bridge, in Preble County, Ohio. At 20.9 feet wide, Cabin Creek Bridge shares with Bennett's Mill Bridge the distinction of being the widest historic covered bridge in Kentucky.

Cabin Creek Bridge is one of seven multiple kingpost truss bridges remaining in Kentucky, unusual in that all of the truss members are doubled, with the end posts quadrupled, although five of the seven multiple kingpost trusses in Kentucky have this double truss member feature. A single Burr arch, consisting of five laminated 4-inch by 4-inch poplar timbers bolted together, was added by Louis Stockton Bower, Sr., during the 1914 restoration. The 115.4-foot long bridge rests on dry, rough-cut limestone abutments, the left downstream or east abutment having a poured concrete facing, which was added in 1981. Three poured concrete pillars supporting a large, crosswise, steel I-beam were added under the bridge, about ten feet from the east abutment, in the 1970s. This large steel I-beam extends about ten feet upstream to support a steel

I-beam buttress brace, which stabilizes the upstream lean of the bridge. Two poured concrete pillar supports were also added under the bridge, at the west end. Transverse oak planking with four-plank-wide wheel treads makes up the floor. Gray weathered, vertical boarding covers the sides and the gables; rusty tin covers the roof, which extends three feet beyond the truss posts, to provide shelter from the elements for the portals. An *11' 2"* diamond-shaped yellow caution sign is mounted above each entrance.

The single-span bridge crosses wide, rock-bottomed Cabin Creek, which flows into the Ohio River, at Springdale. The east/west-aligned bridge has a homestead on the west end and a roadway running parallel to the creek on the east end, which is open downstream to the concrete bypass bridge. Both creek banks are heavily wooded upstream and, on the downstream side, only the west side is wooded. Cabin Creek Bridge was bypassed and closed to motor traffic in 1983 and, unfortunately, abandoned since, resulting in a fair, but deteriorating, condition.

In 1947, Lewis County had four historic covered bridges, but, since 1962, only Cabin Creek Bridge remains. The aging bridge was listed on the National Register of Historic Places on March 26, 1976. Vanceburg is the county seat of Lewis County.

Directions: From Maysville, take SR 10 east to Plumville; go left 1.8 miles on Springdale Road to Cabin Creek Road; go right 1.8 miles to the stop sign on the bypass bridge. The bridge is on the right, with parking available at the portal.

Little Grassy Acres Covered Bridge

Lewis County

Brief Statistics

Type: Non-authentic modern covered bridge. World Guide Covered Bridge Number: KY-68-A. Built in 1987 by Charles H. "Herb" Johnson. One-span bridge crossing Grassy Branch. Stringer bridge 33.0 feet long by 14.4 feet wide, with approximately 12.9-foot wide by 10.0-foot high portal openings. Alternate Name: None known. Located in Vanceburg off SR 59.

Charles H. "Herb" Johnson built Little Grassy Acres Bridge, in 1987, on the driveway to his property, all lumber and timber for the bridge harvested from Johnson's Farm. The 33.0-foot long stringer bridge is supported by two 24-inch steel I-beam stringers, resting on poured concrete abutments, and two piers, constructed from 3-foot by 6-foot concrete blocks, capped with poured concrete. The abutments are below the ground surface, and the piers extend eight feet below ground level. The flooring consists of transverse planking with four-plank-wide wheel treads. Gray weathered, vertical boarding covers the sides and portals, the sides open down the middle, exposing the five 6-inch by 8-inch roof support timbers, which have diagonal braces between them, forming an X. Gleaming tin covers the roof.

The single-span bridge crosses Grassy Branch, which flows into Kinniconick Creek, which flows into the Ohio River, at Garrison. The west-northwest/east-southeast aligned bridge is situated in the open front yard, amid sycamore, elm and maple trees growing along the stream banks. A ford crosses the rocky banks and gravelly bottom of the stream, just downstream of the bridge.

Directions: From Vanceburg, at the intersection of SR 9/SR 10 and SR 59, take SR 59 south 4.2 miles to the bridge, on the right, a short distance back from the highway. Private property: request permission to visit.

R & B Frye Farm Covered Bridge

Lewis County

Brief Statistics

Type: Non-authentic modern covered bridge. World Guide Covered Bridge Number: KY-68-B. Built in 2000 by Rodney Frye. One-span bridge crossing Little Branch. Stringer bridge 30.3 feet long by 13.5 feet wide, with approximately 11.9-foot wide by 13.0-foot high portal openings. Alternate Name: None known. Located in Charters off SR 10.

R & B Frye Farm Bridge was built by Rodney Frye, the stringer bridge portion in 1999 with the cover added in 2000. Dale Dummitt built the abutments and piers. The bridge had been dubbed "The Bridge to Nowhere," as the bridge was completed prior to the construction of the house.

R & B Frye Farm Bridge has a 30.3-foot long covered center span, supported by four steel I-beam stringers, set on two poured concrete piers, with a 19.7-foot long approach ramp on the left downstream or south-southwest end and a 24.2-foot long approach ramp on the right downstream or north-northwest end. The approach ramps are supported by four 8-inch by 16-inch railroad trestle timber stringers and span from the poured concrete abutments to the piers. Thus, the stringer bridge is 74.1 feet long from abutment to abutment, in three spans, but only the center span is covered. The decking inside the bridge and extending to the end of the ramps is made up of transverse railroad ties. Vertical hemlock boarding covers the sides and steeply peaked portals. Tan painted aluminum covers the roof, which is supported along each side by five 4-inch by 6-inch wood posts.

The stringer bridge crosses gravel-bottomed Little Branch, flowing east into Salt Lick Creek, which flows into the Ohio River, at Vanceburg. The bridge is situated on the gravel and crushed stone driveway, elevated near the bridge with large, crushed-rock embankments, amid open fields, well back from State Road 10, at the tree lined stream paralleling the highway. *See R & B Frye Farm Covered Bridge in the color photograph section: C-10.*

Directions: From Vanceburg, go west on SR 10 to just beyond Charters (the junction with SR 989). The bridge is readily visible on the left, set back from the highway. Private property: request permission to visit.

Bagwell Covered Bridge

MARION COUNTY

Brief Statistics

Type: Non-authentic modern covered bridge. World Guide Covered Bridge Number: KY-78-A. Built in 1978 by Ronnie Bagwell. One-span bridge crossing Pontchartrain Creek. Stringer bridge 28.7 feet long by 14.2 feet wide, with approximately 13.1-foot wide by 9.7-foot high portal openings. Alternate Name: None known. Located south of Lebanon near Calvary at 2585 Old Calvary Pike.

Morris Bridge displays a horse and buggy weathervane on the cupola on the rooftop.

Built in 1978 by Ronnie Bagwell, the brother of the owner, Bagwell Bridge is at the entrance to a residence, on a gravel driveway. The 28.7-foot long stringer bridge is supported by four steel I-beam stringers set on poured concrete abutments. There are two steel angle buttress braces along each side to stabilize the cover. Inside the bridge, transverse planking forms the floor. Outside the bridge, light gray painted vertical boarding covers the sides and portals, brown asphalt shingles cover the roof. The sides have three small horizontal openings, under the eaves, to air the bridge. The roof, supported by eight 2-inch by 6-inch planks along each side, extends beyond the flooring, to provide shelter to the entrance. There is a swing gate inside the roadside portal to secure entry.

Aligned west-northwest/east-southeast, the single-span bridge crosses Pontchartrain Creek, running south into Rolling Fork, at Calvary, thence to the Salt River, before entering the Ohio River at West Point, on the Bullitt/Hardin County line. The bridge is fully exposed, with Old Calvary Pike at the left downstream or west-northwest end. The creek has wooded banks on the downstream side.

Directions: From Lebanon, take US 68 west to SR 208; go left 0.5 mile, continuing straight through (SR 208 goes to the right) on Country Club Road 0.6 mile to Old Calvary Pike (the sign says "Road") (SR 2744), on the right; go 2.1 miles to No. 2585, on the left. The bridge is at the entrance to the driveway. Private property: request permission to visit.

Morris Covered Bridge

MARSHALL COUNTY

Brief Statistics

Type: Non-authentic modern covered footbridge. World Guide Covered Bridge Number: KY-79-a. Built in 1979 by Roger Morris. One-span footbridge crossing wet weather stream. Stringer footbridge 34.1 feet long by 9.5 feet wide, with approximately 8.0-foot wide by 8.1-foot high portal openings. Alternate Name: None known. Located in Benton at 2408 Main Street.

Roger Morris built Morris Bridge, in 1979, over a cement-bottomed wet weather stream that bisected his large side yard, to facilitate his crossing the stream on his riding mower. There are no pathways to the south/north-aligned bridge, which sits conspicuously (and picturesquely) visible amid the maintained green grass side yard. The single-span bridge is closed to passage, as it stores a pop-up camper trailer. Morris Bridge, at 9.5 feet wide, is the widest covered footbridge in Kentucky.

Two wood utility pole stringers, set on the ground, support the 34.1-foot long stringer bridge. The floor consists of transverse boards with board wheel treads. Natural weathered vertical boarding covers the sides, portals and cupola, centered on the top of the roof peak. Black asphalt shingles cover the gambrel style roof, with the lower sides flared outward. Vivid white molding trims the rooflines. The roof is supported by four 6-inch by 6-inch wood posts along each side. A horse drawn buggy weathervane caps the cupola.

Directions: From Benton, at the intersection of US 641 and US 641/SR 58 with SR 58 and SR 408, go south 1.2 miles on US 641 (Main Street) to Haltom Drive, on the left. The bridge is in the southeast corner of Main Street and Haltom Drive. Private property: request permission to visit.

Dover Covered Bridge

Mason County

Brief Statistics

Type: Authentic historic covered bridge. World Guide Covered Bridge Number: KY-81-01. Built in 1835 by unknown builder. One-span bridge crossing Lee Creek. Double queenpost truss bridge 70.1 feet long by 16.3 feet wide, with approximately 12.4-foot wide by 11.7-foot high portal openings. Alternate Name: Lee (or Lee's) Creek Bridge. Located in Dover on Lee Creek Road (SR 3113).

Dover Bridge, named after the community of Dover, was built as a toll bridge by an unknown builder, in 1835, making it the second oldest covered bridge in Kentucky, and, for that matter, in the entire southeastern United States. Also referred to as Lee or Lee's Creek Bridge, after the creek beneath, Dover Bridge is one of four historic covered bridges in Kentucky still carrying motor traffic; the other three are Colville Bridge, Goddard Bridge and privately owned Valley Pike Bridge. In 1928, major repairs were made by the Bower Bridge Company (Louis Stockton Bower, Sr.). Ravaged by the elements, the bridge was struck by a tornado, on April 23, 1968, which blew the roof off, and damaged by the record flood of March 1, 1997.

The 70.1-foot long bridge is the only queenpost truss covered bridge remaining in Kentucky, and, a very unusual queenpost truss, at that. It is a double half-height queenpost truss, the horizontal beam extending three panels long (36 feet), with the diagonal braces in 14-foot long end panels. Within the center three panels, under the horizontal beam, is a second queenpost truss, the horizontal beam one panel long (12 feet), with the diagonal braces in 12-foot long end panels. There are double 1½-inch diameter iron rods going through the ends of the horizontal beams in both trusses, running down between the double vertical timber posts, and connected to the crossbeams beneath the lower chord, before the steel I-beams were added, in 1966. Now these rods hang loose below the lower chord and serve no function.

The single-span bridge originally rested on dry, natural limestone abutments, but, during the 1966 restoration by the Kentucky Highway Department, poured concrete was added to the creek side of the abutments to support the two massive steel I-beam stringers, added at that time. The flooring consists of transverse 2-inch by 4-inch pine planks, set on edge, with four-plank-wide wheel treads. Recently installed natural vertical boarding covers the sides and portals, and new tin covers the roof.

Dover Bridge carries tree-lined Lee Creek Road across wide, bedrock-bottomed Lee Creek, which flows into the Ohio River, one mile north of the bridge. The creek has heavily wooded banks near the north/south-aligned bridge. A bridge informational sign is near the bridge, at the junction of Lee Creek Road (State Road 3113) and State Road 8.

Listed on the National Register of Historic Places on March 26, 1976, Dover Bridge, along with privately owned Valley Pike Bridge, are the only historic covered bridges remaining in Mason County. The county seat of Mason County is Maysville.

Directions: From Augusta, take SR 8 east, past Dover, to SR 3113 (Lee Creek Road), on the left—SR 3113 is unmarked, but is the next left after SR 1235—go 0.2 mile to the bridge.

Trip Trap Covered Bridge

Mason County

Brief Statistics

Type: Non-authentic modern covered footbridge. World Guide Covered Bridge Number: KY-81-a. Built in 1976 by Claud "Bud" Highfield and Winona "Winnie" Highfield. Three-span footbridge crossing Sleepy Hollow

Creek. Stringer footbridge 34.1 feet long by 5.3 feet wide, with approximately 4.3-foot wide by 6.2-foot high portal openings. Alternate Name: None known. Located in Maysville off Mason-Lewis Road (SR 10).

Named after the children's fairytale, *The Three Billy Goats Gruff*, and their trip-trapping across the bridge, Trip Trap Footbridge was built, in 1976, by Claud "Bud" Highfield and Winona "Winnie" Highfield, in the front yard of their home. The 5.3-foot wide bridge is the narrowest covered footbridge in Kentucky.

The three-span stringer bridge is supported by two 2-inch by 6-inch stringers, resting on rocks on the ground, and five sets of two 4-inch by 4-inch wood posts, one on each side of the bridge, which extend up to the roof and support the roof, as well as the stringers. The floor is constructed with 2-inch by 4-inch planks laid transversely, with the center span elevated about fifteen inches, giving the bridge a humped appearance. The red painted bridge has open sides, with white painted lattice below the handrails, which run the length of the bridge. Gray asphalt shingles cover the roof. Above each entrance, painted in white, is *6' 6"/TRIP TRAP/BRIDGE.*

The 34.1-foot long bridge spans Sleepy Hollow Creek, which flows under the bridge, through a large steel conduit, to the nearby Ohio River. The east/west-aligned bridge runs parallel to and just off Mason-Lewis Road.

Directions: From Maysville, go east on SR 10 (Mason-Lewis Road) 0.4 mile past the junction with SR 2513, on the left and just after the large cemetery, on the left. The bridge is on the left. Private property: request permission to visit.

Valley Pike Covered Bridge

MASON COUNTY

Brief Statistics

Type: Authentic historic covered bridge. World Guide Covered Bridge Number: KY-81-02. Built in 1864 by unknown builder. One-span bridge crossing tributary to Lee Creek. Kingpost truss bridge 23.2 feet long by 16.5 feet wide, with approximately 14.8-foot wide by 12.4-foot high portal openings. Alternate Names: Bouldin Bridge, Daugherty Bridge. Located east of Fernleaf at 421 Valley Pike.

Valley Pike Bridge is off Valley Pike, from which it derives its name, on the driveway to the Bouldin Farm. In 1864, an unknown builder built Valley Pike Bridge, also known as Bouldin Bridge, after the family who has owned the bridge since its construction. The bridge has also been called Daugherty Bridge. At 23.2 feet long, Valley Pike Bridge is the shortest authentic historic covered bridge in Kentucky. It also is one of four historic covered bridges in Kentucky, one of two in Mason County, still carrying motor traffic, albeit this bridge only carries private traffic.

The single-span bridge has a two-panel single kingpost truss, resting on mortared natural limestone abutments, and is one of seven kingpost truss bridges remaining in Kentucky. In 1972, Louis Stockton "Stock" Bower, Jr., rebuilt the bridge, adding poured concrete caps to the existing abutments, upon which to set six 12-inch steel I-beam stringers, which now support the bridge. Old transverse oak planking makes up the floor and the left downstream or east-southeast end has 2-inch by 2-inch planks, spaced about 3 inches apart, to form a cattle guard, installed in the 1972 reconstruction. Weathered, vertical poplar boarding, most boards original, covers the sides and portals; rusted tin covers the roof.

The old weathered bridge spans a rocky tributary to Lee Creek, which flows north into the Ohio River, at Dover. There are mature trees along the gravel driveway and the narrow creek, upstream of the east-southeast/west-northwest aligned bridge. There are open fields in the four corners around the bridge.

Listed on the National Register of Historic Places on March 26, 1976, Valley Pike Bridge and Dover Bridge are the last historic covered bridges remaining in Mason County.

Directions: From Maysville, the county seat, go west on SR 10, continuing 5.7 miles after SR 10 joins SR 9 to SR 3056, on the right. Go 0.8 mile on SR 3056 to Valley Pike, on the left, following 1.5 miles to the bridge, on the right at No. 421. Private property: request permission to visit.

Hunter Lane Covered Bridge

OWEN COUNTY

Brief Statistics

Type: Non-authentic modern covered bridge. World Guide Covered Bridge Number: KY-94-A. Built in 1990 by Gary Hunter. One-span bridge crossing Hammond Creek. Stringer bridge 37.1 feet long by 12.3 feet wide, with approximately 11.0-foot wide by 9.5-foot high portal openings. Alternate Name: None known. Located southwest of Canby at 170 Pleasant Grove Road.

The Hunter family found that crossing the ford on Hammond Creek to reach their new home, at times of high water, was unsafe; therefore, they decided to build a bridge, giving consideration to a covered bridge. Gary Hunter recalled, from his youth, spent near the former Natlee Covered Bridge (KY-94-01), in Owen County, which was destroyed by fire in 1953, that some of the timbers escaped the fire and were salvaged. Gary Hunter acquired these timbers and used them in the construction of his bridge, in 1990. Seven years later, the great flood of March 1, 1997, washed the bridge off its abutments, carrying it downstream to come to rest against some trees. Fortunately, the Hunters were able to retrieve the bridge intact and reset it on the abutments.

The 37.1-foot long Hunter Lane Bridge is supported by five steel I-beam stringers, set on poured concrete abutments. Transverse planking with two-plank-wide wheel treads comprises the floor; natural vertical boarding covers the sides and the portals; and reddish-brown asphalt shingles cover the roof. Five 8-inch by 8-inch timber posts along each side support the roof. A sign reading *HUNTER LANE* is mounted on the left downstream or northwest portal.

The single-span bridge crosses Hammond Creek, which flows into nearby Eagle Creek, before reaching the Kentucky River, at Worthville. The northwest/southeast-aligned bridge is fully exposed amid the fields and moderately sloped, grassed hillside on the southeast end. A short split-rail fence on the roadside or northwest end guides traffic into the bridge. A poured concrete ford is adjacent to the bridge, on the downstream side.

Directions: From I-75, exit 144 SR 330, near Corinth, go west on SR 330 to SR 607, on the left; follow SR 607 to Pleasant Grove Road, on the right (before New Columbus); follow 4.8 miles to No. 170, on the left. The bridge is on the gravel driveway. Private property: request permission to visit.

Johnson Creek Covered Bridge

Robertson County

Brief Statistics

Type: Authentic historic covered bridge. World Guide Covered Bridge Number: KY-101-01. Built in 1874 by Jacob N. Bower. One-span bridge crossing Johnson Creek. Smith Type 4 truss with Burr arch bridge 114.5 feet long by 19.3 feet wide, with approximately 15.4-foot wide by 12.7-foot high portal openings. Alternate Name: Blue Licks Bridge. Located southwest of Alhambra on old SR 1029.

Johnson Creek Bridge, named after the creek beneath, was built in 1874 by Jacob N. Bower, the father of Louis Stockton Bower, Sr., who repaired this bridge and several other historic covered bridges in Kentucky (including Colville Bridge, also built by his father). Coincidentally, while the author was gathering data at Johnson Creek Bridge, a father arrived with his young son. Now residing in Ohio and visiting friends where he used to live, nearby, he said his own father would bring him to the bridge to play, when he was a boy. Now, he was bringing his son to play at the bridge, and, of course, to carve his initials next to his father's. *CRF.1906* is the oldest carving observed by the author inside the bridge.

The historic bridge has also been known as Blue Licks Bridge, after the nearby community of Blue Licks Spring. Johnson Creek Bridge was bypassed and closed to motor traffic in 1965, the concrete bypass bridge being a comfortable few hundred feet downstream. On September 26, 2001, the Federal Highway Administration approved $643,432, under the National Historic Covered Bridge Preservation Program, for restoration work on Johnson Creek Bridge.

The 114.5-foot long bridge has a twelve-panel Smith Type 4 truss, the only Smith truss covered bridge remaining in Kentucky. During the 1914 restoration, by Louis Stockton Bower, Sr., 1-inch diameter iron rods were added to the truss, connecting the upper chord to the lower chord. At the same time, Mr. Bower added a 4-inch by 20-inch laminated Burr arch, constructed from four 4-inch by 5-inch poplar timbers bolted together, and with ten 1-inch diameter iron rods spaced the length of the arch, tying the arch to the lower chord. The bridge rests on mortared rough-cut limestone abutments, the right downstream or southwest end abutment placed on top of dry limestone blocks. In 1914, poured concrete facings were added under each lower chord, at the abutments, for arch support and to reinforce the abutment supporting the lower chord. A mortared rough-cut limestone pier was added, under the middle of the bridge, at a later date. A new transverse planked floor was installed sometime after 1985, as was the gray weathered, vertical boarding covering the sides and the gables. Rusted tin covers the very low peaked roof. The roof extends about two feet beyond the entrances, providing shelter from the weather.

The single-span bridge crosses Johnson Creek, which flows into the Licking River, west of Piqua. The northeast/southwest-aligned bridge is very open all around, with woods along the downstream creek banks and on the southwest end. The bypassed section of State Road 1029 crosses the bridge, paralleling the downstream creek bank to the far side of the bypass bridge.

The only historic covered bridge in Robertson County, Johnson Creek Bridge was listed on the National Register of Historic Places on September 27, 1976. *See Johnson Creek Covered Bridge in the color photograph section: C-11.*

Directions: From Mt. Olivet, the county seat, take US 62 east through Sardis to SR 1029, on the right; go 4.3 miles to the bridge, on the left.

Mullins Station Covered Bridge

ROCKCASTLE COUNTY

Brief Statistics

Type: Non-authentic historic covered bridge. World Guide Covered Bridge Number: KY-102-A #2. Built in 1997 by N & N Construction. Five-span bridge crossing Roundstone Creek. Stringer bridge 132.1 feet long by 16.4 feet wide, with approximately 15.1-foot wide by 15.0-foot high portal openings. Alternate Name: None known. Located in Mullins off Mullins Station Road at Kentucky Powder Co.

Mullins Station Bridge was built in two stages, the stringer bridge, in 1931, with a cover, added in 1952, which was blown down in a severe storm, in 1997. A new cover was built by N & N Construction of Mount Vernon, Kentucky, later in 1997. The original cover was built by employees of Kentucky Stone Company, the former name of Kentucky Powder Company. Now owned by the Kentucky Powder Company and located at their mines, off Mullins Station Road, the 132.1-foot long bridge has the distinction of being the longest non-authentic motor vehicle covered bridge in Kentucky.

The stringer bridge is supported by five steel I-beam stringers, resting on poured concrete abutments, and three poured concrete piers. Each pier has five upright steel I-beams, set on top of the poured concrete, and a steel I-beam placed crosswise, capping the five steel I-beams. Longitudinal planking is laid across 8-inch by 8-inch timber floor joists. New, gray-painted, vertical aluminum siding covers the sides, gables and roof. The roof is supported by thirty-four 4-inch by 4-inch wood posts along each side. Each side of the bridge has nine buttress braces of steel pipe and steel I-beam construction, added during the 1997 rebuilding, to stabilizing the structure. Steel frame and heavy-duty-wire double swing gates, enclosing the entire entrance opening, secure the left downstream or north-northwest portal.

The five-span bridge crosses Roundstone Creek, flowing into nearby Crooked Creek, which enters the Rockcastle River, just south of Livingston. The bridge is about 65 feet above the heavily wooded, steep banked, rock-bottomed creek. There is a rock escarpment with two large mine tunnels, near the south-southeast end, and gravel roadways on the north-northwest end.

Directions: From I-75, exit 59 US 25 (east of Mt. Vernon), go east 3.2 miles on US 25 to Mullins Station Road (Forest Road 465), on the left (road is unmarked, across from Pine Hill Holiness Church). Follow Mullins Station Road 2.4 miles, crossing the railroad tracks four times, to the bridge, on the right, across from mine tunnels.

Beckham Covered Bridge

SCOTT COUNTY

Brief Statistics

Type: Non-authentic modern covered bridge. World Guide Covered Bridge Number: KY-105-A. Built in 1976 by Richard Beckham. One-span bridge crossing North Rays Fork. Stringer bridge 37.7 feet long by 8.8 feet wide, with approximately 7.7-foot wide by 6.7-foot high portal openings. Alternate Name: None known. Located in Corinth off North Rays Fork Road.

Richard Beckham built Beckham Bridge, in 1976, on the gravel driveway to his property. The 37.7-foot long bridge is the longest single-span non-authentic covered bridge in Kentucky, that carries motor traffic. At 8.8 feet wide, the bridge is also the narrowest non-authentic covered bridge in Kentucky, that carries motor traffic.

The stringer bridge is supported by two steel I-beam stringers, set on earth and large stone abutments. The flooring is composed of longitudinal planking with multi-plank wheel treads. The sides are the aluminum sides of a trailer van. Galvanized metal covers the gabled roof; horizontal lapped boarding covers the gables.

The west/east-aligned bridge crosses North Rays Fork, which joins South Rays Fork nearby, to form Rays Fork, which then joins Little Eagle Creek, to form Eagle Creek, which enters the Kentucky River, at Worthville. A ford crossing is just upstream of the exposed bridge while light woods are on the downstream side.

Directions: From Corinth, at the junction of SR 330 and US 25, go south 1.3 miles on US 25 to North Rays Fork Road, on the right; follow 2.0 miles to the bridge, on the left. Private property: request permission to visit.

Hycliffe Manor Covered Bridge

SHELBY COUNTY

Brief Statistics

Type: Non-authentic modern covered bridge. World Guide Covered Bridge Number: KY-106-A. Built in 1980 by unknown builder. One-span bridge crossing Lincoln Todd Creek. Stringer bridge 30.6 feet long by 17.0 feet wide, with approximately 13.0-foot wide by 10.9-foot high portal openings. Alternate Name: None known. Located south of Conner at 1150 Conner Station Road.

Hycliffe Manor Bridge was built, in 1980, by an unknown builder, on the driveway to the property, set a considerable distance from Conner Station Road. The 17.0-

Beech Fork Bridge, measuring 225.0 feet long, is the longest historic covered bridge in Kentucky.

foot wide bridge is the widest non-authentic covered bridge in Kentucky carrying motor traffic.

Seven stringers, set on poured concrete abutments, support the 30.6-foot long bridge. The stringers consist of three large steel C-beams, two small steel I-beams, and two timbers, the latter placed at the outside. Transverse planking makes up the floor; vertical boarding covers the sides and the portals; and wood shingles cover the roof. Small, opposing, horizontal rectangular window openings are cut into each side. The roof is supported by three 6-inch by 6-inch timber posts along each side. A large wood filigreed embellishment is mounted on the left downstream or west gable. An electric carriage lantern, mounted on a round steel post, is near the west end.

Crossing Lincoln Todd Creek, the single-span bridge is very exposed, amid rolling pastures. There are a few mature trees along the banks of the narrow creek. The long and winding asphalt-paved driveway has black-painted rail fencing along both sides, leading to the bridge. *See Hycliffe Manor Covered Bridge in the color photograph section C-9.*

Directions: From I-64, take exit 28, near Simpsonville; go south on SR 1848, immediately going right on Veechdale Road (SR 1399) 4.2 miles to SR 148 (unmarked); go right 0.6 mile to Conner Station Road, on the right, following 2.4 miles to No. 1150, on the right. The bridge is on the driveway, set back from the road. Private property: request permission to visit.

Beech Fork Covered Bridge

WASHINGTON COUNTY

Brief Statistics

Type: Authentic historic covered bridge. World Guide Covered Bridge Number: KY-115-01. Built in 1865 by L. H. Barnes and William F. Barnes. Two-span bridge crossing Beech Fork. Multiple kingpost truss with Burr arch bridge 225.0 feet long by 18.9 feet wide, with approximately 15.0-foot wide by 10.9-foot high portal openings. Alternate Names: Mooresville Bridge, Mount Zion Bridge. Located north of Mooresville on old SR 458.

Built in 1865 by L. H. Barnes and William F. Barnes, Beech Fork Bridge originally carried the Springfield and Chaplin Turnpike across Beech Fork. Named after the stream beneath, Beech Fork Bridge has also been called Mooresville Bridge, after the nearby community, and Mount Zion Bridge. The 225.0-foot long bridge is the longest historic covered bridge remaining in Kentucky.

Beech Fork Bridge has a twelve-panel multiple kingpost truss with a double Burr arch, in each of its two spans, and is one of seven kingpost truss covered bridges in Kentucky. The truss is constructed with poplar timbers sandwiched between two 4-inch by 20-inch timber arches, bolted together through the truss members. The bridge rests on dry, natural limestone abutments that have been stucco covered, the left downstream or east abutment reinforced at a later date with poured concrete, and a central pier. The pier was constructed with mortared rough-cut limestone, the upstream end crumbling and eventually falling away to the base. In the fall of 1982, a local stone mason rebuilt the upstream end of the pier with large sandstone blocks, which contrast perceptibly with the smaller limestone. The floor is constructed with transverse oak planking with a full width, oak plank wheel tread. Gray weathered, vertical boarding covers the sides and gables; three steel bands were added along the length of the sides, at a later date. The roof is covered with faded brown-painted aluminum. Three small, vertical, rectangular window openings were cut through the siding, on the upstream side, after the bridge was bypassed and closed to motor traffic, in 1975.

Beech Fork flows beneath the two-span bridge, the creek entering Rolling Fork near Boston, then flowing into the Salt River, which empties into the Ohio River, at West Point. The east/west-aligned bridge is open upstream to the concrete bypass bridge, a very comfortable distance away. The wide and deep creek is heavily wooded downstream of the bridge. Steel barriers are at each portal, barring vehicle passage across the bridge on the bypassed section of State Road 458, which offers parking at both ends.

Listed on the National Register of Historic Places on March 26, 1976, Beech Fork Bridge is the only historic covered bridge remaining in Washington County. Springfield is the county seat.

Directions: From the Blue Grass Parkway, west of Chaplin, at exit 34, go south on SR 55 to Mooresville and SR 458, on the left. Follow SR 458 for 2.3 miles to the bridge, on the left. Parking is available at either end of the bridge.

Lincoln Homestead State Park Covered Bridge

WASHINGTON COUNTY

Brief Statistics

Type: Non-authentic modern covered footbridge. World Guide Covered Bridge Number: KY-115-a. Built in 1963 by W. E. Wilkerson and Bill Padgett. One-span footbridge crossing Lincoln Run. Stringer footbridge 31.9 feet long by 6.6 feet wide, with approximately 5.3-foot wide by 6.4-foot high portal openings. Alternate Name: None known. Located north of Springfield in Lincoln Homestead State Park.

W. E. Wilkerson and Bill Padgett, local State Park employees, built Lincoln Homestead State Park Bridge, in 1963, making it the oldest non-authentic covered footbridge in Kentucky. The bridge was built to provide State Park visitors access to the replica of a blacksmith shop, similar to one where Abraham Lincoln's father, Thomas, learned the blacksmith trade.

Mortared natural stone abutments support the four wood utility pole stringers of the 31.9-foot long footbridge. Transverse planks form the flooring; gray weathered, vertical boarding covers the sides and the gables; wood shingles cover the roof. Twelve studs along each side support the roof. A decorative Burr arch is inside the bridge.

The east-southeast/west-northwest aligned bridge has a golf course on the upstream side, with a tee above the east-southeast stream bank. Netting is strung, just upstream of the bridge and blacksmith shop, to shield visitors, the bridge, and the blacksmith shop from errant golf balls. A short distance upstream from the bridge, State Road 438 bisects the golf course. The blacksmith shop is on the west-northwest or right downstream end. Several mature trees are downstream of the bridge, some overhanging the structure. The single-span bridge crosses frequently dry, bedrock-bottomed Lincoln Run, which flows into nearby Beech Fork, then into Rolling Fork, near Boston, eventually reaching the Salt River, before entering the Ohio River, at West Point.

Directions: From the Blue Grass Parkway, west of Chaplin, at exit 34, go south on SR 55 through Mooresville to SR 438, on the left; follow 2.1 miles to Lincoln Homestead State Park and the bridge, on the left.

Maryland

Maryland, at one time, had over fifty-two covered bridges. By 1937, only seventeen historic covered bridges remained, dwindling to eleven, in 1947, and by about one bridge every decade thereafter, until only six remained in 1989. Fortunately, Maryland has been able to preserve those six covered bridges to the present time.

Of the twenty-three covered bridges in Maryland, six are authentic and seventeen are non-authentic; sixteen of the non-authentic bridges are stringer type bridges. Seven of the twenty-three bridges are historic, five authentic and two non-authentic, and were built between 1850 and 1950. The remaining sixteen modern bridges were built between 1965 and 1994. Interestingly, the six authentic bridges are all kingpost truss bridges, comprised of one single kingpost truss, one multiple kingpost truss, and four multiple kingpost truss with Burr arch.

Maryland's twenty-three covered bridges are located in nine of the state's twenty-three counties, those nine counties north and west of Chesapeake Bay. Five of the six authentic covered bridges are located in Cecil and Frederick Counties, with the sixth bridge joining Baltimore and Harford Counties, and five of the six authentic covered bridges still serve the traveler today.

The Betty Ruth Covered Bridge

ALLEGANY COUNTY

Brief Statistics

Type: Non-authentic modern covered bridge. World Guide Covered Bridge Number: MD-01-A. Built in 1993 by Roy Romsburg. One-span bridge crossing Town Creek. Stringer bridge 64.3 feet long by 12.3 feet wide, with approximately 10.7-foot wide by 13.1-foot high portal openings. Alternate Name: Betty Ruth Bridge. Located east of Flintstone at 23200 Romsburg Lane.

Roy Romsburg supervised the construction of the cover for The Betty Ruth Bridge, over the existing stringer bridge, in 1993. The stringer bridge is supported by two large steel I-beam stringers, set on old concrete abutments, with new concrete poured on top and poured concrete wing walls, with broken rock extensions restrained by baling wire. The floor of the stringer bridge has transverse planking with three-plank-wide wheel treads. The red painted cover has vertical boarding on the sides and gables, the sides with eye level openings running the length of the bridge. Light red-brown asphalt shingles cover the exceptionally high, low peaked roof. A sign reading *The Betty Ruth*, in black letters on a white background, is mounted over the right downstream or west-southwest entrance. A short split-rail fence funnels traffic into the single-lane bridge, at each portal. The 64.3-foot long bridge is the longest non-authentic covered motor vehicle bridge in Maryland. The impressive bridge dominates the surrounding open fields, which contain a few scattered mature trees. Moderately wide, shallow, rocky-bottomed Town Creek passes beneath the single-span bridge, flowing southward into the Potomac River, at Town Creek. *See The Betty Ruth Covered Bridge in the color photograph section: C-15.*

Directions: From Flintstone, at I-68 exit 56, Williams Road, go south on Williams Road 0.2 mile to National Pike; go left 1.2 miles to Dry Ridge Road, on the left; follow 0.6 mile to 23200 Romsburg Lane, on the right; follow 0.5 mile to the residence. Private property: request permission to visit.

Jericho Road Covered Bridge

BALTIMORE COUNTY/HARFORD COUNTY

Brief Statistics

Type: Authentic historic covered bridge. World Guide Covered Bridge Number: MD-03-02/MD-12-01.

Built in 1993 at a length of 64.3 feet across Town Creek, The Betty Ruth Bridge is the longest stringer motor vehicle covered bridge in Maryland. This is an upstream view. ***See The Betty Ruth Covered Bridge in the color photograph section: C-15.***

Built in 1865 by Thomas F. Forsyth. One-span bridge crossing Little Gunpowder Falls. Multiple kingpost truss with Burr arch bridge 89.1 feet long by 19.2 feet wide, with approximately 13.0-foot wide by 13.0-foot high portal openings. Alternate Name: Jericho Bridge. Located in Jerusalem on Jericho Road.

Thomas F. Forsyth built Jericho Road Bridge, in 1865, at a cost of $3,125. Also called Jericho Bridge, the old bridge carries Jericho Road over Little Gunpowder Falls. In 1937, a queenpost truss was added to reinforce the Burr arch. Stretching across the entire ten panels, the queenpost truss was constructed by placing a wide timber at each side of the end panel kingpost braces and placing wide timbers horizontally across the eight center panels, sandwiching the tops of each multiple kingpost and running inside the top of the double arch. The queenpost was securely bolted through the kingpost end braces, the tops of each intersection with the multiple kingpost, and the upper part of the Burr arch. Quite some time prior to 1975, steel I-beam stringers were added under the bridge, to facilitate the heavier loads carried by the bridge. Due to unsafe conditions, the bridge was closed in 1980, with restoration work commencing in 1981, to further strengthen the structure, install a new floor, repair the abutments and approaches, and repaint the bridge. Absent for many years, the gables were re-installed in 2002.

The 89.1-foot long bridge has a ten-panel multiple kingpost truss encased between Burr arches, with the queenpost truss added in 1937. Jericho Road Bridge is one of four multiple kingpost truss bridges with a Burr arch remaining in Maryland. The single-lane bridge gains additional support from five steel I-beam stringers, added prior to 1975, that are set on mortared rough-cut stone abutments with wing walls. The flooring is made up of transverse 2-inch by 4-inch planks, set on edge, that were installed during the 1981 restoration. Vertical boarding covers the sides and the gables; cedar shingles cover the roof. The roof extends beyond the entrances, providing extra protection for the interior from inclement weather. The exterior of the bridge wears the weathered and peeling brown paint that was applied in 1981. At the same time, the interior was also painted brown.

The single-span bridge crosses wide, rocky bottomed

LOAD
LIMIT
2 TONS

and boulder strewn Little Gunpowder Falls, which merges with Bird River to form Gunpowder River, below Joppatowne, both rivers estuaries of Chesapeake Bay. Dense woods surround the south/north-aligned Jericho Road Bridge.

With only six authentic covered bridges remaining in Maryland, Jericho Road Bridge is one of the five still open to motorized traffic. It is the only covered bridge joining two counties and the only historic covered bridge remaining in Baltimore and Harford Counties. The old bridge was listed on the National Register of Historic Places on September 13, 1978.

Directions: From Perry Hall, northeast of Baltimore, take US 1 north to Kingsville; go right 1.9 miles on Jerusalem Road to Jericho Road, on the right; go 0.3 mile to the bridge.

Foxcatcher Farms Covered Bridge

CECIL COUNTY

Brief Statistics

Type: Authentic historic covered bridge. World Guide Covered Bridge Number: MD-07-02. Built in 1860 by Ferdinand Wood. One-span bridge crossing Big Elk Creek. Multiple kingpost truss with Burr arch bridge 80.2 feet long by 16.0 feet wide, with approximately 12.0-foot wide by 13.0-foot high portal openings. Alternate Name: Big Elk Bridge, Dupont Bridge, Hills Fording Bridge, Strahorn's Mill Bridge. Located in Fair Hill on Tawes Drive.

In 1860, at a cost of $1,165, Ferdinand Wood built Foxcatcher Farms Bridge, on what was to become the privately owned Dupont Estate, in the area called Foxcatcher Farms. Erik K. Straub, in 1992, rebuilt the bridge, which was originally called Hills Fording Bridge, the name of the bridge site; and also called Strahorns Mill Bridge, after Joseph Strahorns Saw Mill located nearby; Big Elk Bridge, from the creek beneath; and Dupont Bridge. *P.A.P. APRIL 6, 1916*, is the oldest of the few hand carvings found in the bridge by the author. The State of Maryland acquired the Dupont Estate, in 1975, ten years after the death of William Dupont, Jr. At 16.0 feet wide, Foxcatcher Farms Bridge shares with Roddy Road Bridge, in Frederick County, the distinction of being the narrowest authentic historic covered bridge remaining in Maryland.

The 80.2-foot long bridge has a six-panel, single-post, multiple kingpost truss, sandwiched between double, segmented, half-height Burr arches, resting on mortared, rough-cut stone abutments with wing walls, one of four authentic covered bridges in Maryland with the same truss. The vertical tension members of the kingpost slant away from the top of the center kingpost in this bridge. The flooring has transverse planking with three-plank-wide wheel treads. Red painted, horizontal lapped siding covers the sides, open under the eaves, allowing air circulation within the bridge, with a white painted stripe along the uppermost siding board. The red painted portals have horizontal lapped siding in the gables, with vertical boarded pilasters at each side of the entrance, the roof fascia board painted white for contrast. Black weathered wood shingles cover the roof, which has a small wooden spire on the peak over each gable.

Big Elk Creek flows southward beneath the single-span bridge, joining the Elk River, an estuary of Chesapeake Bay. The wide, rock and sediment bottomed creek has dense foliage along the east-southeast bank and at the upstream side of the west-northwest bank. The west-northwest bank on the downstream side is relatively open, with a large mature tree near the bridge. An excellent view of the bridge may be seen from the creek on the downstream side.

Foxcatcher Farms Bridge is one of the five authentic covered bridges in Maryland open to traffic. This bridge and Gilpin's Falls Bridge, now closed, are the only authentic historic covered bridges remaining in Cecil County, for which Elkton is the county seat.

Directions: From Fair Hill, at the intersection of SR 213 and SR 273, go north on SR 273 to Training Center Road at Fair Hill Training Center, on the right. Follow winding Training Center Road 1.8 miles to Tawes Drive, on the left (sign to Nature Center and Covered Bridge); follow 1.2 miles to the bridge. A large parking area is on the left.

Gilpin's Falls Covered Bridge

CECIL COUNTY

Brief Statistics

Type: Authentic historic covered bridge. World Guide Covered Bridge Number: MD-07-01. Built in 1860 by Joseph George Johnson. One-span bridge crossing North East Creek. Multiple kingpost truss with Burr arch bridge 119.6 feet long by 18.1 feet wide, with approximately 13.5-foot wide by 12.1-foot high portal openings. Alternate Names: Bayview Bridge, Gilpin Bridge. Located in Bay View on bypassed section of Northeast Road.

Gilpin's Falls Bridge was built, in December 1860, by Joseph George Johnson of North East, Maryland, at a cost

***Opposite:* The beautifully preserved Foxcatcher Farms Bridge was built in 1860 across Big Elk Creek. Note the spire above the right downstream portal.**

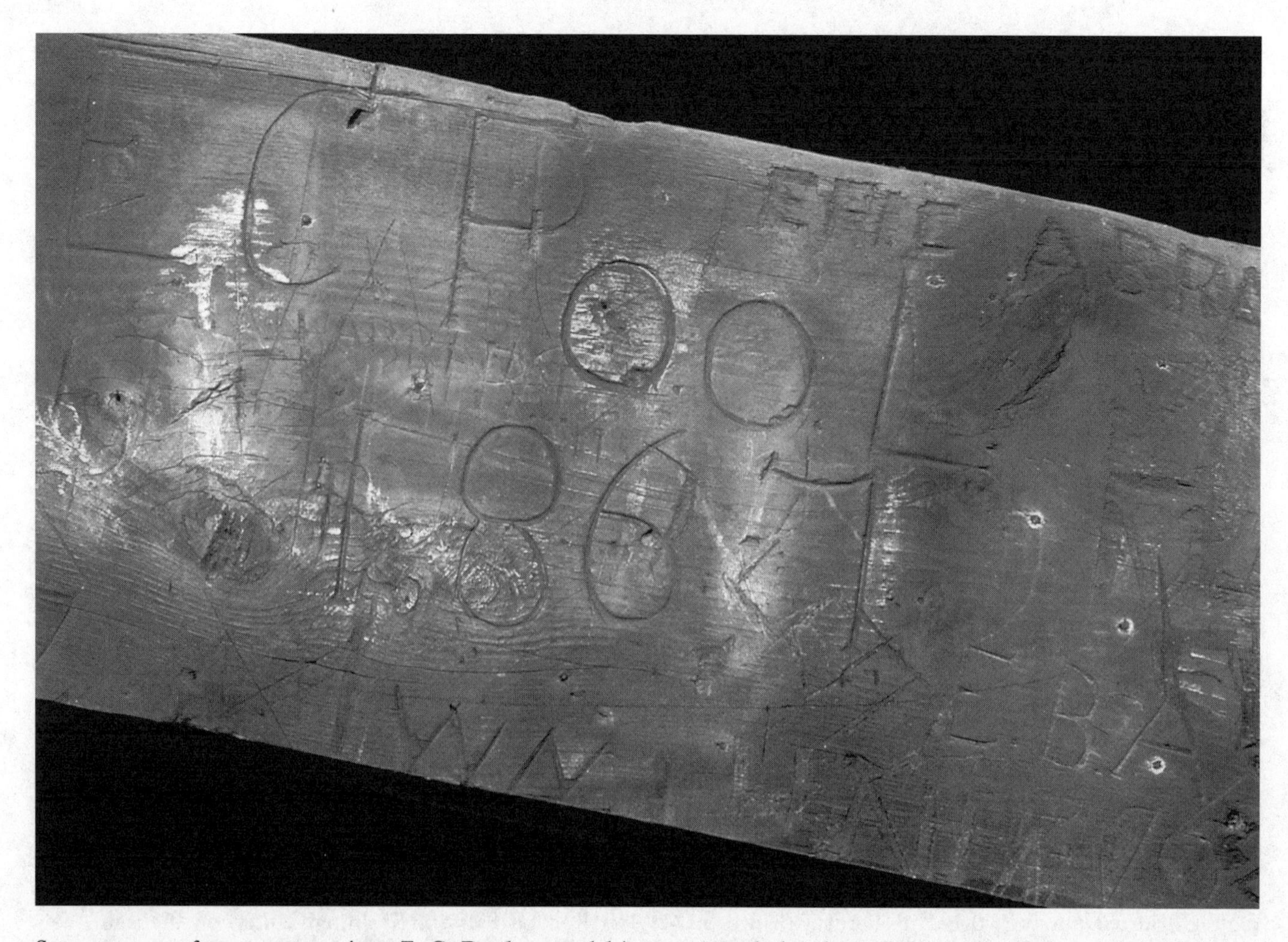

Seven years after construction, E. C. Pool carved his name and the date 1867 on the Burr arch in Gilpin's Falls Bridge. ***See Gilpin's Falls Covered Bridge in the color photograph section: C-13.***

of $2,000, near the site of several mills, the earliest built circa 1735 by Samuel Gilpins. The Gilpins Falls, whence the bridge derived its name, and remnants of the millrace are just downstream of the bridge. The single-lane bridge has also been known as Gilpin Bridge and Bayview Bridge, derived from the local community of Bay View. Gilpin's Falls Bridge is the only authentic covered bridge in Maryland closed to traffic, having been bypassed by the concrete bridge, just upstream, between 1930 and 1937. The old bridge was near collapse when it was restored, in 1959, by the joint efforts of the State of Maryland Roads Commission and the Historical Society of Cecil County. Gilpin's Falls Bridge had extensive repairs, including repainting, in 1990. Owned by Cecil County, the 119.6-foot long bridge is the longest authentic covered bridge remaining in Maryland, which in the early 19th century, had its longest covered bridge measured at an incredible 4,170 feet. (This was the Rock Run Bridge, built in 1817 by Theodore Burr, over the Susquehanna River, joining Cecil and Harford Counties, at Rock Run near Port Deposit. It met its fate in the winter of 1823, when a sleigh with iron runners was dragged through the snow-free interior, causing sparks, when the runners struck nails in the flooring, setting the structure on fire.)

Gilpin's Falls Bridge has a twelve-panel, single-post, multiple kingpost truss sandwiched between double Burr arches, set on dry, natural stone abutments that have a poured concrete facing and stuccoed wing walls. Diagonal planking comprises the flooring; horizontal lapped siding covers the sides and portals; old dark weathered wood shakes cover the roof. Aged and faded red paint covers the bridge, which has several siding boards missing on the downstream side. The sides are open under the eaves, and centered on each side is a long horizontal window opening with a small roof over it.

Aligned west-southwest/east-northeast, the single-span bridge crosses the placid waters of North East Creek, which flows into the North East River, an estuary of Chesapeake Bay, at North East. There are open grassy park areas at both ends of the bridge, the park areas containing scattered mature trees, heavily foliated beyond. State Road 272 and the bypass bridge are close on the upstream side. The bridge is in need of repairs, with a serious sag on the downstream side, and the abutment under the Burr

Set amongst Weeping Willow trees over Carroll Creek, Baker Park Bridge, built in 1965, is especially photogenic in the morning. *This bridge is depicted on the front cover of this book.*

arch, also on the downstream side, broken away. *KEEP OFF/BRIDGE UNSAFE* signs are at each entrance. The oldest legible graffiti found in the bridge is *E.C.PooL/1867*, carved on the Burr arch seven years after the bridge was built; however, an 1865 is discernible on the same arch.

Gilpin's Falls Bridge and Foxcatcher Farms Bridge are the only surviving authentic covered bridges in Cecil County. Elkton is the county seat of Cecil County. *See Gilpin's Falls Covered Bridge in the color photograph section: C-13.*

Directions: From I-95, north of North East, at exit 100 SR 272, go north 1.2 miles on SR 272 to the bridge, on the right.

Baker Park Covered Bridge

FREDERICK COUNTY

Brief Statistics

Type: Non-authentic modern covered footbridge. World Guide Covered Bridge Number: MD-10-a. Built in 1965 by unknown builder. One-span footbridge crossing Carroll Creek. Stringer footbridge 42.4 feet long by 11.6 feet wide, with approximately 6.9-foot wide by 7.7-foot high portal openings. Alternate Name: Frederick Bridge. Located in Frederick in Baker Park.

Built in 1965 by an unknown builder, Baker Park Bridge, also called Frederick Bridge, is located in Baker Park in Frederick. Baker Park Bridge, at 42.4 feet long, is the longest single-span non-authentic covered footbridge in Maryland.

The stringer footbridge is supported by five steel I-beam stringers, set on mortared natural stone abutments. The red painted bridge has a transverse planked floor, horizontal boarded sides and portals, and a gray-black asphalt-shingled roof. The interior of the bridge is painted black. Two long rectangular window openings are centered on each side. Seven 6-inch by 6-inch wood posts along each side support the steeply peaked roof. Low, mortared natural-stone guard walls lead pedestrians to each portal.

The west/east-aligned Baker Park Bridge is in a fully exposed setting, flanked by very large Weeping Willow trees in an expansive city park. The photogenic bridge spans placid Carroll Creek, which flows eastward into the nearby Monocacy River. There is a small dam on the medium-wide creek, downstream of the bridge.

Directions: From downtown Frederick, at the intersection of SR 355 (Market Street) and West Second Street, go 0.9 mile west on West Second Street to Baker Park, on the left. The bridge is beyond the baseball field, in the 700 block of West Second Street.

Holly Hills Country Club No. 1 Covered Bridge

FREDERICK COUNTY

Brief Statistics

Type: Non-authentic modern covered footbridge. World Guide Covered Bridge Number: MD-10-c. Built in 1972 by unknown builder. One-span footbridge crossing Long Branch. Stringer footbridge 17.0 feet long by 11.6 feet wide, with approximately 6.5-foot wide by 6.8-foot high portal openings. Alternate Name: None known. Located in Ijamsville at Holly Hills Country Club at 5502 Mussetter Road.

Holly Hills Country Club has three small covered footbridges on its eighteen-hole golf course, all of similar appearance and construction. Holly Hills Country Club No. 1 Bridge was built, in 1972, by an unknown builder near the 6th tee. The 17.0-foot long footbridge is supported by three steel I-beam stringers, set on poured concrete abutments. The bridge is constructed in a trapezoidal configuration, placed diagonally over the stream. The floor has diagonal planking; the sides and portals have weathered vertical boarding with battens; and the roof has dark weathered wood shakes. The roof and gables extend beyond the entrances, shielding the interior from the weather, the roof supported by three 6-inch by 8-inch wood posts along each side. The sides have two horizontal window openings, non-opposing, on each side. A log bench runs along each side of the interior. Above the left downstream or north-northeast entrance, mounted on the gable, a four-line sign proclaims *You are now entering/"The President's Loop"/President George H. W. Bush/ June 22, 1990.*

The single-span bridge crosses narrow Long Branch, which joins Linganore Creek, just upstream of its confluence with the Monocacy River, east of Frederick. The exposed bridge is situated near the 6th tee, with a small tree at each side and a large mature tree on the upstream side, between the bridge and the 6th tee. *See Holly Hills Country Club No. 1 Covered Bridge in the color photograph section: C-16.*

Directions: From New Market, at I-70 exit 60 SR 75 (Green Valley Road), go north a short distance to Old National Pike. Go left to Mussetter Road, on the left; go 0.2 mile to Ritchie Drive, on the right; go 0.7 mile to Holly Hills Country Club clubhouse, on the left.

Holly Hills Country Club No. 2 Covered Bridge

FREDERICK COUNTY

Brief Statistics

Type: Non-authentic modern covered footbridge. World Guide Covered Bridge Number: MD-10-d. Built in 1972 by unknown builder. One-span footbridge crossing Long Branch. Stringer footbridge 16.2 feet long by 11.5 feet wide, with approximately 6.5-foot wide by 7.5-foot high portal openings. Alternate Name: None known. Located in Ijamsville at Holly Hills Country Club at 5502 Mussetter Road.

Holly Hills Country Club has three small covered footbridges on its eighteen-hole golf course, all of similar appearance and construction. Holly Hills Country Club No. 2 Bridge was built, in 1972, by an unknown builder near the 6th green. The 16.2-foot long stringer footbridge, built in a conventional rectangular configuration, is supported by three steel I-beam stringers, set on poured concrete abutments. The floor is constructed with transverse planking. Weathered vertical boarding with battens covers the sides and portals, the sides with two horizontal window openings, opposing, on each side. Dark weathered wood shakes cover the roof that, along with the gables, extends beyond the entrances for weather protection. The roof is supported by three 6-inch by 8-inch wood posts along each side. The interior has a log bench on each side, running the length of the bridge. Above the right downstream or southwest entrance, mounted on the gable, a five-line sign proclaims *You have now completed/"The toughest hole/in Western Civilization"/Dan Jenkins, Golf Digest/ June 22, 1990.*

Narrow Long Branch passes beneath the single-span bridge, before joining Linganore Creek, just upstream of its confluence with the Monocacy River, east of Frederick. Shrouded in dense foliage, the bridge is situated between the 6th green, at the southwest end, and the 7th tee, at the northeast end.

Directions: From New Market, at I-70 exit 60 SR 75 (Green Valley Road), go north a short distance to Old National Pike. Go left to Mussetter Road, on the left; go 0.2 mile to Ritchie Drive, on the right; go 0.7 mile to Holly Hills Country Club clubhouse, on the left.

Single-span Loy's Station Bridge had this pier added prior to 1937 when steel I-beam stringers were installed, converting the bridge to a two-span stringer bridge.

Holly Hills Country Club No. 3 Covered Bridge

FREDERICK COUNTY

Brief Statistics

Type: Non-authentic modern covered footbridge. World Guide Covered Bridge Number: MD-10-e. Built in 1972 by unknown builder. One-span footbridge crossing Long Branch. Stringer footbridge 16.4 feet long by 11.5 feet wide, with approximately 6.8-foot wide by 7.4-foot high portal openings. Alternate Name: None known. Located in Ijamsville at Holly Hills Country Club at 5502 Mussetter Road.

Holly Hills Country Club has three small covered footbridges on its eighteen-hole golf course, all of similar appearance and construction. Holly Hills Country Club No. 3 Bridge was built, in 1972, by an unknown builder, in the swale below the 17th tee. The 16.4-foot long stringer footbridge is supported by three steel I-beam stringers, set on poured concrete abutments, with low, mortared natural-stone guard walls leading the golfers into the bridge. Built in a trapezoidal configuration, placed diagonally over the stream, the bridge has a diagonal planked floor. Weathered vertical boarding with battens covers the sides and portals. Two opposing, horizontal window openings are on each side and a log bench extends along each side of the interior. Dark weathered wood shakes cover the roof that is supported by three 6-inch by 8-inch wood posts along each side.

Narrow Long Branch passes beneath the single-span bridge into a pond on the downstream side of the bridge, eventually joining Linganore Creek, just upstream of its confluence with the Monocacy River, east of Frederick. The southeast/northwest-aligned bridge is situated at the side of the 17th fairway against a group of mature trees, the nearest tree, a weeping willow, draping its foliage over the bridge. The 17th tee is on top of the hill, at the northwest end. *See Holly Hills County Club No. 3 Covered Bridge in the color photograph section: C-12.*

Directions: From New Market, at I-70 exit 60 SR 75 (Green Valley Road), go north a short distance to Old National Pike. Go left to Mussetter Road, on the left; go 0.2 mile to Ritchie Drive, on the right; go 0.7 mile to Holly Hills Country Club clubhouse, on the left.

Loy's Station Covered Bridge

FREDERICK COUNTY

Brief Statistics

Type: Authentic modern covered bridge. World Guide Covered Bridge Number: MD-10-03 #2. Built in

1994 by unknown builder. Two-span bridge crossing Owens Creek. Multiple kingpost truss bridge 89.4 feet long by 16.4 feet wide, with approximately 12.8-foot wide by 15.0-foot high portal openings. Alternate Names: Loys Bridge, Owens Creek Bridge. Located in Loys on Old Frederick Road.

Loy's Station Bridge was originally built as a one-span, multiple kingpost truss bridge, in 1880, by an unknown builder. Some time prior to 1937, steel I-beam stringers were added, along with the central pier, to strengthen the bridge, converting it to a two-span bridge. Sadly, on June 27, 1991, a stolen pickup truck was parked inside the bridge and set on fire, destroying the bridge. From 1993 to 1994, an unknown builder rebuilt the bridge, the bridge being rededicated on June 25, 1994. Loy's Station Bridge was named after the nearby Western Maryland Railway stop, Loy's Station. The bridge has also been called Loy's Bridge, after the local community, and Owens Creek Bridge, after the creek flowing beneath. Loy's Station Bridge is the only multiple kingpost truss bridge remaining in Maryland.

The 89.4-foot long bridge has a single-span, fourteen panel, single-post, multiple kingpost truss that is now supported by nine steel I-beam stringers, set on mortared cut stone abutments, with the central poured concrete pier creating a two-span structure. A poured concrete cap was added on the abutments during the 1994 reconstruction. The floor consists of transverse planking. The red painted bridge has horizontal lapped siding on the sides and portals. Dark wood shingles cover the roof. Mortared natural stone guard walls, at each portal, funnel traffic into the single-lane bridge.

The two-span stringer bridge crosses wide, rock-bottomed Owens Creek, which flows into the Monocacy River, east of Creagerstown. The southwest/northeast-aligned bridge is relatively exposed, with several mature trees near the bridge. The southwest downstream corner has a field behind the trees, along the creek bank; the other three corners are part of Loys Station Park, which offers picnic pavilions, tables, grills and a portable restroom.

Loy's Station Bridge is one of five authentic covered bridges still carrying motor traffic in Maryland, and three of these are located in Frederick County. The three remaining authentic covered bridges in Frederick County—Loy's Station Bridge, Roddy Road Bridge and Utica Mills Bridge—were jointly listed on the National Register of Historic Places on June 23, 1978, becoming the first historic covered bridges in Maryland to be so listed. Frederick is the county seat of Frederick County.

Directions: From Creagerstown, go north on SR 550 to Old Frederick Road, on the right; go 4.4 miles to the bridge. There is ample parking near the bridge.

Roddy Road Covered Bridge

FREDERICK COUNTY

Brief Statistics

Type: Authentic historic covered bridge. World Guide Covered Bridge Number: MD-10-02. Built circa 1850 by unknown builder. One-span bridge crossing Owens Creek. Kingpost truss bridge 39.2 feet long by 16.0 feet wide, with approximately 12.3-foot wide by 11.5-foot high portal openings. Alternate Name: Roddy Bridge. Located north of Thurmont on Roddy Road.

An unknown builder constructed Roddy Road Bridge, also called Roddy Bridge, circa 1850, to carry Roddy Road across Owens Creek. The old bridge shares the same approximate construction date with nearby Utica Mills Bridge, making them the oldest authentic covered bridges remaining in Maryland. Roddy Road Bridge is also the only authentic covered bridge in Maryland with a kingpost truss. The bridge has the additional distinctions of being the shortest, at 39.2 feet long, and the narrowest, at 16.0 feet wide, authentic covered bridge remaining in Maryland, the latter distinction shared with Foxcatcher Farms Bridge in Cecil County. Steel I-beam stringers were originally added some time prior to 1954, replaced in 1979, and replaced again in November 1994.

Roddy Road Bridge has a two-panel, single-post, kingpost truss supported by nine steel I-beam stringers, set on mortared natural stone abutments. The flooring consists of transverse planking. The red painted bridge has horizontal tongue and groove siding on the sides and portals, the sides open under the eaves to allow air circulation within the bridge. Shiny tin covers the roof. Small clearance signs are mounted above each entrance.

Medium wide, rock-bottomed Owens Creek passes beneath the single-span bridge, on its way to the Monocacy River, east of Creagerstown. The southwest/northeast-aligned bridge is heavily wooded all around, except on the upstream side at the northeast bank, where Roddy Creek Road parallels the creek.

Three of the five authentic covered bridges still carrying motor traffic in Maryland are in Frederick County, and they are the only authentic covered bridges remaining in the county, namely Roddy Road Bridge, Loy's Station Bridge and Utica Mills Bridge. These bridges were listed as a group on the National Register of Historic Places on June 23, 1978, the first covered bridges in Maryland to be listed. The county seat of Frederick County is Frederick.

Directions: From Thurmont, take US 15 north to Roddy Creek Road, on the right; go 0.5 mile to the junction with Roddy Road and the bridge on the right.

Built circa 1850 with a distinctive multiple kingpost truss, encased with a Burr arch, that has the truss vertical tension members slanting away from the top of the center kingpost, Utica Mills Bridge is one of the two oldest authentic covered bridges in Maryland. *See Utica Mills Covered Bridge in the color photograph section: C-13.*

Rudderow Covered Bridge

FREDERICK COUNTY

Brief Statistics

Type: Non-authentic modern covered footbridge. World Guide Covered Bridge Number: MD-10-f. Built in 1972 by Ronnie Zecher and Lee Reichard. Two-span footbridge crossing dry ravine. Stringer footbridge 32.6 feet long by 6.2 feet wide, with approximately 4.3-foot wide by 6.8-foot high portal openings. Alternate Name: None known. Located in Middletown at 4612 Old National Pike (Alternate US 40).

Ronnie Zecher and Lee Reichard built Rudderow Bridge, named after the former owner, in 1972, over a dry ravine atop a hill, in the front yard of the residence. The bridge is nearly concealed by trees and shrubbery dotting the hill.

The two-span stringer footbridge is supported by two stringers, consisting of three 2-inch by 10-inch planks, set on edge, on abutments constructed from mortared rocks on the ground, and one center pier, constructed from 6-inch by 6-inch wood posts, set in a concrete footing and cross-braced with 2-inch by 10-inch planks. Transverse planking makes up the flooring. Vertical boarding covers the sides and portals, the sides open at eye level the length of the bridge. Brown asphalt shingles cover the roof, which is supported by four 4-inch by 6-inch wood posts along each side. The 32.6-foot long bridge is painted red on the exterior and the interior, including the floor. Mortared natural stone forms low guard walls at each end of the northwest/southeast-aligned bridge, except no wall is on the uphill side at the southeast end.

Directions: From Frederick, go west on US 40 to Alternate US 40 (Old National Pike), past I-70 to No. 4612, on the right, shortly after the intersection with Ridge Road. Private property: request permission to visit.

Utica Mills Covered Bridge

FREDERICK COUNTY

Brief Statistics

Type: Authentic historic covered bridge. World Guide Covered Bridge Number: MD-10-01. Built circa

1850 by unknown builder. Two-span bridge crossing Fishing Creek. Multiple kingpost truss with Burr arch bridge 101.4 feet long by 21.0 feet wide, with approximately 15.1-foot wide by 12.2-foot high portal openings. Alternate Name: Utica Bridge. Located in Utica on Utica Road.

Utica Mills Bridge, the former Devilbiss Bridge, was originally built, circa 1850, by an unknown builder, as a two-span bridge, to carry Devilbiss Bridge Road over the Monocacy River, near Devon Farms in Frederick County. The Devilbiss Bridge was severely damaged by the Johnstown Flood on June 1, 1889, losing one of its two spans. That year, the surviving span was disassembled and moved by wagon about two miles to Utica, where it was reassembled to carry Utica Road over Fishing Creek. Also called Utica Bridge, Utica Mills Bridge shares with nearby Roddy Road Bridge the distinction of being the oldest authentic covered bridge remaining in Maryland. At 21.0 feet wide, the bridge is also the widest authentic covered bridge remaining in Maryland. Utica Mills Bridge is one of four examples of a multiple kingpost truss with a Burr arch surviving in Maryland.

The 101.4-foot long bridge has a single-span, ten-panel, single-post, multiple kingpost truss, sandwiched by a Burr arch, unusual in that the vertical tension members slant away from the top of the center kingpost. The bridge is now supported by seven steel I-beam stringers, set on mortared rough-cut stone abutments, and one poured concrete pier, added, to strengthen the bridge, some time prior to 1937. The flooring consists of transverse planking. The red painted bridge has horizontal tongue and groove siding on the sides and portals, the sides open under the eaves, allowing air to circulate in the bridge. Dark weathered wood shingles cover the roof. Low cement-capped, mortared, rough-cut stone guard walls at each end funnel traffic into the single-lane bridge. The bridge was restored in 1997.

The two-span stringer bridge crosses moderately wide Fishing Creek, which flows into the Monocacy River, south of Utica. The north-northwest/south-southeast aligned bridge is fully exposed, with dense woods comfortably downstream of the bridge, an agricultural field upstream, beyond the north-northwest bank, and a residence beyond the south-southeast bank.

Utica Mills Bridge is one of five authentic covered bridges still carrying motor traffic in Maryland, and three of these are located in Frederick County, namely Utica Mills Bridge, Loy's Station Bridge and Roddy Road Bridge. These bridges were listed as a group on the National Register of Historic Places on June 23, 1978, the first covered bridges in Maryland to be listed. Frederick is the county seat of Frederick County. *See Utica Mills Covered Bridge in the color photograph section: C-13.*

Directions: From Frederick, take US 15 north to Old Frederick Road, on the right; go 1.3 miles to Utica Road, on the left; follow 0.3 mile to the bridge.

Casselman Motor Inn Covered Bridge

Garrett County

Brief Statistics

Type: Non-authentic modern covered footbridge. World Guide Covered Bridge Number: MD-11-b. Built c. 1983 by Clifford Maust. Two-span footbridge crossing small unnamed stream. Stringer footbridge 20.4 feet long by 6.4 feet wide, with approximately 3.8-foot wide by 6.8-foot high portal openings. Alternate Name: None known. Located in Grantsville at Casselman Motor Inn off Dorsey Hotel Road.

Clifford Maust, the nephew of the owner, Mrs. Della Miller, built Casselman Motor Inn Bridge sometime between 1982 and 1985, on the property of The Casselman, an 1824 hostelry still serving the traveler today. The stringer footbridge provides passage over a meandering, small unnamed stream between a parking area and the rooms of the motor inn. The 20.4-foot long bridge is fully exposed, with a medium-sized tree upstream. At 6.1 feet wide, Casselman Motor Inn Bridge is the narrowest non-authentic covered footbridge in Maryland.

The two-span bridge is supported by two 2-inch by 10-inch wood planks, acting as stringers, attached to three sets of vertical 4-inch by 4-inch wood posts, set in cement footings and evenly spaced along each side. The 4-inch by 4-inch wood posts extend up to support the dark weathered cedar-shingled roof. The flooring is constructed with 1-inch by 6-inch transverse boards. The sides have 1-inch by 4-inch vertical boards, spaced three inches apart, up to handrail height and open above. The gables are enclosed with vertical boards. A wooden bench is installed inside the bridge along the upstream side. The bridge is constructed with unfinished wood, weathering to pleasing shades of browns. Three wood steps and a handrail provide access to the north-northeast entrance of the bridge; the south-southwest entrance is at ground level.

Directions: From Grantsville, at I-68 exit 19 SR 495 (Bittinger Road), go north to Alternate US 40 (East Main Street); go right 0.3 mile to Dorsey Hotel Road, on the left. The Casselman Motor Inn, the former Dorsey's Hotel, is at the far corner of Dorsey Hotel Road and East Main Street. The bridge is 100 yards down Dorsey Hotel Road, at Ravine Street, on the left.

Cernik Covered Bridge

HARFORD COUNTY

Brief Statistics

Type: Non-authentic modern covered footbridge. World Guide Covered Bridge Number: MD-12-b. Built circa 1990 by Mr. Cernik. One-span footbridge crossing small unnamed stream. Stringer footbridge 16.0 feet long by 8.1 feet wide, with approximately 7.4-foot wide by 7.2-foot high portal openings. Alternate Name: Joshua Bridge. Located in Norrisville at 3017 Duncan Road.

Mr. Cernik built Cernik Bridge, circa 1990, in the side yard of, at that time, his residence. The footbridge has also been called Joshua Bridge, named after a subsequent owner, Joshua Fletcher. At 16.0 feet long, Cernik Bridge is the shortest non-authentic covered footbridge in Maryland.

The stringer footbridge is supported by five 6-inch by 6-inch timber stringers, set on railroad tie abutments. The flooring consists of transverse one-inch thick boarding, extending out the entrances to form approximately five-foot long approach ramps. Vertical tongue and groove paneling, which replaced the lattice in 2001, covers the lower half of the sides, the upper half open except for the 2-inch by 4-inch studding that supports the roof. Old, rusted, corrugated metal covers the low peaked roof.

The single-span bridge crosses a small spring outflow stream that feeds into a small tributary to Deer Creek, at the downstream side of the bridge, nearby Deer Creek flowing eastward into the Susquehanna River, across from Rock Run. The north-northeast/south-southwest aligned bridge is fully exposed in the grassed side yard, with a few mature trees close by. A spring fed pond is at the upstream side and landscape plantings are at the downstream side of the bridge.

Directions: From SR 23, going north toward Norrisville, take the first left (Church Lane) after the junction with SR 136. Go 0.4 mile to Duncan Road, on the right; go 0.6 mile to No. 3017, on the left. Private property: request permission to visit.

Grube Covered Bridge

HARFORD COUNTY

Brief Statistics

Type: Non-authentic modern covered bridge. World Guide Covered Bridge Number: MD-12-A. Built in 1970 by unknown builder. One-span bridge crossing tributary to Thomas Run. Stringer bridge 24.2 feet long by 9.0 feet wide, with approximately 7.6-foot wide by 6.5-foot high portal openings. Alternate Name: None known. Located west of Schucks Corners at 402 Prospect Mill Road.

Grube Bridge was built, in 1970, by an unknown builder at another location, for use as a motor vehicle bridge. The bridge was purchased by the Grubes and moved to its present location on their property and, thenceforth, called Grube Bridge. The bridge is located in the side yard, amongst scattered mature trees; not closed, but not being used. It is becoming overgrown with vines, showing deterioration, and habituated by unfriendly bees, may I add, that take offense to trespassers!

Two wood utility pole stringers, set on concrete and rocks and serving as abutments, support the 24.2-foot long stringer bridge. Transverse planking makes up the floor. The red painted bridge has open sides and no gables. The gray asphalt-papered roof is supported by four 4-inch by 4-inch wood posts along each side, with diagonal braces between the posts and at the outside of the end posts; a horizontal timber runs between the posts above the braces. This side framing slants slightly outward, under the eaves, an unusual feature.

The single-span bridge, aligned east-southeast/west-northwest, crosses a small tributary to Thomas Run, which flows into Deer Creek, near Palmer State Park, continuing to the Susquehanna River, across from Rock Run. The 9.0-foot wide bridge is the narrowest non-authentic covered motor vehicle bridge in Maryland.

Directions: From Bel Air, take SR 22 (Churchville Road) east to Prospect Mill Road, on the left just before Schucks Corners; follow to No. 402, on the left. Private property: request permission to visit.

Burdette Covered Bridge

HOWARD COUNTY

Brief Statistics

Type: Non-authentic historic covered bridge. World Guide Covered Bridge Number: MD-13-A #2. Built c. 1980 by unknown builder. One-span bridge crossing tributary to South Branch Patapsco River. Stringer bridge 24.5 feet long by 15.2 feet wide, with approximately 12.7-foot wide by 9.3-foot high portal openings. Alternate Name: None known. Located in Woodbine off Newport Road.

Hubert P. Burdette built the original Burdette Bridge, in 1942, from the timbers of an abandoned old barn. An unknown builder rebuilt the present Burdette Bridge circa 1980. At 15.2 feet wide, Burdette Bridge is the widest non-authentic covered motor vehicle bridge in Maryland.

Two timber stringers, set on poured concrete abutments, support the 24.5-foot long stringer bridge. These timbers could be old barn timbers, and this part of the structure could date to 1942. The flooring is constructed with transverse planking. Facsimile tongue and groove paneling covers the sides, which have two very large window openings on each side. These window openings, at one time, had glass windows; at present, only one has partial glass remaining. Framed paneling encloses the gables. The interior is covered with plywood paneling. The brown painted bridge has light brown asphalt shingles on the half-hip style roof, which is supported by 2-inch by 4-inch studding.

The single-span bridge carries a dirt road, with grass growing in the center, across a tributary to the nearby South Branch Patapsco River, which merges with the North Branch Patapsco River, at Marriotsville, to give birth to the Patapsco River, which empties into Chesapeake Bay. Heavily foliated woods surround the east-southeast/west-northwest aligned bridge. *See Burdette Covered Bridge in the color photograph section: C-12.*

Directions: From Lisbon, at I-70 exit 73 SR 94 (Woodbine Road), go north on SR 94 to Newport Road, on the left just before the large concrete bridge; go 0.2 mile to a paved road, on the left just before a second bridge. Immediately when you take this left, the pavement ends at a paved turnaround. Continue on the dirt road to the first dirt driveway on the right. Follow to the bridge. Private property: request permission to visit.

Turf Valley Resort Covered Bridge

Howard County

Brief Statistics

Type: Non-authentic modern covered bridge. World Guide Covered Bridge Number: MD-13-B. Built in 1994 by unknown builder. Three-span bridge crossing tributary to Little Patuxent River. Stringer bridge 24.3 feet long by 10.0 feet wide, with approximately 9.0-foot wide by 7.8-foot high portal openings. Alternate Name: Turf Valley Country Club Bridge. Located west of Ellicott City at Turf Valley Resort at 2700 Turf Valley Road.

An unknown builder built Turf Valley Resort Bridge, in 1994, when the North Course was constructed at the Turf Valley Resort and Conference Center, formerly the Turf Valley Country Club. The golfcart bridge spans a heavily foliated tributary to the nearby Little Patuxent River, which enters the Patuxent River near Crofton. The 9th green is at the east-northeast or right downstream end of the bridge.

Turf Valley Resort Bridge is a three-span stringer bridge, measuring 65.3 feet long, with only the 24.3-foot long center span covered. There is a 14.8-foot long span at the west-southwest end and a 26.2-foot long span at the east-northeast end. The bridge is supported by two large timber stringers, with 4-inch by 4-inch steel angle attached to the outside bottom of each timber stringer. The stringers are set on railroad ties, laid on the ground for abutments, and two piers, constructed with two upright railroad trestle timbers with cross-braces, set in poured concrete footings. The flooring consists of transverse planking with plywood wheel treads. The sides are open and vertical paneling encloses the gables. The cedar-shingled roof is supported by five 4-inch by 4-inch wood posts along each side; a handrail with cross-braces below runs between the posts. The outer two spans have a railing, constructed with 4-inch by 4-inch wood posts that have a heavy hemp rope looped along the post tops; a low 2-inch by 6-inch wood plank guardrail runs the length of the three spans.

Directions: From I-70, east of West Friendship at exit 82 US 40, go east on US 40, to Marriotsville Road (first left); follow 0.3 mile. The bridge is visible on the right, on the golf course, in heavy foliage. Or, continue east on US 40 to Turf Valley Road, on the left; go to No. 2700, Turf Valley Resort and Conference Center, on the left. The bridge is on the North Course, at the 9th green.

H. Deets Warfield Covered Bridge

Montgomery County

Brief Statistics

Type: Non-authentic modern covered bridge. World Guide Covered Bridge Number: MD-15-E. Built in 1966 by A. Jacob. Two-span bridge crossing pond. Stringer bridge 44.5 feet long by 12.2 feet wide, with approximately 8.8-foot wide by 9.4-foot high portal openings. Alternate Name: None known. Located in Damascus at 28025 SR 27.

H. Deets Warfield Bridge, named after its former owner, was built, in 1966, by A. Jacob, at the rear of the property over a pond. In May 2002, Mark Rhodes replaced the piers and the flooring on the ramps. H. Deets Warfield Bridge is the oldest non-authentic covered motor vehicle bridge in Maryland.

The 44.5-foot long stringer bridge is supported by four steel I-beam stringers set on three piers, each constructed with two vertical steel I-beams, with a steel I-beam cap. Ramps supported by four smaller steel I-beam stringers span from the end piers to the poured concrete abutments. Transverse planking makes up the floor that extends down the ramps. The red painted bridge, trimmed in white, has vertical boarding on the sides and portals, the sides having four opposing eight-pane glass windows along each side. Weathered wood shingles cover the roof, which is supported by five 4-inch by 6-inch wood posts along each side.

The two-span bridge has the majority of the substantial pond on the east side, with a small portion on the west side. It is heavily wooded around the pond and at the south end of the bridge. A white painted double board railing runs along both sides of the ramps, the north ramp adjoining a three-rail split-rail fence, which runs a short distance along the pond bank. *See H. Deets Warfield Covered Bridge in the color photograph section: C-14.*

Directions: From Damascus, go north on SR 27 to No. 28025, on the right. Follow the private dirt road, bearing to the left, to the bridge. Private property: request permission to visit from any one of the homes along the dirt road.

Lawson King Covered Bridge

MONTGOMERY COUNTY

Brief Statistics

Type: Non-authentic modern covered bridge. World Guide Covered Bridge Number: MD-15-C. Built in 1973 by S. Hawkins. Post supported roof over roadway crossing tributary to Muddy Branch. Bridge 43.3 feet long by 14.8 feet wide, with approximately 13.5-foot wide by 8.0-foot high portal openings. Alternate Name: None known. Located in Gaithersburg off Central Avenue.

Lawson King Bridge was built, in 1973, by S. Hawkins, contracted by W. Lawson King, on the driveway to the Lawson property. The bridge, a post supported roof over the asphalt-covered poured-concrete driveway, allows the passage of a tributary to nearby Muddy Branch, through a three-foot diameter galvanized steel pipe, under the driveway. Muddy Branch flows into the Potomac River. The 43.3-foot long cover has red painted plywood, with white painted battens, spaced 24 inches apart, covering the sides and portals; the portal entrances are trimmed in white. Old weathered wood shingles cover the low peaked roof, which is supported by 1-inch by 6-inch studs on 24-inch centers, the studs doubled at the portals. The roof fascia boards and the bridge interior are painted white. Three 24-inch wide horizontal slot openings are in the bottom of the siding, on each side, to allow the passage of floodwaters. A bridge informational sign is mounted on the south-southwest or right downstream gable.

The bridge has open lawn, with scattered mature trees on the downstream side, one large sycamore tree trunk almost touching the bridge. The bridge is heavily wooded on the upstream side.

Directions: From Gaithersburg, go north on SR 355 (Frederick Road South) to the 500 block, to Central Avenue, on the right. The bridge is at the second driveway on the left. Private property: request permission to visit, at the first or second home on the left side of Central Avenue.

Smokey Glen Farm Covered Bridge

MONTGOMERY COUNTY

Brief Statistics

Type: Non-authentic modern covered bridge. World Guide Covered Bridge Number: MD-15-B. Built in 1976 by Doogie Lowe. One-span bridge crossing tributary to Great Seneca Creek. Stringer bridge 28.3 feet long by 12.7 feet wide, with approximately 11.7-foot wide by 7.4-foot high portal openings. Alternate Name: None known. Located in Gaithersburg at Smokey Glen Farm at 16407 Riffleford Road.

Smokey Glen Farm Bridge was built by Doogie Lowe, in 1976, on the premises of Smokey Glen Farm Barbequers, Inc. The 28.3-foot long stringer bridge is supported by two steel I-beam stringers, with smaller steel I-beam cross-braces, that replaced the oak timber stringers in 1988 or 1989. The stringers are set on poured concrete abutments. The floor is constructed with transverse planks. Naturally weathered vertical boarding covers the sides and gables, the sides open the length of the bridge, at eye level. Old dark weathered wood shingles cover the roof, which is supported by five sets of doubled 2-inch by 6-inch wood plank posts along each side.

A narrow rock and silt bottomed tributary to nearby Great Seneca Creek passes beneath the single-span bridge, continuing on to Seneca Creek and the Potomac River, at Seneca. The southeast/northwest-aligned bridge is in a wooded area of the tributary, with the paved roadway curving to the northwest end of the bridge and coming to a tee at the southeast end. There is an open area with a picnic table downstream of the bridge.

Directions: From north of Gaithersburg, at the junction of SR 355 and SR 124 (Village Avenue), go west on SR 124 to SR 28; go right 0.5 mile to Riffleford Road; go right 0.7 mile to Smokey Glen Farm, at No. 16407, on the right. Private property: request permission to visit.

Valieant Covered Bridge

MONTGOMERY COUNTY

Brief Statistics

Type: Non-authentic modern covered bridge. World Guide Covered Bridge Number: MD-15-D. Built in 1976 by Rockville High School vocational students. One-span bridge crossing tributary to North Branch Rock Creek. Stringer bridge 20.2 feet long by 11.9 feet wide, with approximately 11.1-foot wide by 8.0-foot high portal openings. Alternate Name: None known. Located northeast of Rockville off Muncaster Mill Road (SR 115) at Meadowside Nature Center.

The Rockville High School vocational students built Valieant Bridge, in 1976, in memory of Joan Valieant, a teacher and environmentalist. The Montgomery County-owned bridge was built primarily for foot traffic on the Pioneer Trail and, although it will accommodate small motor vehicles, motor traffic is restricted to Meadowside Nature Center maintenance vehicles. At 20.3 feet long, Valieant Bridge is the shortest non-authentic covered bridge in Maryland accommodating motor vehicles.

The stringer bridge is supported by four wood utility pole stringers, set on the stream banks, with short wood utility pole sections set vertically in the ground, to prevent movement of the two outside stringers. A transverse planked floor extends out the entrances, forming short approaches. Naturally weathered vertical boarding covers the sides and gables, the sides open under the eaves, providing air circulation within the bridge. Dark weathered, green algae covered wood shakes cover the steeply peaked roof. Wood *VALIEANT BRIDGE* signs, with carved letters, are mounted over the entrances.

The single-span bridge crosses a small tributary to nearby North Branch Rock Creek, which flows into Rock Creek, thence entering the Potomac River at Georgetown, District of Columbia. The west-southwest/east-northeast aligned bridge is on a walking trail in dense, mature woods.

Directions: From Rockville, take SR 28 east to SR 115 (Muncaster Mill Road); go left 1.5 miles to Meadowside Nature Center, on the left, continuing to the Nature Center. The bridge is on Pioneer Trail, behind the Center. Trail maps are available at the Nature Center.

Valley Mill Park Covered Bridge

MONTGOMERY COUNTY

Brief Statistics

Type: Non-authentic historic covered footbridge. World Guide Covered Bridge Number: MD-15-a. Built in 1960 by unknown builder. One-span footbridge crossing millrace. Stringer footbridge 32.5 feet long by 8.3 feet wide, with approximately 7.4-foot wide by 7.0-foot high portal openings. Alternate Name: Snowden's Mill Bridge. Located east of Colesville in Valley Mill Park at 1600 East Randolph Road.

An unknown builder built Valley Mill Park Bridge, in 1960, over the old dry millrace to Valley Mill, which ceased operations in 1930. Also known as Snowden's Mill Bridge, the Montgomery County-owned bridge is the oldest non-authentic covered footbridge in Maryland.

The 32.5-foot long stringer footbridge is supported by two wood utility pole stringers, which are shored up in the center of the span by an upright wood utility pole, set in the ground under each stringer. The stringers are resting on mortared brick-fronted poured concrete abutments. The flooring consists of transverse planking; old weathered, red-painted vertical boarding covers the sides and gables, the sides to railing height; old weathered, lichen covered wood shingles cover the roof that, with the gables, extends beyond the entrances, providing additional weather protection to the interior. Electric carriage lamps are hanging inside each portal. A one-foot wide bench is mounted along each side of the interior.

The single-span bridge is exposed for most of the west-northwest end, with a large mature tree, its trunk against the south-southwest side, growing through the roof. The east-southeast end is overgrown with vines. Old mortared stone walls and a brick walkway lead to the east-southeast portal.

Directions: From Rockville, go east on SR 586 to SR 97; go left to Randolph Road; go right to the intersection with SR 650, where Randolph Road becomes East Randolph Road. Continue on East Randolph Road 1.6 miles to No. 1600, Valley Mill Park, on the left. The bridge is on the right, just before the parking area.

Bowie Horse Crossing Covered Bridge

PRINCE GEORGES COUNTY

Brief Statistics

Type: Non-authentic modern covered footbridge. World Guide Covered Bridge Number: MD-16-a. Built in 1989 by unknown builder. Three-span footbridge crossing Race Track Road. Stringer footbridge 113.3 feet long by 25.1 feet wide, with approximately 22.1-foot wide by 12.5-foot high portal openings. Alternate Name: None known. Located in Bowie over Race Track Road.

Bowie Horse Crossing Bridge is most unusual in many ways: uppermost in that only horses and jockeys traverse the bridge; the bridge is a "double barreled" or two-lane bridge; the bridge spans a roadway; and the bridge has a significant hump. Bowie Horse Crossing Bridge was built, in 1989, by an unknown builder, at the Bowie Training Center, over Race Track Road, to enable horses to go to the racetrack and return to the stable area without being

Unique Bowie Horse Crossing Bridge was built as a "double barreled" bridge in 1989 to enable horses to come and go across Race Track Road without being spooked by the traffic. ***See Bowie Horse Crossing Covered Bridge in the color photograph section: C-11.***

spooked by the highway traffic. The unique bridge holds two distinctions in Maryland: first, at 113.3 feet long, it is the longest non-authentic covered footbridge and second, at 25.1 feet wide, it is the widest non-authentic covered footbridge. As a matter of fact, Bowie Horse Crossing Bridge is the longest and the widest non-authentic covered footbridge in the southeastern United States.

The stringer bridge is supported by three large steel I-beam stringers, set on poured concrete abutments, large concrete blocks placed in front of these abutments, and two piers, constructed with three upright steel I-beams, set in footings, capped with a steel I-beam, and braced with three steel angle braces. The stringers are angled upward from the abutments across the piers to a horizontal center section, creating a significant humped effect to the bridge, rivaling the famous Humpback Bridge in Alleghany County, Virginia. The unusual flooring has thick padded matting covering a thick layer of sand, placed on flat corrugated galvanized steel panels set transversely on the stringers. The flooring muffles the noise of horses crossing the bridge. The interior of the bridge has 2-inch by 10-inch planking on the sides and on both sides of the partition running down the center, creating two lanes, all planking covered with thick padded matting. The exterior of the bridge has dark weathered horizontal lapped siding on the sides and gables. Light orange-brown asphalt shingles cover the roof, which is supported by 2-inch by 8-inch studding. Three electric lights are placed above each lane. High wooden portal wing walls block the view of traffic below. The right lane is marked *ENTRANCE*, and the left lane is marked *EXIT*, on the gables.

The portals of the three-span bridge are high above the roadway, with the stables on the southwest end and the runway to the racetrack on the northeast end. *See Bowie Horse Crossing Covered Bridge in the color photograph section: C-11.*

Directions: From Crofton, go south on SR 3 to SR 450 (Annapolis Road); go right 1.2 miles to Race Track Road; go right 2.2 miles to the Bowie Training Center and the bridge over Race Track Road. The portals are on the Bowie Training Center property and access is restricted; however, the sides of the bridge are quite visible from the roadway.

Mississippi

There are no known extant covered bridges in Mississippi. Mississippi certainly had its share of covered bridges dotting the landscape, during the nineteenth century. The last historic covered bridge traceable to the nineteenth century met its fate in the 1950's. The last covered bridge to disappear from Mississippi was the May Bridge, in Lincoln County. This kingpost truss bridge was built in 1966 and was burned and removed in 1990.

Above: Alamuchee Covered Bridge, in Sumter County, Alabama, distinctive with two diamond-shaped windows, is reflected in the small lake on the east side. The 82.3-foot long Civil War bridge was built in 1861. *Below:* Recently renovated Salem-Shotwell Covered Bridge, built in 1900 in Lee County, Alabama, sports a new coat of red paint.

Above: The 51.9-foot long Ashland Covered Bridge, was built circa 1870 in New Castle County, Delaware. This view is from the left downstream end. *Below:* Loockerman Landing Covered Bridge is at the dam and spillway for Mill Pond, which supplies the water to power the Silver Lake Mill at the Delaware Agricultural Museum and Village, in Kent County, Delaware. (Printed by permission of the Delaware Agricultural Museum Association.)

Wooddale Covered Bridge is a 67.3-foot long Town lattice truss bridge built in 1870 over Red Clay Creek in New Castle County, Delaware.

Above: Built in 1979, Clover Leaf Farms Covered Bridge, in Hernando County, is supported by a double kingpost truss, and is the only authentic covered bridge in Florida. *Below:* Modern-structured and picturesque Isla Key Covered Bridge, in Pinellas County, Florida, is 39.7 feet wide, the widest non-authentic covered motor vehicle bridge in the Southeastern United States.

Above: Auchumpkee Creek Covered Bridge was originally built in 1892 in Upson County, Georgia, destroyed by Tropical Storm Alberto in July 1994 and rebuilt in 1997 by Arnold M. Graton Associates. *Below:* The 151.5-foot long Haralson Mill Covered Bridge, in Rockdale County, Georgia, exposes its decorative Town lattice truss as it spans Mill Rock Creek.

Built in 1891, in Early County, Georgia, Coheelee Creek Covered Bridge is the southernmost-remaining historic covered bridge in the United States.

Elder's Mill Covered Bridge was built in 1897 and relocated in 1924 over boulder-strewn Rose Creek in Oconee County, Georgia.

Above: Watson Mill Covered Bridge's 229.0-foot length makes it the longest covered bridge in Georgia. Located in two Georgia counties, the 1885 bridge joins Madison County with Oglethorpe County. *Below:* Ye Olde Gap Covered Bridge, in Fulton County, Georgia, is the sidewalk of a 1950s concrete motor vehicle bridge, enclosed in 1977, with a beige-painted stucco exterior and a reddish-brown asphalt-shingled roof.

Above: Bennett's Mill Covered Bridge was built in 1855 by B. F. Bennett and Pramley Bennett, brothers who operated nearby Bennetts Mill, in Greenup County, Kentucky. *Below:* The long and winding black paved driveway, lined with black rail fencing, leads to Hycliffe Manor Covered Bridge, located in Shelby County, Kentucky.

R & B Frye Farm Covered Bridge, in Lewis County, Kentucky, was dubbed "The Bridge to Nowhere," as the bridge was constructed prior to building the house.

Above: Johnson Creek Covered Bridge was built in 1874 by Jacob N. Bower and restored 40 years later by his son, Louis Stockton Bower, Sr., in Robertson County, Kentucky. *Below:* Bowie Horse Crossing Covered Bridge, built in 1989, in Prince Georges County, Maryland, is restricted for horses and jockeys going across the street to the racetrack.

Above: Burdette Covered Bridge, in Howard County, Maryland, has a distinguishing half-hip style roof covering its 24.5-foot length. *Below:* Trapezoidal Holly Hills Country Club No. 3 Covered Bridge, built in 1972, in Frederick County, Maryland, is set diagonally across Long Branch in a group of mature trees. View is from the upstream side.

Above: Gilpin's Falls Covered Bridge was built, in 1860, by Joseph George Johnson, over North East Creek in Cecil County. The 119.6-foot long bridge is the longest authentic covered bridge in Maryland. *Below:* Utica Mills Covered Bridge, in Frederick County, Maryland, was built circa 1850 over the Monocacy River near Devon Farms, and moved in 1889 to Utica over Fishing Creek.

H. Deets Warfield Covered Bridge was built in 1966, by A. Jacob, over a pond in Montgomery County and is the oldest non-authentic covered bridge in Maryland.

The Betty Ruth Covered Bridge was built in 1993 by Roy Romsburg over Town Creek in Allegany County, Maryland.

Holly Hills Country Club No. 1 Covered Bridge is one of three covered bridges built, in 1972, on the golf course in Frederick County, Maryland.

North Carolina's Bodenheimer Covered Bridge, built in Guilford County in 1966, is a 20.3-foot long stringer bridge which simulates a Town lattice truss.

Above: Chaplin Covered Bridge is the only covered bridge in North Carolina in two counties. The 46.3-foot long bridge spans Elk Creek, the county line between Ashe and Watauga counties. *Below:* Catawba County's Bunker Hill Covered Bridge is the only authentic historic covered bridge remaining in North Carolina. It is an 80.3-foot long Haupt truss bridge.

Above: The 54.3-foot long Pool Creek Covered Bridge is at the entrance to the Bottomless Pools attraction at Lake Lure, in Rutherford County, North Carolina. *Below:* River Island Covered Bridge was built on the private road to River Island, in the Cape Fear River in New Hanover County, North Carolina, to open the island for development.

Above: Highland Farms Covered Bridge, in Polk County, North Carolina, is a marvelously landscaped 42.0-foot long, two-lane post supported roof bridge over Ostin Creek. *Below:* Built in 1898, in Watauga County, North Carolina, Alpine Acres Covered Bridge is the oldest non-authentic covered footbridge in the Southeastern United States.

Above: The 66.6-foot long Caledonia Golf & Fish Club No. 1 Covered Bridge is the longest non-authentic covered footbridge in South Carolina. *Below:* The only historic covered bridge in South Carolina, Campbell Covered Bridge, located in Greenville County, was named after Alexander Lafayette Campbell, who operated a nearby grist mill.

Above: Collins Covered Bridge was built in 1992, in Pickens County, South Carolina, across Town Creek. The latticework in the windows and gables adds appeal to the bridge. *Below:* An upstream view of South Carolina's Hidden Hills No. 1 Covered Bridge, in Pickens County, which has a unique floor constructed from steel railroad track laid transversely.

Above: Veterans Covered Bridge, in Anderson County, South Carolina, is supported by steel Warren trusses that are below the decking. This is a downstream view. *Below:* Significant sag is visible in Bible Covered Bridge in this upstream view, with Little Chucky Creek passing beneath. Bible Covered Bridge is located in Greene County, Tennessee.

Above: Covered Bridge Estates Covered Bridge, in Bedford County, Tennessee, is supported by timber stringers, yet it has a kingpost truss. The stringers and flooring are not connected to the truss or the roof support posts, depriving the truss of any load-carrying role. *Below:* Elizabethton Covered Bridge was built in 1882 by Dr. E. E. Hunter, over the Doe River, in Carter County, Tennessee.

Above: Fairgrounds Park Covered Bridge, in Madison County, at 60 feet, is the longest single-span non-authentic covered motor vehicle bridge in Tennessee. *Below:* Picturesque weeping willow draped Hidden Mountain East Covered Bridge, in Sevier County, Tennessee, welcomes you to the Enchanted Forest.

Above: In Williamson County, Tennessee, the bright red Keeler's Covered Bridge is surrounded by summer greenery. *Below:* Marrowbone Creek Covered Bridge, 70.1 feet long by 10.3 feet wide, is situated in a beautiful setting over rock banked, swiftly moving Marrowbone Creek, in Cheatham County. Surprisingly, this bridge is noted, not for its length, but for being the narrowest non-authentic covered motor vehicle bridge in Tennessee.

Turkey Pen Resort Covered Bridge in Monroe County, Tennessee, was built, in 1982, by Luke Henley, over a tributary to Carringer Creek.

Above: Daniel Ulrich Biedler built Biedler Farm Covered Bridge, in Rockingham County, Virginia, in 1896, and the bridge remained in the Biedler family until 1969. *Below:* Privately-owned C. K. Reynolds Covered Bridge, in Giles County, Virginia, still carries an occasional farm vehicle across Sinking Creek, but more often it allows cows to reach the other side.

Above: Clover Hollow Covered Bridge, in Giles County, Virginia, was built over Sinking Creek in 1916 by James Maurice Puckett and Will Wingo, bypassed in 1949, and closed to traffic in 1964. *Below:* The oldest non-authentic covered motor vehicle bridge in Virginia, Clifford Wood Covered Bridge, was built in Patrick County in 1977, over the Smith River, by Clifford Wood.

Above: Humpback Covered Bridge, in Alleghany County, Virginia, displays a significant "humped" appearance across Dunlap Creek in this upstream view. *Below:* In 1987, Isaac (Ike) W. Ward built a cover over an old pre–1950 state built stringer bridge on his property in Pittsylvania County, Virginia. The bridge is now called Wards Farm Covered Bridge.

Above: The shortest authentic historic covered bridge in West Virginia, 24.4-foot long Laurel Creek Covered Bridge was built in 1911, in Monroe County. This is a downstream view. *Below:* Simpson Creek Covered Bridge, in Harrison County, West Virginia, was built in 1881, by Asa S. Hugill, over Simpson Creek, originally one-half mile upstream, and relocated to its present site in 1899.

Above: Considered one of the best in covered bridge architecture and design, Philippi Covered Bridge (1852), located in Barbour County, is the longest (301.1 feet) historic covered bridge in West Virginia. *Below:* An early nineteenth century two-story log cabin is framed by the portal of Staats Mill Covered Bridge in Jackson County, West Virginia. Originally built in 1887 by Henry T. Hartley over Tug Fork in Statts Mills, the bridge was relocated in 1982 to its present site.

North Carolina

Certainly, North Carolina had a few hundred covered bridges gracing its roadways during the nineteenth and early twentieth centuries. Randolph County alone had sixty covered bridges in the 1930s. Today, only five historic covered bridges remain in North Carolina, and two of those are part of grist or saw mills, one not included in this reference book at the request of the owner.

North Carolina has thirty-six covered bridges, thirty-five included in this reference book, but only two are authentic and only one of these is historic, Bunker Hill Bridge in Catawba County. Of the thirty-four non-authentic covered bridges, four are historic. These thirty-six covered bridges were constructed between circa 1860 and 2001 in twenty-four of the one hundred North Carolina counties. North Carolina's authentic historic covered bridge has the only Haupt truss in the southeastern United States; the authentic modern covered bridge is a multiple kingpost with a Burr arch.

Twenty-nine of North Carolina's covered bridges are located in the western part of the state. All five of the historic covered bridges are closed to motor traffic, and three are privately owned.

Chaplin Covered Bridge

Ashe County/Watauga County

Brief Statistics

Type: Non-authentic modern covered bridge. World Guide Covered Bridge Number: NC-05-A/NC-95-E. Built in 1999 by J. Sam Chaplin. One-span bridge crossing Elk Creek. Stringer bridge 46.3 feet long by 12.3 feet wide, with approximately 10.8-foot wide by 10.5-foot high portal openings. Alternate Name: None known. Located in Todd off Todd Railroad Grade Road on Chaplin Lane.

J. Sam Chaplin built Chaplin Bridge, in 1999, at the entrance to his property on Chaplin Lane. Moderately wide, rocky bottomed Elk Creek, joined by Pine Orchard Creek, just upstream, flows beneath the single-span bridge, entering the South Fork New River a short distance downstream, the South Fork New River merging with the North Fork New River, east of Weavers Ford, to form the New River. The bridge is fully exposed, with some spruce trees on the northeast creek bank, at the downstream side.

Todd Railroad Grade Road passes the bridge near the northeast or right downstream end, running parallel to the creek, which forms the county line. Open fields extend upstream of the bridge to nearby State Highway 194 and away from the southwest end to the hills beyond.

The 46.3-foot long bridge is supported by five steel I-beam stringers, set on poured concrete abutments, which extend to the sides, forming wing walls. A concrete slab forms the floor. Natural vertical boarding with battens covers the sides and the gables, the sides open under the eaves for ventilation and with an opposing, horizontal, rectangular window opening on each side. The red-painted aluminum roof is supported by 2-inch by 4-inch studding along each side. The bridge is stayed along each side with three steel wire-rope guy wires, attached to the sides under the eaves and connected to the ends of steel I-beams, extending out from the sides of the bridge, which are welded to the bottom of the stringers. *See Chaplin Covered Bridge in the color photograph section: C-18.*

Directions: From Boone, go east on US 221/US 421/SR 194. At the junction where SR 194 goes to the left, follow SR 194 to Todd and Todd Railroad Grade Road, on the right. The bridge is within view, on the right. Private property: request permission to visit.

Christmas Cabin Covered Bridge

AVERY COUNTY

Brief Statistics

Type: Non-authentic modern covered footbridge. World Guide Covered Bridge Number: NC-06-b. Built in 1976 by unknown builder. One-span footbridge crossing unnamed stream. Stringer footbridge 19.0 feet long by 5.8 feet wide, with approximately 4.6-foot wide by 6.4-foot high portal openings. Alternate Name: Covered Bridge Crafts Bridge. Located in Banner Elk off Beech Mountain Parkway at Grey Fox Ridge Road.

Built in 1976 by an unknown builder, Christmas Cabin Bridge was renamed, in 1995, when new owners purchased the Christmas gift shop. The footbridge was formerly called Covered Bridge Crafts Bridge, after the former establishment.

Three wood utility pole stringers, set on mortared natural stone footings, support the 19.0-foot long footbridge. Transverse planking makes up the floor. Red-painted vertical boarding covers the sides to rail height, open above to the eaves. Gray-painted vertical boarding covers the gables, which extend beyond the entrances for weather protection. Moss-covered reddish-brown asphalt shingles cover the roof, which is supported along each side by three 4-inch by 4-inch wood posts. Two wood-enclosed buttress braces, on each side, stabilize the bridge.

A small, unnamed mountain stream flows down the mountainside, passing beneath the single-span bridge, set in a shaded mature-tree wooded area in front of the Visions of Sugarplums Christmas Cabin Gift Shop. The bridge provides access from the parking area, along the graveled walkway, to the gift shop. Candy canes line the walkway approaching the footbridge, guarded by lighted toy soldiers and aglow with strings of icicle lights.

Directions: From Banner Elk, at the junction of SR 194 and SR 184, go north 1.8 miles on SR 184 (Beech Mountain Parkway), up Beech Mountain to Grey Fox Ridge Road, on the right. Go right, entering Visions of Sugarplums Christmas Cabin Gift Shop, on the left. The bridge is visible from Beech Mountain Parkway.

Land of Oz Covered Bridge

AVERY COUNTY

Brief Statistics

Type: Non-authentic modern covered footbridge. World Guide Covered Bridge Number: NC-06-a. Built in 1967 by unknown builder. One-span footbridge crossing man-made stream. Stringer footbridge 18.2 feet long by 9.1 feet wide, with approximately 5.0-foot wide by 7.2-foot high portal openings. Alternate Name: None known. Located in Beech Mountain on top of Beech Mountain at Land of Oz.

Built in 1967 by an unknown builder, Land of Oz Bridge was built in the Land of Oz when the attraction, patterned after *The Wizard of Oz*, was built. The attraction opened to the public in 1970, but was closed in 1980. The attraction has been reopened to the public, for only one day in October, each year since.

Timber stringers, set on the ground, support the 18.2-foot long footbridge. Red-painted transverse planking comprises the floor. Green-painted, vertical, facsimile tongue and groove paneling covers the sides and the portals, each side with three large vertical opposing window openings. The gables have arch-cut entrances. Wood benches run along each side of the green-painted interior. Weathered wood shakes cover the steeply-peaked double-sloped roof, the lower slope flaring out over the eaves. The roof, trimmed in red, is supported by six wood posts along each side. A yellow brick "road," winding to and from the bridge, contrasts vividly with the green and red bridge and its surroundings.

Water is pumped from a man-made pond below the bridge to a waterfall above the bridge, flowing under the single-span bridge and returning to the pond. The north-northwest/south-southeast aligned footbridge is placed near the mountain top, along the long, snaking, yellow brick "road," lined with boulders and low stunted trees and shrubs, some tree trunks with ghoulish faces reminiscent of *The Wizard of Oz*.

Directions: From Banner Elk, at the junction of SR 194 and SR 184 go north 3.3 miles on SR 184 (Beech Mountain Parkway), up Beech Mountain, to Ski Loft Road, on the left; follow to Oz Road, on the right, and take to the top of the mountain. Private property: request permission to visit at Emerald Mountain Realty, at 2669 Beech Mountain Parkway, on the left as you proceed up Beech Mountain on SR 184.

Hoggard Grist Mill Covered Bridge

BERTIE COUNTY

Brief Statistics

Type: Non-authentic historic covered bridge. World Guide Covered Bridge Number: NC-08-B. Built in 1928 by Thomas Heckstall. Two-span bridge crossing Hoggard Mill Creek. Stringer bridge 28.4 feet long by 17.9 feet wide, with approximately 13.3-foot wide by 8.6-foot high portal openings. Alternate Name: None known. Located in Butlers off Greens Cross Road.

Hoggard Grist Mill Bridge is a combination grist mill and covered bridge that is only 150 feet east of Hoggard Saw Mill Bridge, another combination mill and covered bridge.

Hoggard Grist Mill Bridge is the upstream part of Hoggard Grist Mill, which is on the earthen dam impounding the waters of Hoggard Mill Creek, to power the grist mill. About 150 feet along the dam, north of the Hoggard Grist Mill, is the Hoggard Saw Mill, which also houses a bridge. The dam, the first combination grist mill and bridge and the first combination saw mill and bridge, were built, in 1736, by Mr. Synnott, under contract to James Castellaw. Originally, the county road to Halifax ran along the dam, passing through these bridges, but following the destruction wreaked by a hurricane in 1928, the county abandoned the road. Following the destruction of the previous combination grist mill and bridge, by the 1928 hurricane, the present combination Hoggard Grist Mill and Hoggard Grist Mill Bridge was built in 1928 by Thomas Heckstall, upon concrete spillways dating to circa 1900. Harry Lewis Thompson replaced the siding in 1972. Hoggard Grist Mill and Hoggard Grist Mill Bridge derive their name from William Hoggard, who purchased the site circa 1820 and operated the mills until 1830.

The north-northeast/south-southwest aligned combination mill and bridge is heavily wooded on the downstream side and, due to a breach in the dam caused by Hurricane Floyd on September 17, 1999, the upstream pond is low, surrounded by a swamp, with many trees scattered about. The two-span bridge crosses the grist mill spillway on Hoggard Mill Creek, which flows into the nearby Cashie River, the river emptying into Albemarle Sound, at Cashoke Landing.

The 28.4-foot long stringer bridge is supported by timber stringers, set on poured concrete abutments (spillway walls), and one pier, constructed with two upright timbers, with timber braces. Transverse planking forms the floor, which has timbers running along the sides for tire bumpers. On the downstream side, the grist mill inner wall is covered with gray weathered vertical boarding with battens, the screws for the spillway gates constructed along this wall. The ends of the mill and bridge are also covered with gray weathered vertical boarding with battens. The upstream side of the bridge is covered to rail height and under the eave with gray weathered vertical boarding without battens, open in-between for the length of the bridge, exposing the roof support posts. The roof over the mill side is covered with what appears to be new tin; the roof over the bridge side is covered with old weathered wood shingles. Eight wood posts, along the upstream side, and the studding or posts in the mill walls support the steeply peaked roof.

Directions: From Windsor, go north on US 13 to Hoggard Mill Road, on the right; go 0.6 mile to Greens Cross Road, on the right; go 0.1 mile to the combination grist mill and bridge, visible on the left. Private property: request permission to visit.

Hoggard Saw Mill Covered Bridge

Bertie County

Brief Statistics

Type: Non-authentic historic covered bridge. World Guide Covered Bridge Number: NC-08-C. Built in 1972 by Harry Lewis Thompson. Three-span bridge crossing Hoggard Mill Creek. Stringer bridge 41.1 feet long by 11.8 feet wide, with approximately 11.8-foot wide by 10.0-foot high portal openings. Alternate Name: None known. Located in Butlers off Greens Cross Road.

The original dam, the first combination grist mill and bridge and the first combination saw mill and bridge, were built, in 1736, by Mr. Synnott, for the owner, James Castellaw. A county road crossed this earthen dam, passing through these bridges on its way to Halifax. Several owners operated the mills, until William Hoggard purchased them circa 1820. The mills were renamed Hoggard Grist Mill and Hoggard Saw Mill and, although William Hoggard operated them only until 1830, the names have stuck with the mills to the present day. Hoggard Saw Mill Bridge is the upstream part of Hoggard Saw Mill, which is on the earthen dam, about 150 feet north of Hoggard Grist Mill and Hoggard Grist Mill Bridge. The dam held back the waters of Hoggard Mill Creek, necessary to power the saw mill, until breached by Hurricane Floyd on September 17, 1999, at the north end of the saw mill, lowering the water level in the pond. In 1928, a hurricane destroyed the mills and bridges, causing the county to abandon the roadway. The present Hoggard Saw Mill and Hoggard Saw Mill Bridge were built, in 1972, by Harry Lewis Thompson, utilizing the brickwork from a previous circa 1780 saw mill.

The north/south-aligned bridge is heavily wooded on both sides, a cypress swamp on the upstream side. The three-span bridge crosses the saw mill spillway on Hoggard Mill Creek, which joins the nearby Cashie River, the river emptying into Albemarle Sound, at Cashoke Landing. The bridge is no longer open to motor traffic.

The 41.1-foot long stringer bridge is supported by eleven 3-inch by 9-inch timber stringers, placed on edge and set on a poured concrete abutment at the north end, a mortared brick abutment at the south end, and two piers, each pier constructed with double mortared brick pillars and upright timbers. The flooring consists of transverse planking. The sides are covered with gray weathered vertical boarding with battens up to rail height and under the eaves, open in-between for the length of the structure on the bridge side and on the mill side. Wood roof support posts run the length of the bridge, between the traffic lane and the saw mill. Gray weathered vertical boarding with battens also covers the ends of the combination mill and bridge. Old weathered wood shakes cover the roof, which is supported on the upstream or bridge side by seven wood posts; the roof is supported on the downstream or mill side by thirteen wood posts and a wood post partition between the saw mill and the bridge lane. The saw mill and the bridge side are separated by this open wood post partition, the saw mill also having smaller open portals at each end.

Directions: From Windsor, go north on US 13 to Hoggard Mill Road, on the right; go 0.6 mile to Greens Cross Road, on the right; go 0.1 mile to the combination saw mill and bridge, behind the grist mill and bridge, on the left. Private property: request permission to visit.

Bakkelund Covered Bridge

Buncombe County

Brief Statistics

Type: Non-authentic modern covered bridge. World Guide Covered Bridge Number: NC-11-A. Built in 1979 by G. Peter Jensen. One-span bridge crossing Broad River. Stringer bridge 39.8 feet long by 12.6 feet wide, with approximately 11.0-foot wide by 12.1-foot high portal openings. Alternate Name: None known. Located south of Black Mountain at 22 Chestnut Hill Road.

G. Peter Jensen built Bakkelund Bridge, in 1979, on the gravel driveway, at the entrance to his property, the name Bakkelund being Danish for hilly land and indicative of the surrounding area. The single-span bridge crosses the headwaters of the Broad River, which merge with the Saluda River in Columbia, South Carolina, to give birth to the Congaree River. Bakkelund Bridge is situated at the base of a high, densely wooded hill on the downstream side, open on the upstream side, with low foliage along the riverbanks. Chestnut Hill Road, formerly Old Fort Road, is at the south end of the south/north-aligned bridge.

The 39.8-foot long stringer bridge is supported by a 40-foot long flatbed trailer, without wheels, set on poured concrete abutments. The flooring consists of transverse planking. Brown-stained vertical tongue and groove boarding covers the sides and the gables, the sides open under the eaves to promote air circulation within the bridge. Brown asphalt shingles cover the low peaked roof, which is supported with 6-inch by 6-inch wood posts at the portals and 2-inch by 6-inch studding along the sides. Electric lights are mounted over the entrances and at the center of the interior. Brown-stained, board rail fencing extends off both sides of the south portal.

Directions: From Black Mountain, at I-40 exit 64, go south 5.2 miles on SR 9 to Chestnut Hill Road, on the right (this is the first right after crossing the Eastern Continental Divide); go 0.5 mile to the bridge, visible on the right at No. 22. Private property: request permission to visit.

Newfound Covered Bridge

BUNCOMBE COUNTY

Brief Statistics

Type: Non-authentic modern covered footbridge. World Guide Covered Bridge Number: NC-11-b. Built in 1980 by Gay M. McPeters. One-span footbridge crossing Newfound Creek. Stringer footbridge 21.1 feet long by 6.4 feet wide, with approximately 5.2-foot wide by 6.5-foot high portal openings. Alternate Name: None known. Located in Newfound at 1419 Newfound Road.

Newfound Bridge was built, in 1980, by Gay M. McPeters, to provide access to the barn in his backyard at the far side of the creek. This is the second covered footbridge Mr. McPeters built at this site. The first covered footbridge, built in 1955, was severely damaged in 1974, when the upstream dam burst and undermined one abutment; then came floods in 1975 and 1976, the latter washing the weakened bridge away.

Three yellow locust log stringers, set on mortared natural stone abutments, support the 21.1-foot long footbridge. The flooring consists of transverse oak planking. The sides and gables of the red painted bridge are open. A six-section box-shaped handrail, aligned on the roof support posts, runs down each side; each section constructed with 2-inch by 6-inch planks forming the box, with 2-inch by 4-inch wood cross-braces within. Light gray asphalt shingles cover the roof, which is supported by seven 2-inch by 4-inch wood posts. An electric light illuminates the interior of the bridge.

The single-span footbridge crosses narrow Newfound Creek, which flows into the French Broad River, northwest of Asheville. The east-southeast/west-northwest aligned bridge is situated in the open, grassed backyard; the deeply eroded creek banks are overgrown with weeds. A concrete walkway leads to the east-southeast portal, and a concrete pad abuts the far end of the bridge, which is flanked by densely foliated shrubs.

Directions: From I-40, north of Canton, take exit 33, Newfound Road, north 5.2 miles to No. 1419, on the left at the corner of McPeters Road. The bridge is visible in the backyard. Private property: request permission to visit.

Bunker Hill Covered Bridge

CATAWBA COUNTY

Brief Statistics

Type: Authentic historic covered bridge. World Guide Covered Bridge Number: NC-18-01. Built in 1895 by Andy J. Ramsour, Eli Kale, George Moller, Cain Bost and Electious Connor. One-span bridge crossing Lyle Creek. Haupt truss bridge 80.3 feet long by 12.2 feet wide, with approximately 9.5-foot wide by 10.8-foot high portal openings. Alternate Name: None known. Located in Claremont on Old Island Ford Road in Connor Park.

In 1894, Catawba County Commissioners called upon local landowners to build and maintain a bridge across Lyle Creek. In response, landowners hired Andy J. Ramsour, keeper of Horse Ford Covered Bridge over the Catawba River at the Caldwell County and Catawba County line, to build the bridge. Andy J. Ramsour selected the Haupt truss and, with locals Eli Kale, George Moller, Cain Bost and Electious Connor, built Bunker Hill Bridge as an uncovered bridge in 1895; the cover was added in 1900. The bridge derived its name from the nearby Bunker Hill Farm, operated by descendants of the Shuford and Lowrance families since the early 1800s. Bunker Hill Bridge is the only surviving Haupt truss covered bridge in the southeastern states and the only authentic historic covered bridge remaining in North Carolina. The bridge had major restoration work done, in 1994, by Arnold M. Graton Associates of Ashland, New Hampshire.

The 80.3-foot long historic bridge is supported by a fourteen-panel Haupt truss, with double-plank oak lattice members, and is set on large rock abutments, which had concrete added at the base of the rocks at a later date. The flooring consists of transverse planking with three-plank-wide wheel treads. Weathered vertical boarding with battens covers the sides and weathered horizontal boarding covers the gables. The shiny tin covered roof was originally covered with wood shingles, in 1900, which were replaced with tin in 1921. The roof extends beyond the entrances, affording protection for the interior from inclement weather. An old painted sign over the right downstream or east-northeast entrance declares *ANY-ONE-Crossing This Bridge/Is Doing So At Their Own Risk R K Bolick*. The date 1895 is carved into the right downstream portal timber on the downstream side; however, it appears carved over initials *WW*, lending suspicion as to when it was carved.

Now closed to motor traffic, the single-span bridge crosses wide, rock and sediment bottomed Lyle Creek, which

Built in 1895, Bunker Hill Bridge has the only Haupt truss in the southeastern United States. ***See Bunker Hill Covered Bridge in the color photograph section: C-18.***

flows into the Catawba River, about three miles downstream. The bridge is in a heavily wooded area with an open area at the west-southwest portal, where several bridge informational signs and plaques are posted. A bridge informational sign is off US Highway 70, at the entrance to the park.

Bunker Hill Bridge is owned by the Catawba County Historical Society, located in Newton, the county seat of Catawba County. The bridge was listed on the National Register of Historic Places on February 26, 1970. *See Bunker Hill Covered Bridge in the color photograph section: C-18.*

Directions: From Claremont, at the intersection of US 70 and North Oxford Street, go 1.9 miles east on US 70 to Connor Park, on the left. The bridge, not visible from the highway, is an easy walk along the trail.

Granny Squirrel Gap Covered Bridge

Cherokee County

Brief Statistics

Type: Non-authentic modern covered bridge. World Guide Covered Bridge Number: NC-20-A. Built in 1973 by unknown builder. Post supported roof over roadway crossing Matheson Creek. Bridge 48.3 feet long by 15.0 feet wide, with approximately 13.6-foot wide by 13.0-foot high portal openings. Alternate Name: None known. Located in Andrews on Granny Squirrel Drive at entrance to Granny Squirrel Gap Subdivision.

Granny Squirrel Gap Bridge was built, in 1973, by an unknown builder. The bridge originally was a cover over a single-span stringer bridge. Sometime following its construction, the stringer bridge was replaced by an asphalt roadway over a pipe, but retaining the cover, thus making the bridge a post supported roof over an asphalt roadway, with a pipe allowing the flow of Matheson Creek to pass underneath. Plywood paneling covers the sides and the gables, the sides with window openings cut in an inverted V at the top, for a total of eleven full openings per side, with a half opening at each end. Brown asphalt shingles cover the roof, which is supported by fourteen double 2-inch by 4-inch wood posts and 2-inch by 4-inch studding along each side. Originally painted white, the 48.3-foot long bridge is now painted red, with white trimmed gable fascia on the exterior and painted red on the interior, except for the original white paint still covering the upper part of the interior.

Located at the entrance to Granny Squirrel Gap Subdivision, on Granny Squirrel Drive, the southeast/northwest-aligned bridge has a white painted sign reading *GRANNY SQUIRREL GAP* over the northwest entrance. Matheson Creek flows beneath the fully exposed bridge into the nearby Valley River, which joins the Hiwassee River at Murphy. A large cedar tree is at each side of the northwest entrance, and a board rail fence runs along Granny Squirrel Drive, on the upstream side.

Directions: From north of Andrews, at the junction of US 19 and US Business 19, go 2.2 miles north on US 19 to Granny Squirrel Drive, and the bridge, visible on the right.

Clyde Cash Covered Bridge

CLEVELAND COUNTY

Brief Statistics

Type: Non-authentic modern covered bridge. World Guide Covered Bridge Number: NC-23-A. Built in 1984 by Marvin Philbeck, Forrest Littlejohn and Carroll Hamrick. One-span bridge crossing tributary to Buffalo Creek. Stringer bridge 15.7 feet long by 12.2 feet wide, with approximately 10.9-foot wide by 6.8-foot high portal openings. Alternate Name: None known. Located in Shelby at Cleveland Community College on Clyde Cash Nature Trail.

Clyde Cash Bridge was built, in the spring of 1984, by the Cleveland Community College maintenance staff, headed by Marvin Philbeck working with Forrest Littlejohn and Carroll Hamrick. Dedication ceremonies were held on April 12, 1984. The bridge was built on the Clyde Cash Nature Trail and, although dimensionally it could accommodate motor vehicles, the bridge was constructed for foot traffic on the nature trail. A sandy-bottomed tributary to nearby Buffalo Creek passes beneath the single-span bridge, Buffalo Creek eventually entering the Broad River, west of Blacksburg, South Carolina. The south/north-aligned bridge is just inside a heavily wooded area along the dirt nature trail. Lacking maintenance, the bridge has large holes in the roof, several inches of washed in soil covering areas of the flooring, and Kudzu rapidly overgrowing the area and the bridge.

The 15.7-foot long stringer bridge is supported by five wood utility pole stringers, now resting on the ground, as the cement blocks and bricks used as footings have washed out. The flooring consists of transverse planking. Weathered vertical boarding covers the sides and the portals. Weathered, moss and vine covered wood shingles cover the low peaked roof, which is supported by triple 2-inch by 4-inch posts at the portals and a single 2-inch by 4-inch post, centered on each side. The right downstream or north entrance is barricaded, preventing vehicle entry, and a dedication sign is mounted on the gable.

Directions: From Shelby, go east on East Marion Street (Business US 74) to the intersection with SR 180 (South Post Road); continue east on East Marion Street 0.4 mile to the last entrance, on the right, to the student parking area for Cleveland Community College. The bridge is in the woods margin, to the left of the entrance road.

Kitty Hawk Woods Covered Bridge

DARE COUNTY

Brief Statistics

Type: Non-authentic modern covered bridge. World Guide Covered Bridge Number: NC-28-A. Built c. 1995 by Endurance Marine Construction Co. Four-span bridge crossing Jean Guite Creek. Stringer bridge 58.1 feet long by 17.3 feet wide, with approximately 14.6-foot wide by 15.1-foot high portal openings. Alternate Name: None known. Located in Kitty Hawk at 4812 The Woods Road.

Built circa 1995 by Endurance Marine Construction Company, Kitty Hawk Woods Bridge is located in North Carolina Coastal Reserve—Kitty Hawk Woods. The four-span bridge crosses wide and deep Jean Guite Creek, a tidal creek flowing between Currituck Sound and Kitty Hawk Bay. The southwest/northeast-aligned bridge is situated on a private asphalt-paved road, among dense trees and shrubbery in a swampy area.

Nine timber stringers, set on abutments and four piers, support the 58.1-foot long stringer bridge. The abutments are constructed with upright wood-utility-pole piles and timbers, with abutted upright wood-utility-pole piles forming the wing walls; each pier is constructed with four braced, upright wood-utility-pole piles. The flooring consists of transverse planking with three-plank-wide wheel treads. Weathered vertical boarding with battens covers the sides to rail height, the sides open above to the eaves, exposing the roof support posts. The sides have rounded-corner cutouts at the bottom, above the two center spans, to provide additional clearance for passing boats. The weathered, steeply peaked gables are covered with vertical boarding with battens, encircled with trim boards. Cedar shingles cover the roof, which is supported along each side by 6-inch by 12-inch wood posts at the portals and six 6-inch by 6-inch wood posts in pairs, creating four panels, with 4-inch by 6-inch X braces. Wood rail guardrails are at each portal.

Directions: From Kitty Hawk, at the junction of US 158 and SR 12, go west 1.2 miles on US 158 to The

Old Salem Pedestrian Bridge has clear acrylic panels protecting the multiple kingpost truss with a Burr arch. The footbridge has a 4.8-foot camber giving the bridge a humpbacked appearance.

Woods Road, on the left; go 0.8 mile to North Carolina Coastal Reserve—Kitty Hawk Woods and gate house on the private road, on the right (No. 4812 on the mailbox). The bridge is 0.4 mile down the private road. Private property: request permission to visit.

Old Salem Pedestrian Covered Bridge

FORSYTH COUNTY

Brief Statistics

Type: Authentic modern covered footbridge. World Guide Covered Bridge Number: NC-34-01. Built in 1998 by Blue Ridge Timberwrights. One-span footbridge crossing Old Salem Road. Multiple kingpost truss with Burr arch bridge 123.6 feet long by 19.5 feet wide, with approximately 11.2-foot wide by 11.7-foot high portal openings. Alternate Name: Old Salem Heritage Bridge. Located in Winston-Salem at Old Salem over Old Salem Road.

The only authentic covered footbridge in Virginia, Old Salem Pedestrian Bridge was built, in December 1998, by Blue Ridge Timberwrights of Christiansburg, Virginia, utilizing southern yellow-pine timbers, mostly salvaged from old buildings and pilings. The footbridge was designed by DCF Engineering of Carey, North Carolina, and built at a cost of $1,500,000. The single-span footbridge was built to provide pedestrian access from the Old Salem historic village and the Museum of Early Southern Decorative Arts across Old Salem Road to the new Old Salem Heritage Center.

The 123.6-foot long truss bridge is supported by double twelve-panel multiple kingpost trusses, encasing a laminated Burr arch, all anchored to poured concrete abutments, which are faced with rough-cut stone. The west/east-aligned bridge has a red-brick-faced concrete stairway at the west end and a massive red-brick-faced concrete ramp and stairways at the east end. The flooring consists of 3-inch by 6-inch tongue and groove planking. The sides are enclosed with dark brown, aluminum framed, 7/16-inch thick, clear acrylic panels, providing a bright daytime light in the interior and visibility of the truss timbers from the outside. The portals have natural horizontal lapped siding covering the gables; the entrances are oval-shaped at the top, with boxed-in, decorative, wide, natural wood pilasters. The roof is covered with dark brown copper sheathing. The roof and gables extend beyond the entrances, providing weather protection for the entrances.

Old Salem Pedestrian Bridge displays a significant humpbacked appearance, due to the 4.8-foot camber built into the timberwork. The bridge interior has electric lighting and black-painted banister-type iron railings running along each side. Black *1998* on a red-brown oval plaque is mounted on the gable above each entrance. Bronze plaques with bridge information are mounted on the brick wall at each portal, the bridge name listed as Old Salem Heritage Bridge, although the bridge is commonly referred to as Old Salem Pedestrian Bridge.

Directions: From Winston-Salem, at I-40, take exit 193, US 52, going north and immediately turning left on Sprague Street, continuing to Main Street; go right (circles right becoming Old Salem Road) to the bridge overhead.

Bodenheimer Covered Bridge

GUILFORD COUNTY

Brief Statistics

Type: Non-authentic modern covered bridge. World Guide Covered Bridge Number: NC-41-A. Built in 1966 by Millard H. Bodenheimer. One-span bridge crossing Richland Creek. Stringer bridge 20.3 feet long by 10.3 feet wide, with approximately 9.3-foot wide by 6.6-foot high portal openings. Alternate Name: None known. Located north of Archdale at end of Lorraine Avenue.

In 1966, Millard H. Bodenheimer, the original owner, built Bodenheimer Bridge over Richland Creek. The stringer bridge has the distinctions of being the shortest, at 20.3 feet long, and the narrowest, at 10.3 feet wide, non-authentic covered motor vehicle bridge in North Carolina.

Bodenheimer Bridge is supported by five wood utility pole stringers, set on mortared stone on top of a natural rock base at the east-southeast end, and a pier, consisting of two poured concrete pillars capped with a timber, on the west-northwest or right downstream end. A ramp at one time spanned from the pier to the far abutment, but all that remain are two of the four wood utility pole stringers; the bridge is barricaded at the pier. The flooring consists of transverse planking; the sides are open, with a gray weathered lattice simulating a Town lattice; the gables are covered with dark weathered vertical boarding. The steeply peaked roof is covered with shiny tin and supported by three 4-inch by 4-inch wood posts along each side. The east-southeast portal has small branches attached to it and two signs mounted on the gable, white letters on dark weathered boards—*WHAT MAKES PIEDMONT GREAT* and *OLD CARTERTOWN.*

The single-span bridge crosses rock-bottomed Richland Creek, which flows eastward about four miles, before entering the Deep River. Closed to motor traffic for several years now, the bridge has a chalet at the upstream east-southeast corner, a short mortared stone pillar and log rail fence along the creek bank joining the two. The bridge is situated in a densely wooded area, thick woods on the far side and a variety of landscape appointments among the trees, on the near side. *See Bodenheimer Covered Bridge in the color photograph section: C-17.*

Directions: From southeast of Archdale, at I-85, exit 111, US 311 (Main Street); go north on US 311 to Baker Road, on the right; go 1.8 miles to Alleghany Street, on the left; go 0.3 mile to Lorraine Avenue, on the left; go 0.2 mile, to the end of the road. The bridge is behind the residence, on the right. Private property: request permission to visit.

Buckhorn Covered Bridge

GUILFORD COUNTY

Brief Statistics

Type: Non-authentic modern covered bridge. World Guide Covered Bridge Number: NC-41-B. Built in 1994 by Dwight R. Sharpe. One-span bridge crossing Buckhorn Creek. Stringer bridge 28.3 feet long by 16.5 feet wide, with approximately 10.7-foot wide by 12.8-foot high portal openings. Alternate Name: None known. Located in Gibsonville at Holly Brooks Subdivision at 7117 Laurel Point Drive.

In 1994, Dwight R. Sharpe built Buckhorn Bridge on the long asphalt-paved driveway to the residence. The single-span bridge crosses rock- and sediment-bottomed Buckhorn Creek, whose waters join nearby Reedy Fork, which flows about four miles eastward to join the Haw River. The driveway is heavily wooded along both sides, with small clearings upstream and downstream of the bridge.

The 28.3-foot long stringer bridge is supported by twelve large timber stringers, set on abutments, constructed with five vertical timbers placed on a horizontal timber and capped with a horizontal timber, horizontal timbers stacked behind the vertical timbers, forming the retaining wall. Steel spreader bars, spanning the creek, hold the vertical timbers in place. Poured concrete wing walls extend out from each side of the abutments. Transverse planking makes up the floor. Brown stained vertical boarding covers the sides and the gables, the gables trimmed in white. Two large horizontal window openings,

Grandover East Course Bridge was built at the fourth tee at Grandover Resort & Conference Center. This 55.5-foot long golfcart bridge has a 354-foot long ramp extending the bridge along the 4th fairway.

filled in with lattice, are opposing along each side. Shiny tin covers the roof, which is supported along each side by 2-inch by 6-inch studding, between multiple 2-inch by 6-inch wood posts at the center and at each end. The roof and gables extend beyond the entrances, to provide protection from inclement weather for the interior. The interior has a raised platform for seats along each side, the upstream side with a bench. The south/north-aligned bridge has a white painted wood sign, lettered in black, *BUCKHORN BRIDGE/1994* mounted on the gable, centered over the south entrance. Plank-rail guardrails funnel traffic into the single-lane bridge.

Directions: From I-85/I-40, south of Gibsonville, take exit 138, SR 61, north to Gibsonville. Continue north 5.9 miles on SR 61 (from where SR 61 turns left and SR 100 continues straight ahead) to Deermont Road, on the right. Follow Deermont Road 0.5 mile into Holly Brooks Subdivision, to Laurel Point Drive; go left 0.1 mile to No. 7117, on the right. The bridge is visible down the long driveway. Private property: request permission to visit.

Grandover East Course Covered Bridge

GUILFORD COUNTY

Brief Statistics

Type: Non-authentic modern covered footbridge. World Guide Covered Bridge Number: NC-41-d. Built in 1997 by Koury Construction. Five-span footbridge crossing tributary to Jenny Branch. Stringer footbridge 55.5 feet long by 12.1 feet wide, with approximately 8.9-foot wide by 9.0-foot high portal openings. Alternate Name: None known. Located southwest of Greensboro at Grandover Resort & Conference Center on the East Course.

Koury Construction built Grandover East Course Bridge, in 1997, on the East Course, beside the 4th tee, at Grandover Resort & Conference Center. The 55.5-foot long stringer footbridge is supported by six 2-inch by 12-inch plank stringers, placed on edge, and attached to six piers, consisting of two upright wood utility pole piles, with a 2-inch by 12-inch plank bolted at the top of each side of the piles. The five short spans are necessary to create the humped appearance of the bridge. A winding cement golfcart path leads from the 4th tee down to the west-southwest portal of the bridge. A 354-foot long winding ramp leads away from the east-northeast portal of the bridge toward the 4th green. The ramp is an extension of the bridge, utilizing the same pier and stringer construction. The flooring in the bridge consists of transverse 2-inch by 8-inch planking, with four 1-inch by 6-inch board-wide wheel treads, the transverse planking continuing to the end of the ramp. The sides are open, with a handrail running between the roof support posts. The handrail consists of a 6-inch by 6-inch timber orientated on the diagonal, forming the top rail, and a diagonal 4-inch by 4-inch timber, forming the lower rail. The gables are covered with vertical tongue and groove planking, a semi-circular cutout centered at the bottom, exposing the roof truss. The steeply peaked roof is covered with weathered and algae-stained wood shingles and is supported along each side by six 8-inch by 8-inch timber posts, placed on the diagonal, decorative bracing at the upper part of the posts. The bridge has short rough-cut-stone guard walls at each portal.

The narrow, sandy bottomed tributary to nearby Jenny Branch flows across the 4th tee area, passing beneath the five-span bridge on its way to join Reddicks Creek, which enters Hickory Creek near its confluence with the Deep River. The west-southwest/east-northeast aligned bridge has the 4th tee upstream, or on the north-northwest side, and dense woods downstream. The bridge curves or bows toward the downstream side.

Directions: From Greensboro, go south on I-85 to exit 120, at Groometown Road; go right 0.1 mile to Grandover Parkway; go left 1.3 miles to Grandover Resort & Conference Center. The bridge is on the East Course.

Grandover West Course Covered Bridge

GUILFORD COUNTY

Brief Statistics

Type: Non-authentic modern covered footbridge. World Guide Covered Bridge Number: NC-41-c. Built in 1997 by Koury Construction. One-span footbridge crossing wet weather stream. Stringer footbridge 37.0 feet long by 15.6 feet wide, with approximately 12.7-foot wide by 9.0-foot high portal openings. Alternate Name: None known. Located southwest of Greensboro at Grandover Resort & Conference Center on the West Course.

Koury Construction built Grandover West Course Bridge, in April 1997, on the West Course, between the 15th tee and the 6th green, at Grandover Resort & Conference Center. A reinforced concrete four-rib floor-panel stringer, set on poured concrete abutments, supports the 37.0-foot long stringer footbridge. The flooring consists of the concrete slab stringer. Light-gray-stained vertical boarding with battens covers the sides to rail height, open above to the eaves, exposing a decorative Town lattice. The steeply peaked gables are also covered with light-gray-stained vertical boarding with battens, encircled with trim boards and with a diamond window opening centered over the entrances. Gray weathered cedar shingles cover the roof, which is supported along each side by five sets of double wood posts that encase the decorative lattice. The roof and gables extend a generous distance beyond the entrances, providing shelter for the entrances from inclement weather. Mortared rough-cut-stone guard walls are at both portals, a bronze bridge informational plaque mounted on the downstream guard wall, at the southeast portal.

The single-span bridge crosses a wet weather stream that most likely runs into Jenny Branch. The northwest/southeast-aligned bridge is fully exposed at the portal ends, but thick foliage, with scattered mature trees along the stream, makes side views difficult. Two cement golfcart paths converge at the southeast portal of the two-golfcart-lane wide bridge, the cement golfcart path winding around the 6th green, after emerging from the bridge. At 15.6 feet wide, Grandover West Course Bridge is the widest non-authentic covered footbridge in North Carolina.

Directions: From Greensboro, go south on I-85 to exit 120, at Groometown Road; go right 0.1 mile to Grandover Parkway; go left 1.3 miles to Grandover Resort & Conference Center. The bridge is on the West Course.

Brannon Forest Covered Bridge

HAYWOOD COUNTY

Brief Statistics

Type: Non-authentic modern covered bridge. World Guide Covered Bridge Number: NC-44-A. Built in 1997 by unknown builder. One-span bridge crossing tributary to Jonathan Creek. Stringer bridge 46.1 feet long by 24.0 feet wide, with approximately 21.9-foot wide by 15.2-foot high portal openings. Alternate Name: None known. Located in Dellwood in Brannon Forest Subdivision on Brannon Forest Drive.

In 1997, an unknown builder built Brannon Forest Bridge on Brannon Forest Drive, in Brannon Forest Subdivision. Thirteen steel I-beam stringers, set on poured concrete abutments, support the two-lane bridge. The 46.1-foot long bridge, aligned west-southwest/east-northeast, has the east-northeast half spanning an unnamed tributary, and the west-southwest half has a concrete slab subfloor, set on the ground. The uniquely-patterned flooring is divided into quarter sections by a transverse plank running the full width of the bridge and two planks running the length of the bridge, butted to the transverse plank, forming a cross. The quarter sections have diagonal planking, with the diagonals radiating out from the center point of the floor. The sides are open, and the gables are covered with brown stained horizontal lapped siding, the board edges not planed, simulating log edges. Red painted aluminum covers the roof, which is supported by five 8-inch by 8-inch timber posts, creating four panels, the two center panels braced, simulating a kingpost, the two outer panels counterbraced. Three timber buttress braces are evenly spaced down each side, to stabilize the bridge. A louvered cupola with a red painted roof, topped with a horse weathervane, is centered on the roof. Electric lights illuminate the interior.

The single-span bridge crosses a narrow wet weather tributary to nearby Jonathan Creek, which joins the Pigeon River, north of the community of Cove Creek. Thick woods, but thinned out near the bridge, with many rhododendrons added, run along Brannon Forest Drive, the entrance road to the subdivision. An ornate sign, inscribed *Brannon Forest/Maggie Valley*, is mounted over the east-northeast entrance and a white board sign, inscribed *Brannon Forest/1997*, is mounted on the west-southwest gable.

Directions: From Dellwood, at the junction of US 19 and US 276, go north 0.5 mile on US 276 to Fox Run Road, on the left; go 0.2 mile to Jayne Cove Road, on the right; follow 1.0 mile to Brannon Forest Subdivision, where the road becomes Brannon Forest Drive. The bridge is at the subdivision entrance.

Hickory Nut Gap Covered Bridge

Henderson County

Brief Statistics

Type: Non-authentic modern covered footbridge. World Guide Covered Bridge Number: NC-45-a. Built in 1980 by unknown builder. One-span footbridge crossing Hickory Branch. Stringer footbridge 20.1 feet long by 8.0 feet wide, with approximately 7.4-foot wide by 7.3-foot high portal openings. Alternate Name: None known. Located in Gerton at former Hickory Nut Gap Motel.

Hickory Nut Gap Bridge was built, in 1980, by an unknown builder, at the former Hickory Nut Gap Motel, amidst scattered mature trees and shrubs on the maintained lawn area between the motel units and the highway. Narrow and rocky Hickory Branch parallels the highway, passing beneath the single-span footbridge on its way into nearby Hickory Creek, which enters the Broad River, at Bat Cave.

Three steel rectangular-tubing stringers, placed on cement blocks on the stream banks, support the 20.1-foot long stringer bridge. The flooring consists of transverse 1-inch boarding. The sides and the gables are open, the sides with 2-inch by 4-inch wood handrails running between the roof support posts, and benches running the length of the bridge. Weathered cedar shingles cover the roof, which is supported along each side by three 4-inch by 4-inch wood posts, braced under the eaves. Interestingly, the west/east-aligned bridge is fastened together with treenails or wooden pegs.

Directions: From Bat Cave, go north on US 74A to Gerton, and the bridge is on the left, just beyond Bearwallow Cemetery Road, on the left. During foliage season, the bridge is difficult to see, due to the trees lining US 74A. Private property: request permission to visit.

Mountain Junction General Store Covered Bridge

Macon County

Brief Statistics

Type: Non-authentic modern covered footbridge. World Guide Covered Bridge Number: NC-57-b. Built in 1990 by Lamar Burnette and Lee Ray Wilburn. One-span footbridge crossing tributary to Mud Creek. Stringer footbridge 12.8 feet long by 6.3 feet wide, with approximately 5.2-foot wide by 7.3-foot high portal openings. Alternate Name: None known. Located in Scaly at Mountain Junction General Store at corner Dillard Road (SR 106) and Old Mud Creek Road.

Lamar Burnette and Lee Ray Wilburn built Mountain Junction General Store Bridge, in 1990, at the Mountain Junction General Store. Located between the store and the café, the single-span footbridge crosses a narrow, sandy-bottomed tributary to nearby Mud Creek, which joins the Tennessee River near Dillard, Georgia. The northeast/southwest-aligned bridge has open fields sprawling to the wooded hills beyond, on the northeast end, and the general store and the café parking areas off Old Mud Creek Road, on the southwest end.

Three wood utility pole stringers, set on the stream banks, support the 12.8-foot long stringer bridge. The flooring consists of transverse 1-inch by 10-inch boarding. The sides are open with small bark-covered log handrails, filled in below with little branches, along each side. The northeast portal is open; the southwest portal is covered with wood shakes, with a log birdhouse mounted below the roof overhang. Very old, weathered wood shakes cover the roof, which is supported by bark-covered log posts at each portal, the posts set into the ground. Old split-rail fencing extends out from the south-west portal to the general store, down-stream, and the café, upstream.

Directions: From Highlands, take SR 106 west to Scaly, continuing to Old Mud Creek Road, on the left (very near Georgia line). The bridge is in the southeast corner, between the Mountain Junction General Store and the Mountain Junction Café.

The Bridge of Madison County Covered Bridge

MADISON COUNTY

Brief Statistics

Type: Non-authentic modern covered footbridge. World Guide Covered Bridge Number: NC-58-a. Built in 1997 by unknown builder. One-span footbridge crossing Spring Creek. Stringer footbridge 30.5 feet long by 8.6 feet wide, with approximately 4.9-foot wide by 8.0-foot high portal openings. Alternate Name: None known. Located in Trust off SR 209.

The Bridge of Madison County was built, in 1997, by an unknown builder, in the landscaped front yard of the residence. The bridge is situated just off State Road 209. Spring Creek swiftly cascades over boulders, under the single-span bridge, on its way to join the French Broad River, at Hot Springs.

The Bridge of Madison County is in Madison County, North Carolina! This is a 30.5-foot long stringer footbridge.

The 30.5-foot long bridge is supported by two sets of triple 2-inch by 10-inch plank stringers, set on mortared natural stone abutments, with short wing walls. The floor is constructed with transverse planking. Red-stained, vertical, facsimile tongue and groove paneling covers the sides and the portals, the sides with a horizontal window opening running the length of the bridge, exposing the roof support posts and braces. A full-radius arch is over each entrance. Cedar shingles cover the steeply peaked roof, which is supported along each side by five 4-inch by 4-inch wood posts with diagonal braces. The roof peak extends slightly beyond the portals; the roof and gable fasciae are scalloped. There are two wooden steps at each portal. The east-southeast/west-northwest aligned bridge has a black-painted metal gate, securing the west-northwest or right downstream entrance.

Directions: From Trust, at the junction of SR 63 and SR 209, go 0.3 mile south on SR 209, to the bridge on the left, at the edge of the highway. Private property: request permission to visit.

Lewis Covered Bridge

MONTGOMERY COUNTY

Brief Statistics

Type: Non-authentic modern covered bridge. World Guide Covered Bridge Number: NC-62-A #2. Built in 2001 by William A. Lewis, Jr. Three-span bridge crossing ponds interconnect. Stringer bridge 28.8 feet long by 11.0 feet wide, with approximately 9.8-foot wide by 8.0-foot high portal openings. Alternate Name: None known. Located in Biscoe at 901 Alternate US 220 (South Main Street).

William A. Lewis, Jr., built the first covered bridge on his driveway, in 1956. This single-span bridge was 14.8 feet long, with approach ramps at both ends. Mr. Lewis felt that the covered bridge needed to be longer and sturdier. In 2001, he, with the help of his family, rebuilt Lewis Bridge, utilizing the previous bridge's piers, stringers and flooring. The new three-span bridge crosses the interconnect between two sizeable man-made ponds, the long graveled driveway leading to the residence, transiting the bridge. The northeast/southwest-aligned bridge is fully exposed in the front yard, amidst scattered mature cedar, pine and maple trees. The large pond on the southeast side has an island.

The 28.8-foot long bridge is supported, under the center span, by five 6-inch by 8-inch timber stringers, placed on edge and set on two piers, each pier constructed with two upright wood utility pole piles, capped with a railroad tie. The outside spans are supported by five stringers: double 2-inch by 4-inch wood lumber, on edge, for the outside stringers, double 2-inch by 8-inch wood lumber, on edge, for the center stringer, and triple 2-inch by 8-inch wood lumber, on edge, for the in-between stringers. The flooring consists of 2-inch by 8-inch planking, rough-cut planking on the original center span and nominal finished planking on the outside spans, which were the ramps on the original bridge. Natural, vertical, rough-cut boarding covers the sides and the gables, the sides to rail height, exposing the roof support posts with bracing. Cedar shingles cover the roof, which is supported along each side by four 4-inch by 6-inch wood posts. Two wood buttress braces, along each side, stabilize the bridge. A horse-and-buggy weathervane is mounted above the northeast gable.

Directions: From Biscoe, at the intersection of Alternate US 220 (North-South Main Street) and SR 24/SR 27 (East-West Main Street), go 1.0 mile south on Alternate US 220 (South Main Street) to No. 901. The bridge is along the long driveway, not visible from the highway. Private property: request permission to visit.

River Island Covered Bridge

NEW HANOVER COUNTY

Brief Statistics

Type: Non-authentic modern covered bridge. World Guide Covered Bridge Number: NC-65-A. Built in 1991 by Sanco Homes. Five-span bridge crossing tidal marsh. Stringer bridge 57.5 feet long by 17.0 feet wide, with approximately 10.9-foot wide by 14.2-foot high portal openings. Alternate Name: None known. Located in Wilmington off River Road on Island Bridge Way.

Sanco Homes built River Island Bridge, in 1991, over the tidal marsh of the Cape Fear River, to gain access to River Island for development. Landscaped with planter boxes leading to the east portal, the fully exposed five-span bridge, on Island Bridge Way, is just off River Road.

The 57.5-foot long stringer bridge is supported by fifteen 6-inch by 14-inch timber stringers, set on an abutment of stacked 6-inch by 14-inch timbers, at the east end, and five piers. Each pier is constructed with three wood utility pole piles, capped with three stacked 6-inch by 14-inch timbers, two on edge on top of one on the flat. A long ramp, of the same construction as the bridge, stretches to the island, from the west portal of the west/east-aligned bridge, and a short ramp is at the east end. Transverse planking makes up the flooring, which extends to the end of the ramps. Each open side has five sets of double 8-inch by 8-inch timber roof support posts, with an extra 8-inch by 8-inch timber roof support post at the portals. Adjacent roof support timbers are joined by an arched timber, near the eave, a brace running from the center of the timber to under the roof. A handrail runs the entire length of the bridge, including the ramps. The gables are also open, with a decorative arch and lattice. Weathered wood shingles cover the steeply peaked roof.

The exterior and the interior of the bridge are painted white; the top handrail plank and the top board on the planter boxes are painted gray-green. A security gate is at the center of the bridge. A lighted, decorative, semi-circular sign, with a windswept tree above *RIVER ISLAND*, is mounted on the east gable. *See River Island Covered Bridge in the color photograph section: C-19.*

Directions: From Wilmington, at the intersection of US 17 and SR 132, go south on SR 132 to Shipyard Boulevard, on the right; follow to the intersection with US 421 (Carolina Beach Road), continuing 0.8 mile past on Shipyard Boulevard to River Road, on the right. Follow River Road 8.2 miles to the bridge, visible at Island Bridge Way, on the right. Island Bridge Way is a private road.

Gait Way Covered Bridge

ORANGE COUNTY

Brief Statistics

Type: Non-authentic modern covered bridge. World Guide Covered Bridge Number: NC-68-A. Built in 1984 by unknown builder. One-span bridge crossing tributary to Collins Creek. Stringer bridge 28.1 feet long by 14.9 feet wide, with approximately 14.0-foot wide by 14.5-foot high portal openings. Alternate Name: None known. Located west of Chapel Hill in Covered Bridge Development on Gait Way.

An unknown builder built Gait Way Bridge, in 1984, on dusty, graveled Gait Way, the entrance road to Covered Bridge Development. Thick woods surround the bridge; however, a small clearing is at the upstream south-southeast corner of the north-northwest/south-southeast aligned bridge. The single-span bridge crosses the rock and sediment bottomed tributary to nearby Collins Creek, which flows into the Haw River, thence continuing through B. Everett Jordan Lake, before merging with the Deep River, south of Haywood, to form the Cape Fear River.

The 28.1-foot long stringer bridge is supported by thirteen 6-inch by 12-inch timber stringers, set on abutments, constructed with a large timber above and below vertical timbers. Upright 4-inch by 6-inch timber piles form the wing walls. The flooring consists of diagonal planking. Gray-green painted horizontal lapped siding covers the sides and the gables. The sides have three small diamond-shaped window openings along each side. Shiny tin covers the roof, which is supported along each side by 2-inch by 4-inch studding. A Covered Bridge Development bulletin board is mounted inside the bridge.

Directions: From Chapel Hill, go west on SR 54 to Dodsons Crossroads; go right 0.8 mile on Dodsons Crossroads to Gait Way, on the left; go 0.3 mile to the bridge.

Albemarle Plantation Covered Bridge

PERQUIMANS COUNTY

Brief Statistics

Type: Non-authentic modern covered bridge. World Guide Covered Bridge Number: NC-72-A. Built in 1986 by unknown builder. Sixteen-span bridge crossing wet weather stream. Stringer bridge 80.6 feet long by 23.8 feet wide, with approximately 21.1-foot wide by 14.8-foot high portal openings. Alternate Name: None known. Located east of Hertford at Albemarle Plantation on Middleton Drive.

Albemarle Plantation Bridge was built by an unknown builder, in 1986, in Albemarle Plantation on Middleton Drive, over a wet weather stream and boggy area, the trees, during foliage season, concealing the nearby golf course. The southeast/northwest-aligned bridge is fully exposed, with Albemarle Boulevard on the northwest end, and homes in the Middleton housing section on the southeast end.

The sixteen-span bridge is supported by stringers of framed panels of double 2-inch by 12-inch lumber, between the fifteen piers and between the abutments and piers. Each abutment and pier consists of seven upright timber piles, with bracing. The abutments have upright, tongue and groove 2-inch by 8-inch lumber forming the wing walls. The flooring consists of a longitudinal plank down the center of the two-lane bridge, with diagonal planking running in opposite directions down each lane, this all on top of a transverse-planked sub-floor. Dark weathered horizontal lapped siding covers the upper third of the sides and the portals, the lower sides open, exposing the roof support posts, which have an approximately 5-foot by 5-foot lattice on the inside. Each roof support post has a wood buttress brace for bridge stabilization. The interior upper third of the sides also has horizontal lapped siding, the ceiling plywood paneled. Gray-black weathered wood shingles cover the roof, which is supported along each side by nine 5-inch by 5-inch wood posts. A large electric carriage lantern is mounted inside each portal of the 80.6-foot long bridge.

Directions: From Edenton, take US 17 north to Hertford, continuing on Bypass US 17 to Harvey Point Road, on the right; go 4.2 miles to Burgess Road, on the right; go 0.7 mile to the tee at Holiday Island Road; go left 1.3 miles to Albemarle Plantation, on the right. Obtain clearance at the gatehouse, then follow Albemarle Boulevard to Middleton Drive and the bridge, on the left.

Caro-Mi Covered Bridge

POLK COUNTY

Brief Statistics

Type: Non-authentic modern covered bridge. World Guide Covered Bridge Number: NC-75-A #2. Built in 1978 by unknown builder. One-span bridge crossing North Pacolet River. Stringer bridge 54.8 feet long by 16.7 feet wide, with approximately 11.9-foot wide by 9.8-foot high portal openings. Alternate Name: None known. Located in Tryon at 3231 US 176 West.

Caro-Mi Bridge was built by an unknown builder, in February 1978, to replace the earlier covered bridge, built in 1962, which was lost in the flood of November 6, 1977. Located at the Caro-Mi Dining Room Restaurant, the single-span bridge crosses the rock- and boulder-strewn North Pacolet River, which merges with the South Pacolet River, southeast of Fingerville, South Carolina, to form the Pacolet River. Aligned south/north, the bridge has the restaurant and dense woods on the south end, and the asphalt paved parking area on the north end. Several trees line the riverbanks.

The 54.8-foot long bridge is supported by two steel I-beam stringers, set on poured concrete abutments. The stringers have steel I-beam crossbeams that extend out the upstream side to support the covered walkway. Transverse planking comprises the flooring on the traffic lane, and longitudinal planking runs the length of the walkway. The downstream side and the partition separating the walkway from the traffic lane have horizontal lapped siding to rail height, open above to the eaves. The walkway has a banister-type railing on the upstream side. The gables are also covered with horizontal lapped siding, the walkway entrance with an arch cut out of the siding. Moss-covered, black asphalt shingles cover the roof, which is supported along each side by seven 6-inch by 6-inch wood posts. All boards on the bridge have edges that are not planed, to simulate log edges. The exterior and the interior of the bridge are painted dark brown, except the gables are painted cream color, and the fasciae are painted white. Electric lights illuminate the interior. *CARO-MI* is on a sign on the north gable, and bridge information plaques are on both gables.

Directions: From Tryon, take US 176 north to the junction with SR 108, continuing 2.8 miles west on US 176 to the Caro-Mi Dining Room at No. 3231, on the left. The bridge crosses the river to the restaurant.

Highland Farms Covered Bridge

Polk County

Brief Statistics

Type: Non-authentic modern covered bridge. World Guide Covered Bridge Number: NC-75-C. Built in 1999 by unknown builder. Post supported roof over roadway crossing Ostin Creek. Bridge 42.0 feet long by 24.3 feet wide, with approximately 22.2-foot wide by 14.3-foot high portal openings. Alternate Name: None known. Located in Mill Spring in Lake Adger Community on Lake Adger Parkway.

Designed by Harrison Design of Landrum, South Carolina, and built, in 1999, by an unknown builder, Highland Farms Bridge is in Lake Adger Community, on Lake Adger Parkway. The 42.0-foot long bridge is a post supported roof over a roadway, which is over two large square cement culverts. Inside the two-lane bridge, the roadway is paved with simulated cobblestone cement. The bridge sides are covered with vertical tongue and groove siding above the four openings between the roof support posts, and the sides are covered with vertical boarding with battens, below the openings, from rail height down. The gables are covered with vertical tongue and groove siding, except that under the peak is open, aligned to a cross-board across the fascia, with a simulated tree in front of the opening. Black asphalt shingles cover the steeply peaked roof, which is supported along each side by five 8-inch by 8-inch timber posts. The entire bridge is stained gray-brown. A black asphalt-shingle-roofed cupola, with six-glass-pane windows on each side and topped with a prancing horse weathervane, is mounted on the roof. Mortared natural stone guard walls, capped with an electric lantern, are at the portals.

Ostin Creek flows beneath the bridge into nearby Lake Adger (Green River). The south-southeast/north-northwest aligned bridge is fully exposed, with thick woods blanketing the hill on the upstream side, and a well-landscaped area with picnic tables and benches among some trees on the downstream side. Landscape planters run the length of the bridge, at both sides. Embankments around the culverts are covered with broken rocks and plantings. *See Highland Farms Covered Bridge in the color photograph section: C-20.*

Directions: From Mill Spring, at the intersection of SR 9 and SR 108, go north 2.3 miles on SR 9 to Garrett Road, on the left (there is a sign here for Lake Adger); go 0.4 mile to Green Hills Road, on the right (Lake Adger Community); go 0.2 mile to Lake Adger Parkway, on the left, and follow 1.6 miles to the bridge.

Pisgah Covered Bridge

Randolph County

Brief Statistics

Type: Non-authentic historic covered bridge. World Guide Covered Bridge Number: NC-76-01.* Built in 1911 by J. J. Welch, Jack Webster and Jeff Webster. Three-span bridge crossing West Fork Little River. Stringer bridge 52.0 feet long by 11.0 feet wide, with approximately 10.1-foot wide by 10.6-foot high portal openings. Alternate Name: Pisgah Community Bridge. Located west of Pisgah off Pisgah Covered Bridge Road on bypassed section.

J. J. Welch (a local farmer) and brothers Jack Webster and Jeff Webster built Pisgah Bridge, at a cost of $40.00, in 1911. Named after the nearby community, the stringer bridge has also been known as Pisgah Community Bridge, the name recorded on the National Register. Pisgah Bridge is the oldest non-authentic covered motor vehicle bridge in North Carolina and, for that matter, in the entire southeastern United States. The bridge has been bypassed and closed to motor traffic for many years.

The 52.0-foot long bridge is supported by four large timber stringers across the center span and five large timber stringers across the two outside spans. The stringers rest on dry, natural river-stone abutments and two dry, natural river-stone piers, the piers having been referred to as stepping stone piers. The flooring consists of longitudinal planking on top of transverse planking. Gray weathered vertical boarding covers the sides and the gables, the sides open under the eaves to promote air circulation within the bridge. Shiny tin covers the steeply peaked roof, which is supported by four 4-inch by 9-inch timber posts with braces. The roof and gables extend beyond the entrances, providing added weather protection for the interior. Four wood buttress braces along each side stabilize the bridge, the buttress braces covered with gray weathered vertical boarding and cedar shingles. A wood guardrail has been added along both sides, inside the bridge. The interior of the bridge has been sandblasted to remove graffiti.

The three-span bridge crosses the rock- and boulder-strewn West Fork Little River, whose waters join the Little River, west of Star, continuing on to the Pee Dee River, near Ingram. The bypass road is near the bridge, on the upstream side. The west-southwest/east-northeast aligned bridge is in an improved park-like area, with boardwalks, trails and picnic tables. The bridge is surrounded by the Uwharrie National Forest and spans a river abounding with large boulders.

In the 1930s, Randolph County had sixty covered bridges; today Pisgah Bridge is the sole survivor. Pisgah Bridge was listed on the National Register of Historic Places on January 20, 1972. Asheboro is the county seat of Randolph County.

Directions: From Asheboro, go south on US 220 to exit 51, at SR 134; go right 0.2 mile to SR 134, on the left; go south 4.5 miles to Burney Road; go right 1.8 miles to Pisgah Covered Bridge Road; go left 0.3 mile and turn right (still Pisgah Covered Bridge Road), continuing an additional 1.6 miles to the bridge, on the left.

**The World Guide Number assigned to this bridge incorrectly refers to an authentic bridge.*

McOwenben Covered Bridge

Rutherford County

Brief Statistics

Type: Non-authentic modern covered bridge. World Guide Covered Bridge Number: NC-81-D. Built in 1997 by Bill McBraer. Two-span bridge crossing Broad River. Stringer bridge 128.2 feet long by 15.2 feet wide, with approximately 12.7-foot wide by 13.9-foot high portal openings. Alternate Name: None known. Located in Uree at Twelve Mile Post Subdivision on McOwenben Pass.

The stringer for McOwenben Bridge was built in 1996; Bill McBraer added the cover in 1997. The covered bridge is on asphalt paved McOwenben Pass, the entrance road to Twelve Mile Post Subdivision. McOwenben Bridge is the longest non-authentic covered motor vehicle bridge in North Carolina.

The 128.2-foot long stringer bridge is supported by four steel I-beam stringers, set on poured concrete abutments, the top part faced with mortared rough-cut stone, as are the planters and square light pillars, and one pier, consisting of two poured concrete pillars capped by a poured reinforced concrete crossbeam. Approximately 9.5 feet at each end of the stringer bridge is uncovered, the uncovered section with a black-painted metal handrail at each side. Transverse planking with three-plank-wide wheel treads comprises the floor. The barn-red painted bridge has vertical boarding on the sides and the gables, the sides with five large cathedral-style window openings, opposing along each side. Gray painted aluminum covers the roof, which is supported by 2-inch by 6-inch studding. The roof and gables extend beyond the entrances, providing added weather protection for the interior. An electric, black-painted metal gate is inside the north-northwest portal. A large bridge name sign is mounted over each entrance.

The two-span bridge crosses the wide, tree-lined, rock and sand bottomed Broad River, which joins the Saluda River in Columbia, South Carolina, to form the Congaree River. The south-southeast/north-northwest aligned bridge is open on the downstream side, but heavily wooded in the upstream south-southeast corner. Open fields extend to the highway, beyond the wide tree line along the riverbank, in the upstream north-northwest corner. A three diagonal-wood-rail fence leads traffic from the highway into the single-lane bridge.

Directions: From Rutherfordton, at the intersection of US 221 and US 64/US 74A, go west 12.3 miles on US 64/US 74A to McOwenben Pass, at Twelve Mile Post Subdivision, on the left. Go 0.1 mile to the bridge.

Pool Creek Covered Bridge

Rutherford County

Brief Statistics

Type: Non-authentic modern covered bridge. World Guide Covered Bridge Number: NC-81-C. Built in 1991 by

Randal Payne with David White and Billy Ellis. One-span bridge crossing Pool Creek. Stringer bridge 54.3 feet long by 18.2 feet wide, with approximately 16.6-foot wide by 14.0-foot high portal openings. Alternate Name: None known. Located in Lake Lure on Bottomless Pools Road at Bottomless Pools.

Randal Payne, with David White and Billy Ellis, built Pool Creek Bridge, in 1991, on Bottomless Pools Road, at the entrance to the Bottomless Pools attraction. The single-span bridge crosses moderately wide, rocky-bottomed Pool Creek, which flows into nearby Lake Lure (Broad River), the Broad River merging with the Saluda River in Columbia, South Carolina, to give birth to the Congaree River. The south-southeast/north-northwest aligned bridge is in a heavily wooded area, with several mature trees near the bridge. In 1996, the bridge was closed, due to a severe sag in the bridge. In 1997, the stringers and abutments supporting the bridge were replaced, and the bridge was reopened.

The 54.3-foot long bridge is supported by four large steel I-beam stringers, set on poured concrete abutments. The floor consists of transverse planking with timber tire bumpers running down each side, forming walkways at each side of the traffic lane. Weathered, brown-stained, vertical boarding with battens covers the sides and the gables, the sides with five opposing, large, cathedral-style, vertical window openings. Weathered cedar shingles, with patches of algae, cover the roof, which is supported along each side by 2-inch by 6-inch studding. The roof overhangs both entrances slightly, the siding extending with the roof, forming a short weather panel. A gate is at the north-northwest entrance. Signs with large, black, wood cutout letters, reading *POOL CREEK/COVERED BRIDGE*, are mounted on each gable. A board rail fence at both ends funnels traffic into the single-lane bridge. *See Pool Creek Covered Bridge in the color photograph section: C-19.*

Directions: From Chimney Rock, going eastbound on US 74A to Lake Lure, cross over the Broad River and take Bottomless Pools Road, on the right, to the Bottomless Pools attraction. The bridge is at the entrance. There is an admission charge to the park, at the far side of the bridge.

Watts Farm Covered Bridge

STOKES COUNTY

Brief Statistics

Type: Non-authentic modern covered bridge. World Guide Covered Bridge Number: NC-85-A. Built in 1984 by George and Sylvia Watts. One-span bridge crossing Watts Creek. Stringer bridge 21.5 feet long by 12.5 feet wide, with approximately 10.5-foot wide by 7.5-foot high portal openings. Alternate Name: None known. Located west of Walnut Cove at 2890 Brook Cove Road.

George and Sylvia Watts built Watts Farm Bridge, in 1984, on the long tree-lined asphalt-paved driveway to their residence. Most of the lumber for the bridge came from the circa 1850 Lawson House, that was on their property. The west-southwest/east-northeast aligned bridge is situated in a densely wooded area immediately surrounding the bridge and on the west-southwest side of the creek. The east-northeast side of the creek has horse paddocks in the upstream corner and an open grassed area in the downstream corner. Narrow, rocky-bottomed Watts Creek passes beneath the single-span bridge, on its way to nearby Town Fork Creek, which flows into the Dan River, at Ceramic.

The 21.5-foot long stringer bridge is supported by three steel I-beam stringers, resting on abutments, constructed with stacked railroad ties, held in place by upright wood utility poles. The flooring consists of transverse planking. Old, weathered, horizontal lapped siding covers the sides, the board edges not planed, to simulate log edges. Horizontal tongue and groove siding covers the gables. Red painted tin covers the low peaked roof, which is supported by five old, hand-hewn timber posts. An electric light illuminates the interior.

Directions: From Winston-Salem, take US 52/SR 8 north, following SR 8 to Germanton, where SR 8 joins SR 65. Continue on SR 8/SR 65 north to where SR 8 splits away to the left; follow SR 8 an additional 0.8 mile to Brook Cove Road, on the right; go 1.1 miles to No. 2890, on the left at Watts Road. The bridge is down the long driveway, at the woods line. Private property: request permission to visit.

Dupont State Forest Covered Bridge

TRANSYLVANIA COUNTY

Brief Statistics

Type: Non-authentic modern covered bridge. World Guide Covered Bridge Number: NC-88-B. Built in 2000 by unknown builder. One-span bridge crossing Little River. Stringer bridge 85.1 feet long by 29.6 feet wide, with approximately 21.4-foot wide by 14.0-foot high portal openings. Alternate Name: None known. Located north of Cedar Mountain in Dupont State Forest on Buck Forest Road.

A developer purchased 2,000 acres, for development in the Dupont State Forest, and, in 2000, built Dupont State Forest Bridge, in a magnificent setting over the

Little River, at the top of High Falls. The development never materialized beyond the covered bridge and some crushed stone roads. The bridge holds two distinctions for a non-authentic covered motor vehicle bridge in North Carolina; first, it is the longest single-span bridge, at 85.1 feet long, and, second, it is the widest bridge, at 29.6 feet wide.

Nine steel I-beam stringers, cross-braced, set on poured concrete abutments with wing walls, support the two-lane stringer bridge. The floor consists of transverse planking. Vertical boarding with battens covers the sides and the portals, the sides with eight large cathedral-style window openings, opposing. Weathered wood shingles cover the roof, which is supported by nine timber posts along each side and between the two traffic lanes and the walkway, on the downstream side. A large timber guardrail runs down each side of the traffic lanes. A handrail extends down the inside of the walkway. The roof has a narrow, braced overhang, aligned on an opening in the gable at the peak, on each end. Each portal has mortared cut-stone faced guard walls, ending in a pillar wired for electric lanterns, which were never installed.

Joining the French Broad River, at Penrose, the Little River flows beneath the single-span bridge and tumbles, cascades, and slides down the smooth rock of spectacular High Falls. The northeast/southwest-aligned bridge is fully exposed, with a crushed-rock road leading to and from the bridge. A breathtaking view of the bridge and falls may be seen from the base of the falls, by taking the hiking trails from the parking area at the Little River bridge.

Directions: From Brevard, take US 276 (Greenville Highway) south to Cedar Mountain, to Cascades Lake Road, on the left; go 2.5 miles to Staton Road, on the right; go 1.5 miles to Buck Forest Road, on the right (just after entrance to AGFA Plant). Buck Forest Road leads into the parking area. The bridge is about a 0.5-mile hike up Buck Forest Road. For scenic hiking and a spectacular view of the bridge from the bottom of the falls, continue on Staton Road 0.9 mile past Buck Forest Road to the parking area, on the left, just over the Little River concrete bridge. Park and cross the road to the far side of the Little River. Take Triple Falls Trail to Triple Falls; continue to High Falls Trail, going left to Pipeline Trail; go left to the bridge (moderately strenuous hike).

Twil-Doo Covered Bridge

TRANSYLVANIA COUNTY

Brief Statistics

Type: Non-authentic modern covered footbridge. World Guide Covered Bridge Number: NC-88-a #2. Built in 1999 by Chris McMinn. One-span footbridge over tributary to French Broad River. Stringer footbridge 11.8 feet long by 5.6 feet wide, with approximately 4.6-foot wide by 6.5-foot high portal openings. Alternate Name: None known. Located in Pisgah Forest at 235 Wilson Road.

In 1998, a tree, felled by an ice storm, destroyed the first Twil-Doo Covered Bridge, built in 1946 by Max Feaster. In 1999, Chris McMinn built the second Twil-Doo Covered Bridge. The single-span footbridge crosses a narrow tributary to the nearby French Broad River. Aligned north-northeast/south-southwest, the bridge is in the nicely landscaped side yard, in summer surrounded by a potpourri of beautiful flowers. Twil-Doo Bridge, at 11.8 feet long and 5.6 feet wide, is the shortest and the narrowest non-authentic covered footbridge in North Carolina.

The 11.8-foot long bridge is supported by three 2-inch by 8-inch lumber stringers, placed on edge, on mortared natural stone abutments. Transverse planking forms the floor. The red painted bridge has vertical boarding with battens, to rail height, on the sides and just below the eaves, open in between, and also in the gables. Three wood posts along each side support the shiny tin roof. Window box planters, overflowing with colorful flowers, run the length of the downstream side. A bench is inside the bridge, on the upstream side.

Directions: From Brevard, go north on US 64/US 276 to the intersection with SR 280, US 276, and US 64. Backtrack 0.1 mile on US 64/US 276 to the first left (Ecusta Road), and follow 1.3 miles to the end, at Old Hendersonville Highway; go right and immediately turn left on Wilson Road; travel 0.5 mile to the bridge, visible on the left, at No. 235. Private property: request permission to visit.

Alpine Acres Covered Bridge

WATAUGA COUNTY

Brief Statistics

Type: Non-authentic historic covered footbridge. World Guide Covered Bridge Number: NC-95-b. Built in 1898 by unknown builder. One-span footbridge crossing Laurel Fork Creek. Stringer footbridge 28.0 feet long by 10.8 feet wide, with approximately 8.0-foot wide by 7.0-foot high portal openings. Alternate Name: None known. Located west of Boone off Old Danner Road at Alpine Acres Road.

Built in 1898 by an unknown builder, Alpine Acres Bridge is the oldest non-authentic covered footbridge in North Carolina and, for that matter, in the entire southeastern United States. Located just off Old Danner Road across from Alpine Acres Road, the north-northwest/south-southeast aligned bridge is now barricaded at the south-southeast portal. The privately owned bridge has two picnic tables and a barbecue grill in the interior, as the old bridge has come to be used for cookouts. Situated in a densely wooded area, the single-span bridge crosses rocky-bottomed Laurel Fork Creek, which empties into the nearby Watauga River.

Four log stringers that rest on dry natural stone abutments support the 28.0-foot long stringer footbridge. The flooring consists of transverse planking placed across the stringers. The sides of the north-northwest half of the bridge are covered with horizontal lapped siding, with two vertical window openings, with red-painted wood shutters, opposing along each side; the south-southeast end is open, with a cross-braced handrail between the roof support posts. The south-southeast gable and the north-northwest portal are also covered with horizontal lapped siding. The horizontal lapped siding boards have edges that are not planed, to simulate log edges, and, at one time, were painted red, the very noticeably faded paint now mostly washed away. Very old, deteriorated wood shingles cover the steeply peaked roof, which is supported along each side by five 4-inch by 4-inch wood posts. *See Alpine Acres Covered Bridge in the color photograph section: C-20.*

Directions: From Boone, at the intersection of US 321 and US 221/SR 105, go south 4.4 miles on SR 105 to Old Danner Road, on the right; go 0.3 mile to the bridge, on the left, just beyond Alpine Acres Road, on the right. Private property: request permission to visit.

Boone Greenway Covered Bridge

WATAUGA COUNTY

Brief Statistics

Type: Non-authentic modern covered footbridge. World Guide Covered Bridge Number: NC-95-d. Built in 1991 by Greene Construction. One-span footbridge crossing South Fork New River. Stringer footbridge 72.0 feet long by 12.3 feet wide, with approximately 10.0-foot wide by 9.0-foot high portal openings. Alternate Name: None known. Located in Boone on Lee & Vivian Reynolds Greenway behind Appalachian State University.

Boone Greenway Bridge was built, in the spring of 1991, by Greene Construction, with the concrete work performed by Watauga Ready Mix. Dedication ceremonies were held on June 11, 1991. Owned by the Town of Boone, the footbridge is located on the Lee & Vivian Reynolds Greenway walking trail, behind Appalachian State University. The north-northeast/south-southwest aligned bridge is in an open area, with several trees at the portals. Moderately wide, gravel bottomed South Fork New River swiftly flows beneath the single-span bridge, merging with the North Fork New River, east of Weavers Ford, to form the New River. The 72.0-foot long Boone Greenway Bridge is the longest non-authentic covered footbridge in North Carolina.

The stringer bridge is supported by a stack of 43 laminated 2-inch by 10-inch lumber, running along each side—also forms the sides up to rail height—with five cross-braces, each consisting of a stack of 16 laminated 2-inch by 4-inch lumber, evenly spaced down the length of the bridge, and four longitudinal braces, each consisting of a stack of 12 laminated 2-inch by 4-inch lumber, evenly spaced side-to-side between the cross-braces. The laminated stringers are set on massive poured concrete abutments that also serve as approach ramps into the bridge. The floor is transverse planked between the ramps. The sides consist of the laminated stringers, up to rail height, capped by a low banister-style handrail, open above to the eaves. Black weathered cedar shingles cover the roof, which is supported along each side by seven 6-inch by 6-inch timber posts. A *LEE & VIVIAN REYNOLDS GREENWAY* sign is at the south-southwest portal.

Directions: From Boone, go south on US 221/US 321 (Blowing Rock Road) to Shadowline Road, on the left; go 0.3 mile to the traffic light, continuing straight across into Appalachian State University, bearing right and going down into the parking lot. Park in the far right corner. Proceed onto the asphalt paved walking trail, going left to the bridge.

Sleepy Hollow Covered Bridge

WATAUGA COUNTY

Brief Statistics

Type: Non-authentic modern covered bridge. World Guide Covered Bridge Number: NC-95-C. Built in 1971 by unknown builder. Two-span bridge crossing Watauga River. Stringer bridge 60.6 feet long by 13.0 feet wide, with approximately 10.5-foot wide by 12.8-foot high portal openings. Alternate Name: None known. Located in Foscoe in Sleepy Hollow Subdivision on Sleepy Hollow Road.

In 1971, an unknown builder built Sleepy Hollow Bridge in Sleepy Hollow Subdivision, on Sleepy Hollow Road over the Watauga River, which flows into the South Fork Holston River (Boone Lake), at Spurgeon, Tennessee.

Three steel I-beam stringers and double massive timber stringers, under each side wall, support the 60.6-foot long stringer bridge. The bridge sits on mortared rough-cut stone abutments and one mortared rough-cut stone pier. The floor consists of transverse planking with three-plank-wide wheel treads. The red painted two-span bridge has vertical boarding covering the sides, the gables and the five buttress braces, evenly spaced down each side. The sides have one framed central viewing window opening and eight window openings under the eaves, in three evenly spaced pairs and one at each end, all openings opposing. Weathered cedar shingles cover the buttress braces and the roof, the roof supported along each side by five wood posts, with inverted V braces between the posts. The roof extends a short distance beyond the portals, which have a small weather panel.

The south-southeast/north-northwest aligned bridge has dense woods at the south-southeast end and heavily wooded riverbanks. The river cascades over bedrock and boulders, passing beneath the bridge into an open and wide riverbed downstream.

Directions: From Boone, at the intersection of US 321 and US 221/SR 105, go south 6.7 miles on SR 105 to the west side of Foscoe, to Sleepy Hollow Road, on the left, the entrance road to Sleepy Hollow Subdivision. Go 0.2 mile to the bridge.

South Carolina

About nine historic covered bridges survived in South Carolina, in 1950; today, only one remains, Campbell Bridge in Greenville County, a Howe truss bridge, built in 1909. Campbell Bridge has been the solitary survivor since 1982, when Lower Gassaway Bridge collapsed into Twelvemile Creek. *See Campbell Covered Bridge in the color photograph section: C-21.*

Of the twenty-five covered bridges in South Carolina, only one is authentic. Fifteen are footbridges, and seventeen are located within a twenty-five mile radius of Greenville, in Greenville County and three adjoining counties: Laurens, Pickens and Spartanburg. The twenty-four non-authentic modern bridges were constructed in the 21-year span from 1980 to 2000. Covered bridges stand in nine of the forty-six counties in South Carolina.

Mineral Spring Park Covered Bridge

ANDERSON COUNTY

Brief Statistics

Type: Non-authentic modern covered footbridge. World Guide Covered Bridge Number: SC-04-a. Built in 1986 by City of Williamston. One-span footbridge crossing Big Creek. Stringer footbridge 30.4 feet long by 6.5 feet wide, with approximately 5.5-foot wide by 7.6-foot high portal openings. Alternate Name: None known. Located in Williamston in Mineral Spring Park.

Mineral Spring Park Bridge is a footbridge, built in 1986, by a local landscaper for the City of Williamston, in downtown Williamston, in Mineral Spring Park. Another covered footbridge, Veterans Bridge, is in Veterans Park, which adjoins Mineral Spring Park to the south, beyond the railroad tracks. Mineral Spring Park Bridge spans Big Creek, which flows southerly into the Saluda River, southeast of Williamston, the river forming the eastern boundary between Anderson County and adjacent Greenville County.

The stringer footbridge is supported by two steel I-beam stringers, set on mortared rough-cut stone abutments. Longitudinal 2-inch by 6-inch planks form the floor, which abuts the poured concrete sidewalk at the east-northeast or right downstream portal and the poured concrete approach at the west-southwest portal. The sides are open, with a decorative railing along both sides, the upstream side having a covered alcove extending four feet out, with a wood bench centered on the inside length, allowing egress to the alcove at each end. Black asphalt shingles cover the roof, which overhangs the portals, the roof tapering to a point at the peak. A sawtooth-edged plank is mounted, on edge, the length of the roof peak, with a small wood spire surmounted above each gable. The gables are enclosed with a narrow wood lattice enhancing the ornate portals. Six 5-inch by 5-inch weathered wood posts along each side support the roof.

The single-span bridge is situated in a park with abundant mature trees crossing a moderately wide muddy creek. State Road 20 is upstream of the unobstructed bridge, which has a picnic pavilion along the upstream side of the concrete walkway, leading from the bridge to the park road, on the east-northeast end.

Directions: In downtown Williamston, the bridge is located in Mineral Spring Park, off SR 20, on the south side, at the concrete bridge over Big Creek.

Veterans Covered Bridge

ANDERSON COUNTY

Brief Statistics

Type: Authentic modern covered footbridge. World Guide Covered Bridge Number: SC-04-b.* Built in 1997 by unknown builder. One-span footbridge crossing Big Creek. Warren truss footbridge 40.3 feet long by 5.0 feet wide, with approximately 4.4-foot wide by 6.6-foot high portal openings. Alternate Name: None known. Located in Williamston in Veterans Park.

Veterans Bridge is the only Warren truss covered bridge in South Carolina, unique in that the steel trusses are below the decking, rather than in the walls, and was built, in 1997, by an unknown builder, in downtown Williamston in Veterans Park, a South Carolina Army National Guard park. In Mineral Spring Park, adjacent to Veterans Park, is another covered footbridge, which is also named after a park, this one called Mineral Spring Park Bridge. Veterans Bridge crosses Big Creek a short distance downstream from Mineral Spring Park Bridge.

The 40.3-foot long footbridge is supported by three, steel, thirteen-panel Warren trusses, placed under the flooring, which are set on abutments built from wood utility poles, placed upright in the ground and aligned transversely to the bridge. The steel trusses are welded construction with double 2-inch by 2-inch steel angle chords sandwiching 1-inch diameter steel rod diagonal members, making thirteen V sections (panels), the three trusses stabilized transversely with 1-inch diameter steel rods at the upper chords and lower chords, at each end. Transverse 2-inch by 6-inch planks, set on the steel trusses, constitute the floor or decking, which extends down the ramp, on the east-northeast or right downstream end. The open sides have railings, with banister-type posts along both sides and extending down the ramp; under the eaves is a narrow wood lattice. The gables are enclosed with vertical, facsimile tongue and groove paneling, the bottom of the panels cut in the form of an arch over the entrances. The gray, asphalt-shingled roof is supported along each side by seven 4-inch by 4-inch wood posts, the end posts stabilized at each portal by 2-inch by 4-inch wood buttress braces. All wood in the bridge is coated with a brown stain. The bridge was dedicated on August 24, 1997, and an engraved plastic dedication sign is mounted on the west-southwest or left downstream gable. A park sign is near the west-southwest portal.

The single-span footbridge is fully exposed, in a park, with a few small trees nearby. Flowing beneath the bridge is a moderately wide, muddy creek, running southerly into the Saluda River, southeast of Williamston. Just upstream of the bridge is a concrete railroad bridge, with the tracks elevated on embankments. An asphalt paved walkway leads to the bridge and into the park; a concrete picnic table and benches are on the left, just beyond the bridge. *See Veterans Covered Bridge in the color photograph section: C-23.*

Directions: From downtown Williamston, going southbound on SR 20, enter Mineral Spring Park on the left, just before the concrete bridge, where parking is available. Veterans Bridge is toward the south, beyond the railroad tracks, in adjoining Veterans Park.

**Assigned number is for a non-authentic footbridge. Although a truss footbridge, this footbridge has a steel truss below the flooring.*

Falcon Island Covered Bridge

CHARLESTON COUNTY

Brief Statistics

Type: Non-authentic modern covered footbridge. World Guide Covered Bridge Number: SC-10-a. Built in 1996 by unknown builder. One-span footbridge crossing Blue Heron Pond. Stringer footbridge 13.1 feet long by 9.0 feet wide, with approximately 5.6-foot wide by 7.2-foot high portal openings. Alternate Name: None known. Located in The Preserve on Kiawah Island.

Owned by the Kiawah Island Community Association, Inc., Falcon Island Footbridge is located near the eastern end of Kiawah Island, on a sub-island called Falcon Island. The stringer footbridge carries The Preserve Hiking Trail across a finger of Blue Heron Pond. The bridge was built, in 1996, by an unknown builder.

Falcon Island Bridge consists of a 13.1-foot long covered center span, resting on piers, with a 25.2-foot long north approach ramp and a 24.6-foot long south approach ramp, each ramp resting on an abutment and the covered section pier, with an additional supporting pier at the midpoint of each ramp. The abutments and piers are constructed from two pile-driven, upright, wood utility poles joined with two 2-inch by 12-inch lateral supports, bolted through the top of each pole, one support on each side. The two central piers are stabilized with 2-inch by 6-inch X cross-braces, one on each side of the poles. The transverse 2-inch by 6-inch floor planks are placed on top of five 2-inch by 12-inch wood stringers, set on edge and spanning the lateral supports. The abutment poles and pier poles extend to rail height along the ramps and extend to roof height at the two central piers, to support the roof. Handrails joining the poles are filled in with a rustic, bark-covered tree limb lattice below the railings. The sides of the cover are open, with sitting benches along each side. The steeply peaked roof is covered with gray weathered wood shingles and has a dormer with a louvered opening, on each side of the roof. Wood shingles cover the gables of the dormers and the gables over the entrances.

The picturesque bridge is highly visible over the elongated arm of Blue Heron Pond, the pond with a dense mixture of palms, pine trees and deciduous trees at each side, the hiking trail meandering off both ends of the bridge into the cool shade of the woods. Landscape plantings along the pond banks enhance the scene.

Directions: From US 17, west of Charleston, take Main Road (SSR 20) south through Johns Island (Main Road becomes Bohicket Road) to Kiawah Island. Kiawah Island is a gated community, requiring the issuance of a pass at the security gate. Follow Kiawah Island Parkway, the entrance road, to Governor's Drive (past second security gate), on the right; take the first left after Sweet Gum Lane; follow to the tee at Blue Heron Pond Road; go right, to the bridge that is on the right.

Caledonia Golf & Fish Club No. 1 Covered Bridge

GEORGETOWN COUNTY

Brief Statistics

Type: Non-authentic modern covered footbridge. World Guide Covered Bridge Number: SC-22-a. Built in 1993 by unknown builder. Eleven-span footbridge crossing man-made channel. Stringer footbridge 66.6 feet long by 12.8 feet wide, with approximately 10.9-foot wide by 8.4-foot high portal openings. Alternate Name: None known. Located in Pawleys Island at Caledonia Golf & Fish Club at 369 Caledonia Drive.

Caledonia Golf & Fish Club is located on the site of the original Caledonia Plantation, founded in the 18th century by Dr. Robert Nesbit. Caledonia Golf & Fish Club was constructed in 1993, officially opening in January 1994, and has two covered golfcart bridges. Caledonia Golf & Fish Club No. 1 Bridge was built, in 1993, during the construction of the golf course, by an unknown builder, crossing one of the man-made channels, with the 8th green near the south-southwest side of the bridge.

The stringer footbridge is supported by eleven framed panels of ten 2-inch by 10-inch wood timbers, on edge, forming the stringers, which are set on the abutments and ten piers. The abutments and piers are constructed from three pile-driven, upright, wood utility poles, joined by two 2-inch by 10-inch lateral supports, bolted through the top of each pole, one support on each side. The cover extends over the central nine spans for 66.6 feet, leaving the two end spans uncovered. Transverse planking makes up the floor, the sides are open, and white-painted horizontal lapped siding encloses the gables. Ten black-painted 6-inch by 6-inch wood posts support the black asphalt-shingled roof, the roof painted white on the underside. Black painted three-rail handrails, constructed from diagonally orientated 4-inch by 4-inch wood rails, are placed along each side, between the roof supports, and extend to the ends of the uncovered spans.

Aligned west-northwest/east-southeast, the eleven-span bridge is fully exposed on the golf course, crossing a landscaped, wide channel. The north-northeast side on the east-southwest end is filled in with thick foliage from trees and shrubbery. *See Caledonia Golf & Fish Club No. 1 Covered Bridge in the color photograph section: C-21.*

Directions: From Georgetown, go north on US 17 toward the community of Pawleys Island, going left on Kings River Road 2.2 miles to Caledonia Drive, on the left. Follow to Caledonia Golf & Fish Club clubhouse and parking. If Kings River Road is missed, continue on US 17 north to Beaumont Drive, on the left just after the Ford Dealership, on the left. Follow to the tee at Kings River Road, and go right 1.0 mile to Caledonia Drive, on the left.

Caledonia Golf & Fish Club No. 2 Covered Bridge

GEORGETOWN COUNTY

Brief Statistics

Type: Non-authentic modern covered footbridge. World Guide Covered Bridge Number: SC-22-b. Built in 1993 by unknown builder. One-span footbridge crossing dry wash. Stringer footbridge 14.8 feet long by 10.8 feet wide, with approximately 8.9-foot wide by 8.4-foot high portal openings. Alternate Name: None known. Located in Pawleys Island at Caledonia Golf & Fish Club at 369 Caledonia Drive.

Caledonia Golf & Fish Club No. 2 Bridge, the Club deriving its name from being on the former Caledonia Plantation site, was built by an unknown builder, in 1993, during the golf course construction. The east/west-aligned golfcart bridge crosses a dry wash on a tree line between the 7th and 8th fairways, the encroaching trees overhanging the bridge, giving the illusion that the bridge is a tunnel through the trees.

The four-span bridge, covered for 14.8 feet over the two center spans, is supported by four framed panels of eight 2-inch by 10-inch wood timbers, on edge, plus two additional 2-inch by 10-inch wood timbers at the outside edges, forming the stringers, which are set on the abutments and three piers. The abutments and piers are constructed from two pile-driven, upright, wood utility poles joined by two 2-inch by 10-inch lateral supports, bolted through the top of each pole, one support on each side. Transverse planking makes up the floor, the sides are open, and white-painted horizontal lapped siding encloses the gables. Three black-painted 6-inch by 6-inch wood posts support the black asphalt-shingled roof. Black-painted three-rail handrails, constructed from diagonally orientated 4-inch by 4-inch wood rails, are placed along each side, between the roof supports, and extend to the ends of the uncovered spans.

Campbell Bridge, built in 1909, has a four-panel Howe truss and is the only historic covered bridge remaining in South Carolina. ***See Campbell Covered Bridge in the color photograph section: C-21.***

Directions: From Georgetown, go north on US 17 toward the community of Pawleys Island, going left on Kings River Road 2.2 miles to Caledonia Drive, on the left. Follow to Caledonia Golf & Fish Club clubhouse and parking. If Kings River Road is missed, continue on US 17 north to Beaumont Drive, on the left just after the Ford Dealership, on the left. Follow to the tee at Kings River Road, and go right 1.0 mile to Caledonia Drive, on the left.

Campbell Covered Bridge

GREENVILLE COUNTY

Brief Statistics

Type: Authentic historic covered bridge. World Guide Covered Bridge Number: SC-23-02. Built in 1909 by Charles Irwin Willis. One-span bridge crossing Beaverdam Creek. Howe truss bridge 37.7 feet long by 14.8 feet wide, with approximately 11.6-foot wide by 11.7-foot high portal openings. Alternate Name: Campbell's Bridge. Located in Gowensville on Campbell Covered Bridge Road.

Since the collapse of Lower Gassaway Bridge, in Pickens County, in 1982, Campbell or Campbell's Bridge is the sole survivor of many historic covered bridges scattered around South Carolina. Built in 1909 by Charles Irwin Willis, to replace an earlier bridge washed away by a flood, on August 24, 1908, the historic bridge, after many years of service, was closed to motor traffic in 1981. Campbell Bridge was named after Alexander Lafayette "Fate" Campbell, who operated a grist mill near the bridge site. The 37.7-foot long bridge was restored, in 1964, by Greenville County with the cooperation of the Crescent Community Club, and again, in 1990, by Cunningham-Waters Construction Company of Greer, South Carolina. It was again repaired and improved, in 2001, by Casey Crew and Boy Scout Troop 155 of Landrum, South Carolina.

Campbell Bridge is supported by a four-panel Howe truss, utilizing double timber braces and counterbraces with double iron rods, set on mortared natural stone abutments, capped with concrete. The floor has transverse planking with no wheel treads, except both portals have short planks running longitudinally a short distance into the bridge. Vertical boarding with battens covers the sides and gables and both sides have three buttress braces, boxed in by vertical boarding with battens, on the sides of the braces, and old weathered wood shingles, on the ends of the braces. The bridge exterior and interior is painted red, with an old rusty tin roof, which has the fascia trimmed in white. The steeply peaked roof, along with the gables, extends beyond the entrances, affording protection for the interior from the weather. Mounted on each gable is a white lettered on blue background sign, giving the bridge name, construction date and restoration information. Low, timber guardrails are on each side, at both ends of the bridge.

The south-southwest/north-northeast aligned bridge crosses Beaverdam Creek in a single span. The bedrock bottomed creek flows southeasterly into nearby Middle Tyger River, which joins the North Tyger River, near Roebuck, continuing the southeasterly flow before combining with the South Tyger River, east of Woodruff, to form the Tyger River. The creek banks are densely wooded and have thick underbrush along both sides, requiring periodic clearing near the bridge. A dirt road runs perpendicular to the bridge, near the right downstream or north-northeast portal. Campbell Covered Bridge Road ends at the left downstream portal of the bridge, forming a small cul-de-sac suitable for parking. *See Campbell Covered Bridge in the color photograph section: C-21.*

Directions: From I-26, take the exit at SR 14, going west through Landrum and Gowensville to SR 414, on the right. Follow SR 414 to Pleasant Hill Road, on the left; go 0.4 mile to Campbell Covered Bridge Road, on the right; go 0.2 mile to the bridge.

Kendall Few Covered Bridge

GREENVILLE COUNTY

Brief Statistics

Type: Non-authentic modern covered footbridge. World Guide Covered Bridge Number: SC-23-b. Built in 1984 by local rock mason. Four-span footbridge crossing Stillhouse Branch. Stringer footbridge 22.5 feet long by 6.1 feet wide, with approximately 5.0-foot wide by 7.5-foot high portal openings. Alternate Name: None known. Located in Greer at Stillhouse Ridge Drive off Old Spartanburg Road.

J. Kendall Few designed Kendall Few Bridge, built on the Few property, in 1984, by a local rock mason. The stringer footbridge is supported by three 5-inch by 5-inch timber stringers, resting on horizontal timber abutments and three double 5-inch by 5-inch wood post piers. The four-span bridge has three spans covered for a length of 22.5 feet, the right downstream or north-northwest span remaining open. The floor has transverse planking; the sides and gables are open, with 5-inch by 5-inch weathered timber handrails along both sides, between the roof supports and extending the length of the uncovered span. The unique left downstream end has two portals, one out each side of the bridge, at 90-degree angles to the usual portal on the end, which is blocked off with a handrail between the roof end posts. The roof, covered with weathered, green-algae-covered wood shakes, is supported along each side by four 5-inch by 5-inch wood posts.

Kendall Few Footbridge, joined by several other types of footbridges on the property, crosses Stillhouse Branch, which flows into the nearby Enoree River. Stillhouse Branch has been lined with concrete and natural stone near the bridges, the entire setting in thick, mature deciduous woods, with walking trails abounding. A log cabin and other old buildings are near the right downstream end of the bridge. At one time, the area around the structures appears to have been a delightful, well kept woodland garden, but, since abandoned, the bridge now shows signs of deterioration.

Directions: From Greer, go south on SR 14 to Hammets Bridge Road (SSR 94), on the right. Follow 2.5 miles (becomes Old Spartanburg Road) to Stillhouse Ridge Drive, on the right (1.0 mile after the junction with Suber Road). The bridge is on a trail in the woods. Private property: request permission to visit.

Klickity Klack Covered Bridge

GREENVILLE COUNTY

Brief Statistics

Type: Non-authentic modern covered bridge. World Guide Covered Bridge Number: SC-23-E. Built in 2000 by Donald C. Spann and Troy Coffey. Four-span bridge crossing drainage ravine. Stringer bridge 47.9 feet long by 16.0 feet wide, with approximately 14.5-foot wide by 11.9-foot high portal openings. Alternate Name: None known. Located west of Gowensville at 918 Highway 11.

Donald C. Spann modeled Klickity Klack Bridge after Flume Bridge in Franconia Notch, New Hampshire. Built in 2000 by Donald C. Spann and Troy Coffey, the bridge, although new, utilized old wood for the stringers, roof supports, roof timbers and wheel treads. The 47.9-foot long bridge is supported by four 9-inch by 12-inch timber stringers, set on poured concrete abutments and three piers, each pier constructed with three poured concrete tapered pillars on a concrete footing. The flooring has transverse planking, with one-plank-wide wheel treads, and a walkway on the upstream side, separated from the traffic lane by a row of 6-inch by 8-inch timbers, set on edge, running the length of the bridge. The roof is supported by five 6-inch by 8-inch timbers with diagonal

The 47.9-foot long Klickity Klack Bridge was built in 2000 at LookAway Farm.

braces between the timbers, giving the appearance of two side-by-side kingpost trusses. The open sides have a two-plank-high railing, along the inside of each side, which is attached to old, weathered, roof support timbers. The gables are enclosed with horizontal facsimile tongue and groove paneling, painted light brown, with arched trimming, painted white. Red painted aluminum covers the roof, which has a small cupola centered on the peak. The roof fascia on the sides is painted white. An ornate bridge name sign is mounted on the left downstream or south-southeast gable.

The four-span bridge crosses a drainage ravine, which parallels the white, rail fence lined, asphalt-paved driveway, on its way to a retention pond. The stringer bridge is just off the driveway at State Road 11 and carries a side road into the dense woods, on the far side of the ravine. The nicely exposed bridge presents a pretty sight to the motorist traveling westbound on State Road 11.

Directions: From Gowensville, at the intersection of SR 11 and SR 14, go west 2.4 miles on SR 11 to the bridge, on the right, at LookAway Farm. The bridge is visible from the highway. Private property: request permission to visit.

Marshall Covered Bridge

GREENVILLE COUNTY

Brief Statistics

Type: Non-authentic modern covered bridge. World Guide Covered Bridge Number: SC-23-A. Built in 1985 by unknown builder. Three-span bridge crossing tributary to Reedy River. Stringer bridge 48.0 feet long by 26.5 feet wide, with approximately 23.6-foot wide by 13.8-foot high portal openings. Alternate Name: None known. Located in Greenville at Brookside Forest Subdivision off Sylvan Drive.

Marshall Bridge was built, in 1985, by an unknown builder at the entrance to Brookside Forest Subdivision. The two-lane bridge is the widest non-authentic covered motor vehicle bridge in South Carolina, at 26.5 feet wide, and also is the oldest non-authentic covered motor vehicle bridge in the state.

Six steel I-beam stringers, set on mortared brick abutments with mortared brick wingwalls and two mortared brick piers, support the 48.0-foot long Marshall Bridge. The floor has longitudinal planking, the sides and gables have vertical boarding, and the roof has weathered wood shakes. Three opposing, long horizontal openings are along each side, with the downstream side having an external walkway. Similar smaller openings are in the gables. The roof is supported by four 6-inch by 12-inch wood posts on each side; a glass-windowed cupola, topped by a rooster weathervane, is centered on the steeply peaked roof. Inside the bridge, three electric lanterns hang from the peak, providing interior lighting. The brown painted bridge has an electric security gate at the right downstream or north-northwest end.

The three-span bridge crosses a tributary to nearby Reedy River, which flows through Lake Greenwood, near Waterloo, and into the Saluda River. There is a mortared brick spillway at the downstream side of the bridge, with the small impoundment backed up under the bridge and on the upstream side. Dense woods surround the bridge, with a home in the downstream north-northwest corner.

Directions: From I-95, south of Greenville, take exit 46A (Augusta Road), going north 0.9 mile to Douglas Drive, on the right; go 0.1 mile to Granada Drive, on the right; go 0.1 mile to Sylvan Drive, on the left; go 0.1 mile to the bridge, on the right, at the entrance to Brookside Forest Subdivision.

Spearman Covered Bridge

GREENVILLE COUNTY

Brief Statistics

Type: Non-authentic modern covered footbridge. World Guide Covered Bridge Number: SC-23-.* Built in 1985 by Gene Spearman. One-span footbridge crossing unnamed creek. Stringer footbridge 10.0 feet long by 4.0 feet wide, with approximately 3.4-foot wide by 6.8-foot high portal openings. Alternate Name: None known. Located in Taylors in Avondale Forest Subdivision at 11 Armsdale Drive.

Spearman Bridge was built, in 1985, by Gene Spearman, across a small, unnamed creek at the side of, at that time, the Spearman property. The single-span stringer footbridge is supported by three 2-inch by 8-inch wood stringers attached to abutments, constructed from two 4-inch by 4-inch wood posts set in concrete footings. The flooring is transverse 2-inch by 6-inch planks; the sides and gables are open; the uncovered roof framing is supported by two 4-inch by 4-inch wood posts at each end. A 2-inch by 4-inch railing extends along each side, between the roof supports.

The 10.0-foot long bridge shares with Maston Homes Bridge the distinction of being the shortest non-authentic covered footbridge in South Carolina. The southeast/northwest-aligned bridge is at the edge of the side lawn, spanning a wooded creek leading to a neighbor's property. The current owners plan to replace the missing roof covering.

Directions: From Greenville, take US 29 north to East Lee Road, on the right, across from the Hampton Village Shopping Center. Go 2.5 miles on East Lee Road (makes left turn at 1.5 miles) to its tee at Taylors Road; go left 0.2 mile to Avondale Forest Subdivision, on the right, taking the second entrance on Drewry Road; follow 0.3 mile to Armsdale Drive, on the right; go 0.1 mile to No. 11, on the left. Private property: request permission to visit.

**No number assigned, as the bridge does not meet the minimum length established by the National Society for the Preservation of Covered Bridges.*

Troy Hoskins Covered Bridge

GREENVILLE COUNTY

Brief Statistics

Type: Non-authentic modern covered footbridge. World Guide Covered Bridge Number: SC-23-d. Built in 1984 by Troy Hoskins. One-span footbridge crossing drainage ravine. Stringer footbridge 15.7 feet long by 3.1 feet wide, with approximately 2.5-foot wide by 6.6-foot high portal openings. Alternate Name. None known. Located in Taylors at 301 Cardinal Drive.

In 1984, Troy Hoskins built Troy Hoskins Bridge across a drainage ravine in his backyard. At 3.1 feet wide, his bridge is the narrowest non-authentic covered footbridge in South Carolina. Two wood utility pole stringers, set on the ground, support the 15.7-foot long stringer footbridge. The floor is made with transverse 2-inch by 6-inch wood planks. The open sides have a railing along each side running between the roof supports; spaced apart, vertical 1-inch by 3-inch boards fill in below the handrail. The flat roof is covered with weathered transverse boarding and is supported by three 3-inch by 4-inch landscape-log posts on each side, with short diagonal braces flanking each post, just under the roof.

The single-span bridge is fully exposed, with scattered mature trees nearby. A narrow, high-banked drainage stream, with a bedrock bottom, flows under the south-southwest/north-northeast aligned bridge.

Directions: From Greenville, take US 29 north to East Lee Road, on the right, across from the Hampton Village Shopping Center. Follow East Lee Road 0.3 mile to Cardinal Drive, on the right; go 0.5 mile to No. 301, on the left, at the dip in the road. Private property: request permission to visit.

Medlin Covered Bridge

LAURENS COUNTY

Brief Statistics

Type: Non-authentic modern covered bridge. World Guide Covered Bridge Number: SC-30-A. Built in 1999 by Johnny Skinner. One-span bridge crossing tributary to Little Durbin Creek. Stringer bridge 17.3 feet long by 10.3 feet wide, with approximately 9.9-foot wide by 8.0-foot high portal openings. Alternate Name: None known. Located west of Woodruff off Medlin Drive.

Johnny Skinner, with some help from Ray Medlin, the owner's brother, built Medlin Bridge, in 1999, in time for the owner's wedding, held at the bridge, which was a wedding present for his wife. Medlin Bridge has the distinctions of being the shortest non-authentic motor vehicle covered bridge, at 17.3 feet long, and the narrowest non-authentic motor vehicle covered bridge, at 10.8 feet wide, in South Carolina.

The stringer bridge is supported by nine 8-inch by 8-inch timber stringers, set on abutments, consisting of a 10-inch by 10-inch timber laid on concrete poured on the ground. The bridge was modeled after a historic bridge and has a decorative kingpost truss. Transverse 2-inch by 6-inch tongue and groove planking makes up the floor. The sides are open, with a handrail between the truss members of the decorative truss, the handrail constructed with 2-inch by 6-inch planks, capped with transverse 12-inch lengths of the floor planking. Naturally finished vertical boarding with battens encloses the gables, and red painted aluminum covers the roof, which is supported by three 6-inch by 6-inch wood posts along each side. Three buttress braces along each side stabilize the roof posts. A wood sign, with *MEDLIN* burned into it, is mounted on the center rafter, inside the bridge.

A narrow, high banked and sandy bottomed tributary to Little Durbin Creek flows beneath the single-span bridge into Little Durbin Creek, about 50 yards downstream, the Little Durbin entering nearby Durbin Creek, which joins the Enoree River, south of Woodruff. The west-northwest/east-southeast aligned bridge is surrounded by thick, old growth woods, cleared of all brush and understory a comfortable distance around the bridge.

Directions: From Simpsonville, go northeast on SR 417 to Bethany Road (SSR 315), on the right; go 3.2 miles to the end, at SR 418; go left 2.2 miles to Knighton Chapel Road, on the right; go 1.9 miles to Meadowlark Lane, on the right; go 0.1 mile to Bluebird Drive, on the right; go 0.3 mile to Medlin Drive, on the right; go to the first dirt road, on the left, and follow to the bridge. Private property: request permission to visit at MBI General Contractors, 3333 Knighton Chapel Road.

B. K. Jones Covered Bridge

LEXINGTON COUNTY

Brief Statistics

Type: Non-authentic modern covered bridge. World Guide Covered Bridge Number: SC-32-A. Built in 1993 by Bill K. Jones. One-span bridge crossing Lin Creek. Stringer bridge 19.6 feet long by 11.0 feet wide, with approximately 9.0-foot wide by 8.4-foot high portal openings. Alternate Name: None known. Located north of Lexington at 221 Lincreek Drive.

Bill K. Jones, with the assistance of his son, built B. K. Jones Bridge, in 1993. The 19.6-foot long stringer bridge is supported by five wood utility pole stringers and one timber stringer—the second stringer in from the upstream side—set on wood utility pole abutments, both poles laid on the ground, perpendicular to the bridge. Transverse, varying-width planking forms the floor. Vertical boarding covers the gables and the sides under the eaves, the lower part of the sides open. The leaf and algae covered tin roof is supported by four wood utility pole posts, one in each corner.

The single-span bridge crosses Lin Creek, a narrow creek meandering through the woods into nearby Rawls Creek, which flows into the Saluda River, south of Irmo, thence into the Broad River, at Columbia. Two small spring-fed streams converge to form Lin Creek, just upstream of the bridge. Surrounded by moderately heavy woods, the south-southwest/north-northeast aligned bridge carries the secondary gravel driveway to the main driveway, leading to the residence.

Directions: From Lexington, go north on SR 6 (North Lake Drive), taking the first right (Lincreek Drive) after crossing the Lake Murray Dam. Go 0.2 mile to No. 221, on the right (just after Lake Murray Church). Private property: request permission to visit.

Carpenter Creek Covered Bridge

PICKENS COUNTY

Brief Statistics

Type: Non-authentic modern covered bridge. World Guide Covered Bridge Number: SC-39-D. Built in 1993 by Harold W. McConnell. One-span bridge crossing Carpenter Creek. Stringer bridge 53.5 feet long by 14.3 feet wide, with approximately 9.8-foot wide by 11.0-foot high portal openings. Alternate Name: None known. Located in Dacusville at 1348 Pace Bridge Road.

Carpenter Creek Bridge derives its name from the easterly flowing creek running beneath the bridge, which was built by Harold W. McConnell and local volunteers, in 1993, and was dedicated to the Pioneer Farm Day Show. The single-span 53.5-foot long bridge is the longest non-authentic covered bridge that accommodates motor vehicles in South Carolina. In fact, the bridge exceeds the 37.7-foot length of authentic historic Campbell Bridge, in Greenville County.

The stringer bridge is supported by two large steel I-beam stringers, set on a large steel I-beam, laid transversely on each creek bank. Transverse planking with two-plank-wide wheel treads comprises the floor. The sides, slightly tapered in toward the roof, are covered by weathered, natural, vertical boarding with battens and have a single, horizontal, rectangular window opening centered on each side. The portals are enclosed with weathered, natural, horizontal lapped boarding, with the entrances trimmed with 1-inch by 3-inch boards. The shiny tin roof is supported by 2-inch by 6-inch wood studding along each side. A large circular saw blade, hand-painted with a farm scene, is mounted in the gable of the right downstream or north portal. An engraved brass bridge informational sign is mounted on the right side of the same north portal. A short split rail fence funnels traffic into the north portal of the bridge. Old metal advertising signs and South Carolina license tags are mounted inside the bridge.

Carpenter Creek Bridge is situated in Robinson Field, along a dirt road, spanning tree-lined, moderately wide Carpenter Creek, which cascades over rocks on the upstream side of the bridge. The rock and sand bottom creek flows into the nearby South Saluda River, a few miles north of its confluence with the North Saluda River, giving birth to the Saluda River. There is a ford on the upstream side of the bridge. The open park-like field is at the north end of the bridge; a cleared hillside with scattered trees is at the south end.

Directions: From Dacusville, go north about 2 miles on SR 186 (Earls Bridge Road) to Pace Bridge Road, on the left; go 0.5 mile to Farmall Lane and Robinson Field, on the left. The bridge is 0.2 mile into the field, on the left. The bridge is not visible from the roadway.

Collins Covered Bridge

Pickens County

Brief Statistics

Type: Non-authentic modern covered bridge. World Guide Covered Bridge Number: SC-39-C. Built in 1992 by C. Roy Collins. Two-span bridge crossing Town Creek. Stringer bridge 48.7 feet long by 11.3 feet wide, with approximately 9.1-foot wide by 9.8-foot high portal openings. Alternate Names: C. Roy Collins Bridge, Roy Collins Bridge. Located in Pickens at Jaycee Park.

C. Roy Collins built Collins Bridge, also called C. Roy Collins Bridge and Roy Collins Bridge, in 1992. It was dedicated on May 16, 1992. Although primarily for foot traffic, the bridge will accommodate small motor vehicles.

The 48.7-foot long bridge has two large steel I-beam stringers, with a smaller steel member between, stretched across the abutments and a central pier, the abutments and pier constructed with three upright wood utility poles, topped with a railroad tie crossbeam. The pier has X braces, and horizontal railroad ties are in back of the abutment, at the right downstream or south end. The floor has transverse planking with no wheel treads. Weathered vertical boarding with wide battens covers the sides, which have two opposing, long horizontal, rectangular window openings, filled in with a narrow wood lattice, on each side. Similar latticework encloses the gables. The red-painted aluminum roof is supported by eight 4-inch by 6-inch wood posts and, with the gables, extends beyond the entrances, providing shelter from inclement weather. Behind both gables, the upper part of the end posts is enclosed with vertical boarding, cut at the bottom to form an arch over the entrance. *COLLINS BRIDGE* signs, in the style of public street signs, white letters on a green background, framed in white, are mounted on the gable above each entrance.

The two-span bridge crosses heavily foliated Town Creek, flowing westerly into nearby Twelvemile Creek, which empties into Hartwell Lake, north of Clemson. Ball fields and soccer fields of Jaycee Park are at both ends of the bridge. A picnic pavilion is near the south end, in the upstream or east corner. *See Collins Covered Bridge in the color photograph section: C-22.*

Directions: From Pickens, at the intersection of SR 183 (West Main Street) and US 178 (Ann Street), go north 0.5 mile on US 178 to Jones Avenue, on the left; go 0.1 mile to Jaycee Park, on the right. The bridge is visible across the field, on the right.

Crowe Creek Covered Bridge

Pickens County

Brief Statistics

Type: Non-authentic modern covered footbridge. World Guide Covered Bridge Number: SC-39-f. Built in 1980 by Cecil W. Durham. One-span footbridge crossing tributary to Crowe Creek. Stringer footbridge 21.8 feet long by 5.2 feet wide, with approximately 4.4-foot wide by 7.3-foot high portal openings. Alternate Name: None known. Located in Six Mile at 426 Crowe Creek Road.

Named after the nearby creek, Crowe Creek Bridge was built, in 1980, by Cecil W. Durham, in his front yard. The 21.8-foot long stringer footbridge is supported by three 6-inch by 6-inch timber stringers, mortised together at the center of the bridge and tensioned with a chain and turnbuckle beneath the bridge, to give the bottom of the bridge a moderate upward incline toward the center. The stringers are set on poured concrete abutments. Transverse 1-inch by 6-inch boards form the floor. The sides have vertical boarding under the eaves and a narrow lattice up to rail height and open between, maintaining the upward incline toward the center. Vertical boarding encloses the gables. The light brown painted bridge has dark brown asphalt shingles on the roof, which is supported on each side with 4-inch by 4-inch wood posts at the ends and in the center, with 2-inch by 4-inch wood posts between them.

A concrete walkway leads pedestrians through the single-span bridge, from the side driveway to the front of the nicely landscaped residence. Surrounded by scattered mature trees, northwest/southeast-aligned Crowe Creek Bridge spans a narrow, rock and silt bottomed tributary, flowing into nearby Crowe Creek, which empties into nearby Lake Keowee.

Directions: From Pickens, go west on SR 183 to SR 133 (Crowe Creek Road); go right to No. 426, on the right. The bridge is visible from the highway. Private property: request permission to visit.

Hidden Hills No. 1 Covered Bridge

PICKENS COUNTY

Brief Statistics

Type: Non-authentic modern covered bridge. World Guide Covered Bridge Number: SC-39-E. Built in 1993 by Harold W. McConnell. One-span bridge crossing Hawks Creek. Stringer bridge 26.5 feet long by 14.9 feet wide, with approximately 10.5-foot wide by 10.2-foot high portal openings. Alternate Name: McConnell Bridge. Located northwest of Dacusville.

Named after its location, Hidden Hills No.1 Bridge was built by Harold W. McConnell, the stringer bridge constructed in May 1976, with the cover added in 1993. Also called McConnell Bridge, the motor vehicle bridge was joined by a second bridge on the property, a stringer footbridge, built in 2000.

Two large steel I-beam stringers, resting on abutments of poured concrete on top of random rocks, support the 26.5-foot long stringer bridge. The unique floor is constructed with transverse steel railroad track attached to the steel I-beam stringers. Three-plank-wide wheel treads are laid over the railroad tracks, the asphalt-paved driveway extending a short distance into the bridge, abutting the railroad tracks. Cedar shingles, made on the premises, cover the sides and portals. Old discolored tin covers the roof, which is held up by three upright wood utility poles along each side. A *HIDDEN HILLS* sign hangs from the left downstream or northeast entrance.

The single-span bridge crosses Hawks Creek, flowing northerly into the nearby Oolenoy River, just west of its confluence with the South Saluda River, which joins the North Saluda River, to form the Saluda River, northeast of Dacusville. Open grassed areas are at the northeast end of the bridge, which has a ford on the upstream side and a water wheel on the downstream side. The moderately wide, rock-strewn creek has large crushed rock lined banks, downstream of the bridge. A cabin is at the southwest or right downstream end. The entire area has an abundance of trees. *See Hidden Hills No. 1 Covered Bridge in the color photograph section: C-22.*

Directions: The bridge is on private property. At Dacusville, make telephone contact with Harold McConnell for permission to visit and directions to the bridge. He is well known.

Hidden Hills No. 2 Covered Bridge

PICKENS COUNTY

Brief Statistics

Type: Non-authentic modern covered footbridge. World Guide Covered Bridge Number: SC-39-g. Built in 2000 by Harold W. McConnell. One-span footbridge crossing small tributary to Hawks Creek. Stringer footbridge 13.6 feet long by 8.4 feet wide, with approximately 7.1-foot wide by 7.3-foot high portal openings. Alternate Name: None known. Located northwest of Dacusville.

Harold W. McConnell built Hidden Hills No. 2 Bridge, the second covered bridge constructed on his property, in 2000. Three 4-inch by 10-inch timber stringers, set in poured concrete footings, support the 13.6-foot long stringer footbridge. Transverse 2-inch by 8-inch planks form the floor. Cedar shingles, made on the premises, cover the sides and portals. A major portion of the sides is open, with a handrail spanning the opening and vertical posts, spaced approximately one foot apart, supporting the handrail. Horizontal lapped boarding covers the roof, which is supported by four 2-inch by 6-inch wood posts along each side.

The single-span bridge crosses a small unnamed tributary to Hawks Creek, located just downstream from the bridge, which flows into the nearby Oolenoy River, just west of its confluence with the South Saluda River, which joins the North Saluda River, to form the Saluda River, northeast of Dacusville. The east-southeast/north-northwest aligned bridge is in a moderately wooded area, with the upstream west-northwest bank of the narrow mud bottomed tributary rock lined.

Directions: The bridge is on private property. At Dacusville, make telephone contact with Harold McConnell for permission to visit and directions to the bridge. He is well known.

The Kenny Fisher Covered Bridge

PICKENS COUNTY

Brief Statistics

Type: Non-authentic modern covered footbridge. World Guide Covered Bridge Number: SC-39-b. Built in 1986 by Willy Morgan. One-span footbridge crossing Bets Aiken Branch. Stringer footbridge 28.2 feet long by 4.4 feet wide, with approximately 2.7-foot wide by 6.6-foot high portal openings. Alternate Name: Kenny Fisher Bridge. Located west of Sunset at 1640 Cleo Chapman Highway.

Willy Morgan built The Kenny Fisher Bridge, in September 1986, across Bets Aiken Branch, so his grandson, 3-year old Kenny Fisher, would not have to walk the dangerous 100 yards along the side of curving Cleo Chapman Highway, to visit his grandparents, but could instead take a shortcut, crossing Bets Aiken Branch. Following the sale of Willy Morgan's property for development, The Kenny Fisher Bridge was moved, by Brezeale Shoreline Erosion Control, on May 10, 2001, the 100 yards along Cleo Chapman Highway, to his daughter's property, and placed over the headwaters of Bets Aiken Branch.

Two wood utility pole stringers, set on a railroad tie laid transversely at each end, support the 28.2-foot long stringer footbridge. Cement blocks and shims, set under each stringer near the ends, lend additional support to the stringers. The flooring is constructed with transverse 2-inch by 6-inch planks. Weathered vertical boarding with battens covers the sides and gables. Shiny tin covers the low peaked roof, which is supported on both sides by seven 2-inch by 4-inch studs with diagonal braces, replicating a six-panel multiple kingpost truss, and double 2-inch by 4-inch studs forming the end posts.

The west/east-aligned single-span bridge is now fully exposed, in the open, grassed front yard, just off the gravel driveway. Tiny Bets Aiken Branch passes under the bridge, into nearby Eastatoe Creek, entering Lake Keowee, near Keowee-Toxaway State Park.

Directions: From northeast of Sunset, at the intersection of US 178 and SR 11, go southwest 4.8 miles on SR 11 to Roy F. Jones Highway, on the right. Follow 2.5 miles to the junction with Cleo Chapman Highway (appears that Cleo Chapman Highway goes to the right only, but in actuality Roy F. Jones Highway ends here and Cleo Chapman Highway goes straight through and to the right). Go straight through 0.4 mile to No. 1640, on the right. Private property: request permission to visit.

Covered Bridge Farm Covered Bridge

SPARTANBURG COUNTY

Brief Statistics

Type: Non-authentic modern covered bridge. World Guide Covered Bridge Number: SC-42-B. Built in 1994 by Jack Belue. One-span bridge crossing Motlow Creek. Stringer bridge 42.0 feet long by 24.4 feet wide, with approximately 22.4-foot wide by 14.0-foot high portal openings. Alternate Name: Motlow Creek Bridge. Located in Motlow Creek on Stablegate Drive.

Covered Bridge Farm Bridge was built, in 1994, by Jack Belue, with the concrete work done by Metromont Prestress Company, in the former Covered Bridge Farm Subdivision, now known as Motlow Creek Farms Subdivision. The bridge has also been called Motlow Creek Bridge.

The 42.0-foot long stringer bridge is supported by concrete slab stringers, set on poured concrete abutments with short poured concrete wingwalls. The two-lane bridge has an asphalt paved roadbed for a floor. The sides have vertical boarding with battens, up to railing height, and are open, up to vertical tongue and groove boarding, under the eaves. Vertical tongue and groove boarding also encloses the gables. Brownish asphalt shingles cover the roof and the horse weathervane topped cupola, centered on the peak, the cupola having glass windows on four sides. Five 8-inch by 8-inch wood posts along each side support the steeply peaked roof. The bridge wood is blackened from the wood preservative. Mortared rough-cut stone guard walls are at both ends of the bridge, with electric lanterns mounted at the ends of the walls.

The single-span bridge crosses Motlow Creek, which flows into the nearby North Tyger River. Aligned south-southeast/north-northwest, the bridge has tree-lined creek banks at both sides and open horse paddocks at both ends.

Directions: From Campobello, go west on SR 11 to New Cut Road, on the left just before the Spartanburg/Greenville County line; go 2.0 miles to Motlow Creek

Equestrian Center and Stablegate Drive, on the right; follow Stablegate Drive 0.2 mile to the bridge.

Cushman Covered Bridge

SPARTANBURG COUNTY

Brief Statistics

Type: Non-authentic modern covered bridge. World Guide Covered Bridge Number: SC-42-A. Built in 1991 by Larry S. Cushman. One-span bridge crossing Abner Creek. Stringer bridge 39.0 feet long by 12.3 feet wide, with approximately 11.1-foot wide by 9.3-foot high portal openings. Alternate Name: None known. Located south of Duncan off Mayfield Road.

Larry S. Cushman built Cushman Bridge, in 1991, over Abner Creek. The 39.0-foot long stringer bridge is supported by three steel I-beam stringers, set on a poured concrete and timber abutment, on the left downstream or south-southwest end, and an abutment constructed with four upright wood utility poles, capped with a timber crossbeam, on the right downstream end. The flooring is constructed with transverse planks. The sides and gables are covered with weathered vertical boarding, the sides having three opposing, vertical, rectangular window openings on each side. The old rusty, tin-covered roof is supported by six 3-inch by 6-inch wood posts along each side. The peaked roof and gables extend beyond the end posts, providing protection for the entrances from inclement weather. Four buttress braces, boxed in with weathered vertical boarding, are evenly spaced down each side.

Moderately wide, sandy bottomed Abner Creek placidly flows east-southeastward under the single-span bridge and into the Enoree River, southeast of Pelham. Dense woods obstruct a view of the bridge on the upstream and downstream sides; a barn is on the south-southwest end; the north-northeast portal peeks out of the woods, separated from the large pasture by a fence and cattle gate.

Directions: From I-85, south of Greer, take exit 60 at SR 101, going south 0.3 mile to Leonard Road, on the right; go 0.8 mile to Mayfield Road; go left 0.3 mile to the bridge, on the left at the end of the pasture, partially visible from the roadway. Private property: request permission to visit.

Maston Homes Covered Bridge

SPARTANBURG COUNTY

Brief Statistics

Type: Non-authentic modern covered footbridge. World Guide Covered Bridge Number: SC-42-.* Built in 1998 by Mike Stone. One-span footbridge crossing dry wash. Stringer footbridge 10.0 feet long by 5.9 feet wide, with approximately 5.1-foot wide by 8.9-foot high portal openings. Alternate Name: None known. Located in Greer at 2239 Highway 101 South.

Mike Stone built Maston Homes Bridge, in 1998, along the concrete walkway leading to model homes. The red and white single-span footbridge is fully exposed and highly visible, crossing a dry wash in the landscaped lawn area, just off the highway. Maston Homes Footbridge, at 10.0 feet long, shares the distinction of being the shortest covered footbridge in South Carolina with Spearman Footbridge, in Greenville County.

Six 2-inch by 8-inch wood plank stringers, set on edge, support the stringer bridge. Lacking abutments, the four 4-inch by 6-inch wood corner posts supporting the roof extend into the ground, anchoring the north-northwest/ south-southeast aligned bridge. Transverse 2-inch by 6-inch wood planks form the floor. The open sides have a handrail, secured with closely spaced banister type posts, extending between the roof corner posts. The gables are open. The bright red painted aluminum roof caps the white painted bridge.

Directions: From Greer, take SR 101 south, crossing over the I-85 overpass to Maston Homes at No. 2239, on the left. The bridge is visible in front of the model homes, just off the highway.

**No number assigned, as the bridge does not meet the minimum length established by the National Society for the Preservation of Covered Bridges.*

The Stillhouse Covered Bridge

SPARTANBURG COUNTY

Brief Statistics

Type: Non-authentic modern covered footbridge. World Guide Covered Bridge Number: SC-42-c. Built in 1980 by John Peake. Two-span footbridge crossing Tims Creek. Stringer footbridge 48.2 feet long by 10.1 feet wide, with approximately 6.2-foot wide by 7.1-foot high portal openings. Alternate Name: Walnut Grove Seafood Restaurant Bridge. Located in Moore in southwest corner US 221 and Stillhouse Road.

This covered footbridge was built, in the summer of 1980, by John Peake and called Walnut Grove Seafood Restaurant Bridge. The restaurant had been closed for some time, but reopened under new management in 2002, changing the name of the restaurant to The Stillhouse Restaurant and the name of the bridge to The Stillhouse Bridge.

The 48.2-foot long footbridge is supported by two steel I-beam stringers, resting on poured concrete footings on the creek banks and poured concrete pillars for a pier, one pillar under each stringer. The steel I-beam stringers are dated 7/18/80 near the weld joint at the pier. The flooring is plywood, secured to thirty-seven 2-inch by 8-inch wood-plank floor joists, set on edge on the stringers. Weathered vertical boarding with battens covers the sides and portals. Five vertical, single-pane glass windows are on each side, and double, glass swinging doors are at each entrance. The interior of the bridge is air conditioned, with a carpeted floor and paneled walls, to provide a waiting room for the restaurant. A shiny tin roof, supported by studding, completes the bridge.

The west-northwest/east-southeast aligned two-span bridge crosses narrow and sandy bottomed Tims Creek, flowing southerly into the very close by North Tyger River, which joins the South Tyger River east of Woodruff to form the Tyger River. The stringer bridge provides access from the parking lot to the restaurant, doubling as a waiting room. There are several trees along the creek, downstream of the bridge, and bamboo stands flank the west-northwest portal.

Directions: From Spartanburg, go south on I-26 to exit 28, at US 221; go left 0.1 mile on US 221 to Stillhouse Road, the first road on the right. The bridge is at The Stillhouse Restaurant, on the right.

McDuffie Pedestrian Crossover Covered Bridge

Sumter County

Brief Statistics

Type: Authentic modern covered footbridge. World Guide Covered Bridge Number: SC-43-b. Built in 1995 by unknown builder. One-span footbridge crossing West Liberty Street. Stringer and Pratt truss footbridge 73.5 feet long by 9.2 feet wide, with approximately 7.9-foot wide by 8.8-foot high portal openings. Alternate Name: McDuffie Overpass Bridge. Located in Sumter at Swan Lake Iris Gardens.

An unknown builder built McDuffie Pedestrian Crossover Bridge, also known as McDuffie Overpass Bridge, in the spring of 1995. The footbridge, dedicated in May 1995, received its name in tribute to James C. McDuffie, a local benefactor of Swan Lake Iris Gardens, where the bridge is located. The 73.5-foot long single-span bridge is the longest non-authentic covered footbridge in South Carolina.

McDuffie Pedestrian Crossover Bridge appears to be equally supported by two 4-inch by 6-inch steel rectangular tubing stringers, with 2-inch by 6-inch steel channel cross-braces, and a ten-panel Pratt truss, constructed from square steel tubing. The significantly cambered bridge is supported by the brick-faced concrete block stairwell towers at each end of the bridge. The flooring is transverse planking; the sides have weathered, natural, vertical boarding up to railing height, with steel chain-link fencing continuing up to the roof. Weathered, slanting vertical boards with battens cover each side of the peaked roof. Large towers, housing the stairwell and an elevator, are at both ends, with the roof and sides a continuation of the bridge roof and siding. Electric lighting illuminates the interior of the bridge and towers.

The city-owned bridge spans West Liberty Street, allowing pedestrians to cross the busy highway from the Bland Gardens, on the north-northeast end, to the Heath Gardens and Swan Lake of the Swan Lake Iris Gardens park, at the south-southwest end. The Swan Lake Iris Gardens offers—in addition to the beautiful gardens, lake and swans—picnic shelters, tennis courts, restrooms and another covered footbridge, Swan Lake Bridge.

Directions: From Sumter, at the junction of US 76/US 378 and SR 120, go south on SR 120 (Alice Drive) to West Liberty Street (SR 763); go left 0.3 mile to Swan Lake Iris Gardens parking lot, on the right. The bridge crosses West Liberty Street, just beyond the entrance to the parking lot.

Swan Lake Covered Bridge

Sumter County

Brief Statistics

Type: Non-authentic modern covered footbridge. World Guide Covered Bridge Number: SC-43-a. Built in 1991 by City of Sumter. Five-span footbridge crossing Swan Lake channel. Stringer footbridge 56.0 feet long by 14.0 feet wide, with approximately 13.4-foot wide by 10.1-foot high portal openings. Alternate Name: None known. Located in Sumter at Swan Lake Iris Gardens.

The City of Sumter built Swan Lake Bridge, in 1991, over a channel of Swan Lake in Swan Lake Iris Gardens. At 14.0 feet wide, Swan Lake Bridge is the widest non-authentic covered footbridge in South Carolina. The city-owned footbridge is supported by four 9-inch by 9-inch timber stringers, set on poured concrete abutments and four piers, each pier constructed from four upright wood utility poles capped with a timber crossbeam. The flooring consists of transverse 2-inch by 8-inch planks. The open sides have weathered, natural, horizontal lapped boarding under the eaves and in the gables, a bench on the channel side and a handrail on the lake side, running the 56.0-foot length of the bridge. The brown asphalt-shingled roof is supported by eight 4-inch by 4-inch wood posts along each

Swan Lake Bridge is reflected in the lake channel that creates Camellia Island. The 56.0-foot long bridge was built in 1991.

side, creating seven open panels, 2-inch by 4-inch wood X braces in the four end panels, two at each end.

The five-span bridge provides access to the gardens on Camellia Island, a small island in Swan Lake. The fully exposed bridge, aligned north-northwest/south-southeast, is situated in a beautifully landscaped setting, abundant with tall pines, cypress and many species of deciduous trees and irises, thousands of beautiful irises during the late spring blooming season.

Directions: From Sumter, at the junction of US 76/US 378 and SR 120, go south on SR 120 (Alice Drive) to West Liberty Street (SR 763); go left 0.3 mile to the Swan Lake Iris Gardens parking lot, on the right. The bridge is at the far side of Swan Lake, circling to the left, and is not visible from the highway.

Tennessee

Tennessee certainly had hundreds of covered bridges throughout its past. Today, the historic covered bridges number five. Winona Bridge, in Scott County, was the last historic covered bridge lost, and that was in 1982.

The total of all covered bridges in Tennessee is thirty-six; of these, five are authentic, leaving thirty-one non-authentic, five of the thirty-six being historic (four authentic and one non-authentic, the non-authentic being Clifford Holder Bridge). These bridges were built between 1875 and 2001 in twenty of the ninety-five counties in Tennessee. Tennessee's five authentic covered bridges display four different truss types, namely, queenpost, kingpost, Howe and Pratt. Three historic covered bridges are still open to motor traffic—Elizabethton Bridge (1882), Harrisburg Bridge (1875) and private Clifford Holder Bridge (1919).

Several of Tennessee's covered bridges are not readily available for public view, but those that are offer a variety of styles and sizes to be enjoyed by all, hopefully for many years to come.

Covered Bridge Estates Covered Bridge

Bedford County

Brief Statistics

Type: Non-authentic modern covered bridge. World Guide Covered Bridge Number: TN-02-A. Built in 1997 by Spencer Turrentine. One-span bridge crossing Hatchett Creek. Stringer bridge 22.0 feet long by 14.8 feet wide, with approximately 13.0-foot wide by 11.3-foot high portal openings. Alternate Name: None known. Located in Bell Buckle on Emily Lane in Covered Bridge Estates.

Spencer Turrentine built Covered Bridge Estates Bridge, in the spring of 1997, for Covered Bridge Estates developer Venson Hawkins. The 22.0-foot long bridge carries Emily Lane, the entrance road to the subdivision, across Hatchett Creek, whose waters flow into nearby Wartrace Creek, then into Garrison Fork, south of Wartrace, before entering the Duck River.

Covered Bridge Estates Bridge is supported by seven 8-inch by 10-inch timber stringers, set on large mortared rough-cut stone block abutments, whose wing walls extend up to form guard walls at both ends of the bridge. The flooring consists of transverse planking placed directly on the stringers. The bridge has a single kingpost truss; however, this truss does not carry any load crossing the bridge, as the stringers and flooring are not connected to the truss or roof support posts. Rather, the truss is decorative, as are the wooden pegs used on the bridge timbers. Gray weathered vertical boarding covers the sides, which have opposing long horizontal window openings, exposing the upper part of the kingpost truss. The gables have vertical boarding, which has not yet weathered, due to the roof extending beyond the portals, sheltering the gables and providing additional protection for the interior from inclement weather. Dark wood shingles cover the roof, which is supported by three 8-inch by 8-inch timber posts along each side, the center post being the vertical tension member in the kingpost truss.

The rock and gravel bottomed creek passes beneath the sturdy single-span bridge, mature trees scattered along the banks. Maintained grass lawns are at the northeast or left downstream end; Happy Valley Road and landscape plantings are at the southwest end. *See Covered Bridge Estates Covered Bridge in the color photograph section: C-24.*

Directions: From Beechgrove, at I-24 SR 64 exit, go west on SR 64 to SR 82; go right to Bell Buckle to SR 269; go right 2.4 miles to Happy Valley Road, on the right; follow 0.6 mile to Emily Lane and the bridge, visible on the left.

Elizabethton Bridge (1882) crosses the Doe River upstream of the spillway dam. ***See color section: C-24.***

Elizabethton Covered Bridge

CARTER COUNTY

Brief Statistics

Type: Authentic historic covered bridge. World Guide Covered Bridge Number: TN-10-01. Built in 1882 by Dr. E. E. Hunter. One-span bridge crossing Doe River. Howe truss bridge 137.0 feet long by 20.0 feet wide, with approximately 15.4-foot wide by 12.8-foot high portal openings. Alternate Name: Doe River Bridge. Located in Elizabethton on Third Street.

Two iron barred windows were added to each side of Elizabethton Bridge at a later date. ***See Elizabethton Covered Bridge in the color photograph section: C-24.***

White clad Elizabethton Bridge was built by Dr. E. E. Hunter, early in 1882, in an area then known as Sycamore Shoals. Supervised by George Lindamood, the construction costs to Carter County were $3,000.00 for the bridge and $300.00 for the abutments. The county-owned bridge has also been called Doe River Bridge, after the placidly flowing river beneath, the lone surviving bridge along the Doe River, in Carter County, following the great flood in May 1901, when the river was not so placid. Elizabethton Bridge holds the authentic historic covered bridge distinctions of being the longest single-span bridge, at 137.0 feet long, and the widest, at 20.0 feet wide. Of the five authentic historic covered bridges in Tennessee, only two remain open to motor traffic: Elizabethton Bridge and Harrisburg Bridge, in Sevier County.

The only remaining Howe truss covered bridge in Tennessee, Elizabethton Bridge is supported by a fourteen-panel long Howe truss that rests on mortared, large, rough-cut stone abutments. The flooring has longitudinal planking, with unique resin-covered steel plate wheel treads. The sides are clad with white-painted, horizontal lapped siding, and cedar shingles, weathered to a black color, cover the hip roof that extends beyond the entrances, to provide additional shelter to the interior from inclement weather. A walkway, separated from the traffic lane by a railing, runs along the upstream side of the brown painted interior. Interior electric lights were installed down the center of the bridge at a later date. Two opposing horizontal rectangular window openings, with vertical iron bars, were added to the sides when the bridge siding was last replaced.

The east/west-aligned bridge spans the Doe River, which flows north into the nearby Watauga River, then enters the South Fork Holston River, whose confluence with the North Fork Holston River, west of Kingsport, forms the Holston River. The lengthy bridge is in full view over the wide river, which has a bank-to-bank spillway dam, creating a double waterfall on the downstream side. A public park adjoins the river on the east end, on both sides of Third Street. Riverside Drive runs along the west riverbank, separated by a maintained grassed area. A splendid view of the historic bridge and the spillway waterfalls is afforded from the bridge on Elk Avenue, just downstream of the dam.

Elizabethton Bridge was listed on the National Register of Historic Places on March 14, 1973, under the Elizabethton Historic District listing. This is the only historic covered bridge remaining in Carter County.

Directions: From Elizabethton, the county seat of Carter County, at the junction of US 19E/SR 321/SR 37 and SR 91, go south 0.2 mile on US 19E/SR 321/SR 37 to East Elk Avenue; go right 0.1 mile to the circle with East Main Street, going left on East Main Street to Third Street, on the right, following 0.1 mile to the bridge.

Marrowbone Creek Covered Bridge

CHEATHAM COUNTY

Brief Statistics

Type: Non-authentic modern covered bridge. World Guide Covered Bridge Number: TN-11-A. Built in 1991 by Raymond Benkley Construction and Carl West. Three span bridge crossing Marrowbone Creek. Stringer bridge 70.1 feet long by 10.3 feet wide, with approximately 9.6-foot wide by 9.0-foot high portal openings. Alternate Name: None known. Located in Ashland City at 1495 Big Marrowbone Road.

Marrowbone Creek Bridge was constructed, in 1991, the stringer portion and the piers built by Carl West and the cover built by Raymond Benkley Construction. The bridge sits high above the swiftly flowing waters of Marrowbone Creek, on their way to the Cumberland River, at Sulphur Springs. Marrowbone Creek Bridge, at 10.3 feet wide, is the narrowest non-authentic covered motor vehicle bridge in Tennessee.

The 70.1-foot long stringer bridge is supported by two steel I-beam stringers, set on the ground at the northeast end and on the bedrock cliff at the southwest end, and two piers. The piers are constructed with an upright steel I-beam, set in a circular steel-encased poured concrete footing and capped with a horizontal steel I-beam, which is braced from the ends to the center of the upright with lengths of steel I-beam. The flooring consists of transverse planking placed directly on the steel I-beam stringers. Weathered orange-brown to black, vertical boarding covers the sides and portals, the sides open under the eaves. Horizontal window openings, opposing on each side, are centered over the creek below, to take advantage of the beautiful view. The roof, supported with 2-inch by 4-inch studding, is covered with dark weathered wood shakes. A small water wheel is mounted on the upstream side of the bridge, near the southwest end.

The three-span bridge has a wide bedrock- and gravel bottomed creek running beneath, which has a shale cliff on the southwest downstream bank and an unmortared large-rock retaining wall on the northeast downstream bank; dense woods run along both banks. Paddocks are at the northeast end; the driveway and outbuildings are at the southwest end. The southwest end also has gardens, behind split rail fences extending to the portal. *See Marrowbone Creek Covered Bridge in the color photograph section: C-26.*

Directions: From Nashville, take I-40 west to the SR 155 exit; go right to exit 24 at SR 12; go left 10.0 miles to Little Marrowbone Road, on the right; follow 1.2 miles to Big Marrowbone Road, on the left; go 1.9 miles to No. 1495, on the left. The bridge is not visible from the roadway. Private property: request permission to visit at the gate.

Heatherhurst Country Club Covered Bridge

CUMBERLAND COUNTY

Brief Statistics

Type: Non-authentic modern covered footbridge. World Guide Covered Bridge Number: TN-18-A. Built in

1990 by unknown builder. One-span footbridge crossing Otter Creek. Stringer footbridge 43.3 feet long by 12.4 feet wide, with approximately 11.2-foot wide by 9.1-foot high portal openings. Alternate Name: None known. Located in Fairfield Glade at Fairfield Glade Resort at Heatherhurst Country Club.

Built in 1990 by an unknown builder, Heatherhurst Country Club Footbridge has the honors of being the longest, at 43.3 feet, and the widest, at 12.4 feet, non-authentic covered footbridge in Tennessee. The single-span bridge crosses Otter Creek, whose waters flow into the Obed River, at the Morgan County line, before entering the Emory River and continuing to the Tennessee River, at Kingston. The privately owned bridge is set on poured concrete abutments that are faced with natural stone mortared at the back, so the mortar does not show. Supported by a reinforced concrete slab stringer, the gray-brown painted bridge has facsimile vertical tongue and groove paneling on the exterior and interior of the sides, to railing height, with an exposed lattice, constructed of 4-inch by 4-inch lumber, up to the eaves. The gables are closed in with gray-brown painted diagonal boarding, following the roofline, meeting at the peak. Reddish brown asphalt shingles cover the steeply peaked roof, which is supported by six 4-inch by 4-inch wood posts along each side. Flowering planters, hanging under the side eaves at each portal, add an attractive touch.

The south-southeast/north-northwest aligned bridge is situated between the 4th-hole fairway and green of the Brae Course, over a rock-bottomed creek with high embankments, which are densely wooded. The fairway is at the north-northwest end, with a wide golfcart path passing by the other end.

Directions: From I-40, northeast of Crossville, take the exit at SR 101 (Peavine Road), going north to the four-way stop at Fairfield Glade Resort. Go left, following the signs to Heatherhurst Country Club. The bridge is on the Brae Course, at the 4th hole.

Emerson E. Parks Covered Bridge

DYER COUNTY

Brief Statistics

Type: Authentic historic covered bridge. World Guide Covered Bridge Number: TN-23-01 (formerly TN-66-01). Built in 1904 by Emerson E. Parks. One-span bridge crossing dry land. Modified kingpost truss bridge 33.1 feet long by 12.7 feet wide, with approximately 10.9-foot wide by 7.8-foot high portal openings. Alternate Names: Parks Bridge, Parks Farm Bridge. Located in Trimble at Parks Plaza (downtown park).

Originally built, in 1904, by Emerson E. Parks on his private property in Obion County, over the new Obion River Drainage Canal between US 51 and old SR 3, where it was known as Parks Farm Bridge and Parks Bridge, the single-span bridge was relocated, in 1997, about one mile away, to the park in the center of Trimble, and renamed Emerson E. Parks Bridge. The original bridge spanned two concrete piers, with approach ramps extending to the concrete abutments, the ramps being covered at a later date, effectively creating a three-span 61-foot long bridge. Mr. Parks gave the bridge, abandoned since 1928, to the Town of Trimble. Local volunteers dismantled the bridge in July 1996, leaving the truss and structural timbers on the piers while stacking the siding, extensions and roof materials at the site, in anticipation of the arrival of a crane and flatbed truck that was to move the bridge to Trimble. The equipment never arrived. Unfortunately, in November of that year, the Obion River flooded the area, washing away some of the dismantled lumber, the truss and decking surviving. In 1997, the equipment arrived and moved the bridge and remaining materials to Trimble. In the meantime, local volunteers found a barn nearby that was built around the turn of the 20th century, affording the necessary replacement materials to complete the bridge, the owner donating the barn in exchange for its removal. Barry Whitehouse, a local carpenter, reconstructed the bridge in Trimble, in time for its dedication on December 3, 1997. The reconstructed bridge is only the original 33.1-foot long truss span that went from pier to pier.

Emerson E. Parks Bridge enjoys the distinctions for authentic historic covered bridges in Tennessee, of being the shortest, at 33.1 feet, and the narrowest, at 12.7 feet. It also is the only kingpost truss covered bridge remaining in Tennessee.

Set on poured concrete abutments over dry land, Emerson E. Parks Bridge is supported by a three-quarter-height single kingpost truss, modified by the replacement of the vertical tension truss member with an iron rod. The flooring is transverse planking, the sides and portals have brown weathered, horizontal lapped siding, and the roof has dark weathered wood shingles. An external, central wood timber buttress brace on each side stabilizes the kingpost truss.

Closed to vehicular traffic, the north-northwest/south-southeast aligned bridge resides in a nicely landscaped park, in the center of the Town of Trimble. Emerson E. Parks Bridge was listed on the National Register of Historic Places on November 27, 1978, under the name of Parks Covered Bridge, while at its previous location in Obion County. It is the only historic covered bridge in Dyer County.

Directions: From Kenton, at the intersection of US 45W/SR 5 and SR 89, take SR 89 west to SR 105; go right to Trimble. The bridge is at the north end of Parks Plaza (an extension of Main Street), on the left.

Bible Covered Bridge

GREENE COUNTY

Brief Statistics

Type: Authentic historic covered bridge. World Guide Covered Bridge Number: TN-30-01. Built in 1922 by unknown builder. One-span bridge crossing Little Chucky Creek. Queenpost/kingpost combination truss bridge 57.0 feet long by 12.9 feet wide, with approximately 11.5-foot wide by 9.7-foot high portal openings. Alternate Name: Chucky Bridge. Located east of Warrensburg on old Denver Bible Lane.

Bible Bridge was named after the Christian Bible family that settled the area in 1783. Also called Chucky Bridge, after the creek flowing beneath, the single-span bridge was built, in 1922, by an unknown builder. Bypassed and closed to motor traffic, the decrepit bridge displays extreme sag as it carries old Denver Bible Lane across Little Chucky Creek. Sadly, arson attempts are visible and severe timber rot is evident. Bible Bridge and Harrisburg Bridge are the only two queenpost truss bridges remaining in Tennessee.

The 57.0-foot long bridge rests on a poured concrete abutment, at the right downstream or south-southwest end, and a pier, constructed with upright timbers on a poured concrete base, on the left downstream or north-northeast end, the poured concrete abutment located at the end of the 10.6-foot long ramp, which, interestingly, is only at this end. There is a timber support at the right downstream end, which appears to have been added a long time ago, although it may have been part of the original construction. The bridge is supported by a half-height queenpost/kingpost combination truss, modified by the addition of iron rod tension members at each end of the horizontal member and a kingpost truss centered under the horizontal member. The kingpost truss has an iron rod in place of the normally used vertical timber tension member. The modified queenpost/kingpost combination truss is off center toward the left downstream portal, an unusual feature. Could the bridge have been shortened, creating the need for the ramp and pier at the north-northwest end? The reddish brown painted bridge has transverse planking for the floor, vertical boarding on the sides and portals, 2-inch by 3-inch boards set diagonally and spaced apart under the eaves, and red-brown shingles on the roof. The roof extends beyond the gables, providing additional weather shelter for the entrances. There are three opposing, horizontal, rectangular window openings on each side, each with a peaked gable above. A sign above the right downstream portal reads *THE/BIBLE BRIDGE/Cross this bridge at a walk*. The oldest dated graffiti found inside the bridge is 1923, one year after the 1922 construction date.

The Greene County-owned bridge is in an open setting, with the Denver Bible Lane bypass bridge very close, on the upstream side, SR 349 at the right downstream portal and bypassed old Denver Bible Lane culminating in a parking area, at the left downstream portal. Slow moving, silted Little Chucky Creek flows westerly into the Nolichucky River, at Warrensburg, thence into the French Broad River. Bible Bridge is the only historic covered bridge in Greene County. Greenville is the county seat. *See Bible Covered Bridge in the color photograph section: C-23.*

Directions: From east of Morristown, at I-81 exit 15 SR 340 (Fish Hatchery Road), go south on SR 340 through Warrensburg to SR 349 (Warrensburg Road), on the left; go 1.7 miles to the bridge, on the left at Denver Bible Lane.

Horse Creek Park Covered Bridge

GREENE COUNTY

Brief Statistics

Type: Non-authentic modern covered footbridge. World Guide Covered Bridge Number: TN-30-a. Built in 1995 by US Forest Service. One-span footbridge crossing tributary to Horse Creek. Stringer footbridge 20.7 feet long by 9.0 feet wide, with approximately 7.1-foot wide by 8.0-foot high portal openings. Alternate Name: None known. Located in Cherokee National Forest at Horse Creek Park.

Horse Creek Park Bridge was built, in 1995, by Marvin Harrison, Ray Brookshire and Clyde Hensley, all three employed by the US Forest Service, in Horse Creek Park in the Cherokee National Forest. Three 6-inch by 12-inch timber stringers, set on edge, support the 20.7-foot long stringer footbridge. The abutments are constructed with three 8-inch by 8-inch timbers stacked horizontally upon mortared river stone, at the southeast end, and one 8-inch by 8-inch timber placed horizontally upon mortared river stone, at the northwest end. The abutments have wing walls, consisting of stacked 6-inch by 6-inch timbers. The flooring consists of transverse 4-inch by 6-inch planking. The sides are open, with handrails between the four 6-inch by 6-inch timber roof supporting posts along each side. The gables are enclosed with vertical boarding, with the roof support brace beam over each entrance cut to form an arch over the entrances. Moss and lichen covered wood shingles enclose the roof. All lumber used in the construction of the bridge is pressure treated.

Still carrying an occasional vehicle across Moyer Branch, Clifford Holder Bridge is a private bridge built in 1919.

The single-span footbridge crosses a small tributary, at its confluence with Horse Creek, which flows into the Nolichucky River, southeast of Tusculum. Surrounded by the dense Cherokee National Forest, the bridge is at the edge of an opening that used to be a swimming pool, through which Horse Creek flowed. Following a freshet in 2001, the swimming pool was washed out and filled in with river stone and gravel from Horse Creek.

Directions: From Greeneville, go east on SR 107 to the first right (0.2 mile) after the junction with SR 351 North (Chucky Road), on the left. This right is Horse Creek Park Road; follow 2.7 miles to Horse Creek Park, in the Cherokee National Forest. The bridge is in the first picnic area, on the left just after the day use pay station. Park in the parking lot and walk a short distance down the path, to Horse Creek and the bridge. Admission.

Clifford Holder Covered Bridge

HAMBLEN COUNTY

Brief Statistics

Type: Non-authentic historic covered bridge. World Guide Covered Bridge Number: TN-32-A. Built in 1919 by Clifford Holder. One-span bridge crossing Moyer Branch. Stringer bridge 26.9 feet long by 16.3 feet wide, with approximately 9.7-foot wide by 7.9-foot high portal openings. Alternate Name: Holder Bridge. Located northwest of Morristown at 3475 Cherokee Drive.

Clifford Holder built Clifford Holder Bridge, also called Holder Bridge, in 1919, near the entrance to his driveway. This single-span bridge is the oldest non-authentic covered motor vehicle bridge in Tennessee.

The stringer bridge rests on unmortared stone abutments, supported by four massive timber stringers, cement block supports having been added under the outside stringers. A transverse planked floor, vertical boarded sides, weathered to a pleasing old gray color, and a rusty metal-clad roof enclose the 26.9-foot long bridge. The portals are open, except for a vertical boarded panel on the right side of the left downstream end.

Still allowing the occasional motor vehicle to clatter through, Clifford Holder Bridge is in an open field, in the front yard spanning Moyer Branch, a small creek feeding into nearby Cherokee Lake, formed by the damming of the Holston River. The north-northeast/south-southwest aligned bridge has a large cedar tree, on the downstream side, and a sparsely foliated deciduous tree, on the upstream side.

Directions: From Morristown, at the intersection of US 11E and SR 343, go north 0.7 mile on SR 343 to

Cherokee Drive, on the left; go 4.0 miles to the bridge, visible on the left, at 3475 Cherokee Drive. Private property: request permission to visit.

Hedrick Crossing Covered Bridge

Hamilton County

Brief Statistics

Type: Non-authentic modern covered bridge. World Guide Covered Bridge Number: TN-33-D. Built in 2000 by Steve Hedrick. Post supported roof over roadway crossing tributary to Chestnut Creek. Bridge 18.2 feet long by 10.4 feet wide, with approximately 9.4-foot wide by 11.3-foot high portal openings. Alternate Name: None known. Located in Apison at 3510 Prospect Church Road.

Steve Hedrick built Hedrick Crossing Bridge, in 2000, as a post supported roof over the river-stone driveway to his residence, where a culvert allows the tributary to Chestnut Creek to pass beneath the bridge. A transverse planked floor is laid on longitudinal timbers, set on the ground. Vertical boarding covers the sides and the gables, the sides with two large, opposing, horizontal window openings on each side. The roof rafters and trusses are in place, but the bridge lacks a roof covering, allowing the wood to evenly weather. A roof covering will be added in the near future. Three rough-cut 6-inch by 6-inch wood posts support the roof framing.

The west-southwest/east-northeast aligned bridge is in the grassed front yard, built parallel to a densely wooded area, upstream. Just downstream, a spillway dams the tributary, the waters continuing into nearby Chestnut Creek, which joins Wolftever Creek, east of Collegedale, before entering the Tennessee River, at Chickamauga Lake. On the downstream side of the 18.2-foot long bridge are a large man-made pond, a waterwheel and scattered mature trees. An old wagon has been fittingly placed beside the west-southwest portal.

Directions: From I-75, in Chattanooga, take the East Brainerd Road exit (exit 3), going east 9.7 miles to Prospect Church Road, on the left. The bridge is visible in the far corner, at No. 3510. Private property: request permission to visit.

Mill Run Covered Bridge

Hamilton County

Brief Statistics

Type: Non-authentic modern covered bridge. World Guide Covered Bridge Number: TN-33-C. Built in 1978 by John Howard Stakely, Robert C. Stakely and Raymond E. Stakely. Two-span bridge crossing Wolftever Creek. Stringer bridge 59.9 feet long by 20.3 feet wide, with approximately 18.8-foot wide by 11.0-foot high portal openings. Alternate Name: None known. Located in Ooltewah in Mill Run Subdivision on Mitchell Mill Road.

Mill Run Bridge was built, in 1978, by John Howard Stakely and his two sons, Robert C. Stakely and Raymond E. Stakely, at the entrance to Mill Run Subdivision.

The two-span bridge is supported by seven steel I-beam stringers, set on poured concrete abutments and one poured concrete pier. The flooring consists of transverse planking with four-plank-wide wheel treads. The gray-painted vertical board siding on the sides and portals is from a 50-year old local tobacco barn, the sides open at eye level for the length of the 59.9-foot long bridge. The tin on the roof was installed in 1999, replacing the original shakes. The low peaked roof is supported by seven wood posts along each side, the posts acquired from the same old tobacco barn. Timbers stacked three high run the length of the interior, separating the walkways along each side from the single traffic lane. A large electric carriage lamp is mounted on each gable; a central electric light hangs from the interior roof rafters.

Aligned south/north, Mill Run Bridge spans Wolftever Creek, which flows into the Tennessee River, at Chickamauga Lake. The wide creek is heavily wooded on the north side, and lightly wooded on the south side with the bridge nestled among the verdant trees, as it carries Mitchell Mill Road traffic into the subdivision.

Directions: From Chattanooga, take I-75 north to the US 11/US 64 (Lee Highway) exit; go left to Mountain View Road, just under the overpass. Go right 0.2 mile to Snow Hill Road, on the left; go 0.2 mile to Mill Run Subdivision and Mitchell Mill Road, on the left. This is a gated community; the bridge is visible, just beyond the gate.

Narrowbridge Covered Bridge

Hamilton County

Brief Statistics

Type: Non-authentic modern covered bridge. World Guide Covered Bridge Number: TN-33-A. Built in 1965 by unknown builder. Two-span bridge crossing Mackey Branch. Stringer bridge 31.5 feet long by 17.9 feet wide, with approximately 16.5-foot wide by 14.0-foot high portal openings. Alternate Name: None known. Located in Chattanooga at 1420 Jenkins Road.

Narrowbridge Bridge was built, in 1965, by an unknown builder, on private property that at one time was a restaurant. The two-span bridge crosses Mackey Branch, a tributary to nearby South Chickamauga Creek, which flows into the Tennessee River.

A reinforced concrete slab stringer, set on poured concrete abutments and one poured concrete pier, supports the stringer bridge. The upper sides and gables have gray weathered, vertical boarding; the lower 6.7 feet of the sides are open, with a railing between the nine 2-inch by 6-inch roof-supporting posts along each side. Black asphalt shingles cover the low peaked roof. A centrally located, rafter-mounted electric light is in the interior; a security gate has been installed in the left downstream or roadside entrance.

The 31.5-foot long bridge, aligned north-northwest/south-southeast, carries the private driveway across a tree and thicket lined brook. Open areas are at both ends of the bridge.

Directions: From Chattanooga, go north on I-75 to exit 3, SR 320, East Brainard Road. Go east 1.6 miles to Jenkins Road, on the left. The bridge is in the far corner of this junction, at No. 1420. Private property: request permission to visit.

Tony Kennedy Covered Bridge

HAMILTON COUNTY

Brief Statistics

Type: Non-authentic modern covered bridge. World Guide Covered Bridge Number: TN-33-E. Built in 1987 by Ardis and Soles Construction. One-span bridge crossing Ryall Springs. Stringer bridge 28.3 feet long by 12.1 feet wide, with approximately 10.8-foot wide by 12.4-foot high portal openings. Alternate Name: None known. Located in Chattanooga at 1495 Morris Hill Road.

Ardis and Soles Construction built Tony Kennedy Bridge, in 1987, along the driveway to Mr. Kennedy's residence. The 28.3-foot long stringer bridge is supported by two steel I-beam stringers, set on poured concrete abutments. The flooring is constructed with 4-inch by 6-inch planks, placed on edge, with five-plank-wide wheel treads. The sides and the portals are covered with gray weathered vertical boarding, the sides up to rail height and under the eaves, open the length of the bridge in-between. The exceptionally tall covered bridge has old weathered wood shingles covering the steeply peaked roof, which is supported along each side by five wood posts. An electric carriage lantern is mounted at each side of the east-southeast entrance.

Narrow, rock- and gravel-bottomed Ryall Springs passes beneath the single-span bridge, on its way to nearby Mackey Branch, before entering South Chickamauga Creek, which winds its way across Chattanooga into the Tennessee River. Aligned east-southeast/west-northwest, Tony Kennedy Bridge is situated in the nicely landscaped front yard, a short distance down the asphalt paved driveway that passes between mortared stone pillars at Morris Hill Road.

Directions: From Chattanooga, at I-75, take exit 3, SR 320 (East Brainerd Road), east 3.0 miles to Morris Hill Road, on the left; go 0.4 mile to No. 1495, on the left. The bridge is visible from the roadway. Private property: request permission to visit.

David Crockett State Park Covered Bridge

LAWRENCE COUNTY

Brief Statistics

Type: Non-authentic modern covered bridge. World Guide Covered Bridge Number: TN-50-A #2. Built in 1999 by Jeff Kelly Construction. Post supported roof over roadway crossing Shoal Creek. Bridge 58.3 feet long by 22.9 feet wide, with approximately 19.9-foot wide by 14.0-foot high portal openings. Alternate Name: Shoal Creek Bridge. Located in David Crockett State Day Use Park.

David Crockett State Park Bridge, also called Shoal Creek Bridge, after the creek flowing beneath, was built in 1999 by Jeff Kelly Construction, to replace the original 55-foot long covered bridge, built in 1959, that was washed away in the 1998 flood. The current bridge is a post supported roof over a roadway that allows the waters of Shoal Creek to pass through a pipe under the roadbed. The two-lane bridge, at 22.9 feet wide, is the widest non-authentic covered motor vehicle bridge in Tennessee.

The 58.3-foot long bridge is set on the concrete roadway, supported by six wood posts along each side. The interior and exterior of the sides and the exterior of the portals are covered with vertical boarding, the sides open at eye level for the length of the bridge. Blue painted aluminum covers the roof. There is an uncovered board-floored walkway, with a baluster handrail, on the upstream side. Double wide 8-inch by 8-inch timbers form a bumper guard along both sides of the interior.

The southwest/northeast aligned bridge has a pond on the upstream side, created by the backflow of the creek, and thick woods on the downstream side. The still surface of the pond affords the photographer beautiful reflections of the bridge against the verdant woods background. Shoal Creek flows beneath the roadway on its way to joining the Tennessee River, in Alabama.

Directions: From Lawrenceburg, take US 64/SR 15 west, following the signs to David Crockett State Day Use

David Crockett State Park Bridge is reflected in the still waters of Shoal Creek.

Park, on the right. The bridge is in the park, on the perimeter road.

Salt Lick Creek Covered Bridge

MACON COUNTY

Brief Statistics

Type: Non-authentic modern covered bridge. World Guide Covered Bridge Number: TN-56-B. Built in 1973 by unknown builder. One-span bridge crossing Salt Lick Creek. Stringer bridge 27.4 feet long by 21.6 feet wide, with approximately 18.3-foot wide by 11.0-foot high portal openings. Alternate Name: None known. Located in Red Boiling Springs on Valley View Road.

Salt Lick Creek Bridge shares the same history and construction style as its companion, Village Bridge. Built in 1973, by an unknown builder, Salt Lick Creek Bridge replaced a non-covered bridge that was washed away by the devastating flood of June 24, 1969. The rain started to fall late in the evening of June 23; six hours later, nine inches of rain had fallen, flooding the entire Salt Lick Valley.

Seven steel I-beam stringers, resting on poured concrete abutments, support the single-span bridge. Although wide enough for two lanes of traffic, the transverse planked floor has single-lane wheel treads, consisting of five-plank-wide runners, with an additional two planks laid longitudinally in-between. The exterior and the interior of the lower sides, up to rail height, and the exterior of the gables have weathered vertical boarding. The sides are open 3.4 feet, between the rail height lower section and a boxed-in secondary roof, which a 9-inch open area separates from a continuation of the vertical boarding to the eaves. The main roof and secondary roofs are wood shingled, the shingles having weathered to a dark gray-brown. Steel guardrails run along each side for the interior length of the 27.4-foot long bridge, extending out both portals.

Salt Lick Creek Bridge aligns north-northeast/south-southwest as it crosses Salt Lick Creek, which flows northerly into the Barren River, in Monroe County. The narrow rocky-bottomed creek has scattered trees along its banks, with a park, on the south-southwest side, stretching to State Road 151.

Directions: From Red Boiling Springs, at the junction of SR 52 and SR 151, go south on SR 151, past the library to Valley View Road, on the left. The bridge is 0.1 mile. Parking is available in the park, at the bridge.

Village Covered Bridge

MACON COUNTY

Brief Statistics

Type: Non-authentic modern covered bridge. World Guide Covered Bridge Number: TN-56-A. Built in 1973 by unknown builder. Two-span bridge crossing Salt Lick Creek. Stringer bridge 58.2 feet long by 21.4 feet wide, with approximately 18.5-foot wide, by 11.6-foot high portal openings. Alternate Name: None known. Located in Red Boiling Springs on Church Street.

Sharing the same history and construction style as its companion, Salt Lick Creek Bridge, Village Bridge was built, in 1973, by an unknown builder, replacing a non-covered bridge that was washed away by the devastating flood of June 24, 1969. The rain started to fall late in the evening of June 23; six hours later, nine inches of rain had fallen, flooding the entire Salt Lick Valley.

The two-span bridge is supported by seven steel I-beam stringers, set on poured concrete abutments and one poured concrete pier. The floor consists of transverse planking with four-plank-wide wheel treads at each side of a centered five-plank-wide wheel tread. The exterior and the interior of the lower sides, up to rail height, and the exterior of the gables have gray weathered vertical boarding. The sides are open 3.9 feet, between the rail height lower section and a boxed-in secondary roof, which a 9-inch open area separates from a continuation of the vertical boarding to the eaves. Dark gray-brown weathered, wood shingles cover the main roof and secondary roofs. Steel guardrails run along each side for the interior length of the 58.2-foot long bridge, extending out both portals.

Village Bridge spans gravel-bottomed Salt Lick Creek, which flows northerly into the Barren River, in Monroe County. The southeast/northwest-aligned bridge has scattered trees on the northwest side of the creek, with open grassed areas on the opposite side and a park upstream.

Directions: From downtown Red Boiling Springs on SR 52, go east to Church Street; go right to the bridge.

Fairgrounds Park Covered Bridge

MADISON COUNTY

Brief Statistics

Type: Non-authentic modern covered bridge. World Guide Covered Bridge Number: TN-57-A. Built in 1998 by City of Jackson. One-span bridge crossing Anderson Branch. Stringer bridge 60.0 feet long by 18.0 feet wide, with approximately 17.0-foot wide by 12.6-foot high portal openings. Alternate Name: None known. Located in Jackson in Fairgrounds Park at 800 South Highland Avenue (US 45 South).

Fairgrounds Park Bridge was built, in 1998, by Bobby Gregson and Raymond Trolinger, for the City of Jackson at Fairgrounds Park. At 60.0 feet long, Fairgrounds Park Bridge has the distinction of being the longest single-span non-authentic covered motor vehicle bridge in Tennessee.

Two massive 12-inch by 36-inch steel I-beam stringers, set on poured concrete abutments, support the two-lane stringer bridge. The flooring consists of transverse planking. The sides are open, with banister style handrails, constructed with 2-inch by 4-inch lumber between the roof supports. The gables are enclosed with vertical facsimile tongue and groove paneling. A red painted aluminum roof is supported by nine 6-inch by 6-inch wood posts along each side. On each side are two 4-inch by 4-inch wood buttress braces, anchored to upright wood utility poles and attached to the third roof support in from the west-northwest portal and the second roof support in from the east-southeast portal. The bridge is painted white on the exterior and the interior, contrasting nicely with the red roof.

The bridge spans the tree-lined banks of moderately wide sandy-bottomed Anderson Branch, which joins the nearby South Fork Forked Deer River. The bridge is fully exposed, with a large field at the east-southeast or right downstream end and a roadway, with pull-in parking and buildings beyond, at the west-northwest end. *See Fairgrounds Park Covered Bridge in the color photograph section: C-25.*

Directions: From Jackson, at the intersection of US 70/SR 1 and US 45, go south on US 45, about 1.0 mile, to Fairgrounds Park, at 800 South Highland Avenue (US 45 South), on the left. The bridge is in the park, behind the large building, and is visible from the highway. Ample parking is available at the bridge.

Turkey Pen Resort Covered Bridge

MONROE COUNTY

Brief Statistics

Type: Non-authentic modern covered bridge. World Guide Covered Bridge Number: TN-62-A. Built in 1982 by Luke Henley. One-span bridge crossing wet weather stream. Stringer bridge 27.6 feet long by 12.8 feet wide, with approximately 11.0-foot wide by 9.3-foot high portal openings. Alternate Name: None known. Located in Citico Beach at Turkey Pen Resort.

Luke Henley built Turkey Pen Resort Bridge, in 1982, over a wet weather stream, a tributary to Carringer Creek, which flows into the Little Tennessee River. The 27.6-foot long stringer bridge has four log stringers resting on a log that is set in the ground. The floor has diagonal planking with four-plank- wide wheel treads. The diagonal flooring is unusual, in that only full-length planks were used, leaving the two corners that would have used shorter lengths as bare ground. The sides have dark weathered vertical boarding up to railing height, open above. The steeply peaked gables also have dark weathered vertical boarding, with *TURKEY PEN RESORT*, made from small branches, attached to the boarding. Covered with red painted aluminum, the roof is supported by eight 6-inch by 6-inch wood posts along each side.

The single-span bridge, aligned west-southwest/east-northeast, is situated in the open front lawn area of the resort, with thick mature woods on the downstream side, the red roof standing out vividly against the deep green background of mixed coniferous and deciduous trees. *See Turkey Pen Resort Covered Bridge in the color photograph section: C-27.*

Directions: From Vonore, at the junction of US 411/SR 33 and SR 360 (Citico Road), take SR 360 east 7.2 miles, going straight just before the SR 360 bridge (Citico Road continues straight, while SR 360 makes a sharp right turn across the bridge). Continue another 7.1 miles on Citico Road to Mt. Pleasant Road, on the left, crossing the concrete bridge and following Mt. Pleasant Road past the Chilhowee Dam, on the Little Tennessee River, to Turkey Pen Resort, on the right. The bridge is toward the end of the circular drive into the resort.

Port Royal Covered Bridge

Montgomery County

Brief Statistics

Type: Authentic modern covered bridge. World Guide Covered Bridge Number: TN-63-01 #2. Built in 1977 by Burkhardt & Horn. Two-span bridge crossing Red River. Pratt truss with Burr arch bridge approximately 192 feet long by 13.8 feet wide, with approximately 10.0-foot wide by 10.0-foot high portal openings. Alternate Name: None known. Located in Port Royal State Historic Park.

The previous Port Royal Covered Bridge, World Guide No. TN-63-01 #1, was built, in 1904, by J. C. McMillan, as a two-span, 270-foot long, Howe truss with a Burr arch. This bridge, neglected, deteriorated until it collapsed into the Red River, in April 1972. The current Port Royal Covered Bridge is a three-quarter-size replica of this previous bridge, which was built in 1977 by Burkhardt & Horn of Memphis, Tennessee. Sadly, the bridge was struck by large floating debris during the flood of June 12, 1998, severely damaging the center pier, which partially collapsed, dropping the center of the bridge into the flood-stream. An heroic attempt was made to save the bridge by sawing out the partially submerged siding, to prevent the bridge from being washed away, by allowing the passage of the raging current through the bridge. The attempt saved the right downstream span; the left span was washed downstream, but later saved and dragged up the left embankment. The right span was subsequently dragged off the center pier, to rest over the approach ramp. Thankfully plans are underway to restore the bridge.

The previous Port Royal Bridge, named after the nearby community, carried the old Port Royal Road across the Red River, before the upstream concrete and steel bridge bypassed it in 1954. The Red River joins the Cumberland River, at Clarksville, the county seat. The 192-foot long bridge is the longest authentic covered bridge remaining in Tennessee.

Port Royal Bridge was held high above the Red River on three mortared cut stone piers, backfilled with earth and rubble, with ramps extending to the poured concrete abutments. The ramps are supported by multiple timber supports. The only Pratt truss covered bridge in Tennessee and the only Pratt truss in a historic covered bridge in the southeastern United States, the two-span bridge has an all wood eight- panel Pratt truss with a Burr arch in each span. The floor has transverse planking with no wheel treads, as the bypassed bridge was intended only for foot traffic. The sides have vertical boarding; the portals have vertical boarding with battens. The roof is wood shingled. Each span has a long rectangular window opening on each side.

Aligned south-southeast/north-northwest, the bridge is situated in a wooded area of the Port Royal State Historic Park, highly visible across the wide river. Nearby in the park is an interesting open iron Pratt truss bridge, built in 1890, by the Converse Bridge Company, called Sulphur Fork Bridge, now only open to foot traffic.

Directions: East of Clarksville, on I-24, take exit 11, SR 76; go east 3.4 miles to Old Clarksville-Springfield Road, on the left; follow about 2 miles to the tee at SR 238 (Port Royal Road), in Port Royal State Historic Park. The bridge is on the left; parking is available, on the right, behind the museum. No admission.

Mulberry Creek Covered Bridge

Moore County

Brief Statistics

Type: Non-authentic modern covered bridge. World Guide Covered Bridge Number: TN-64-A. Built in 2001 by Don Spencer. One-span bridge crossing East Fork

Mulberry Creek. Stringer bridge 49.9 feet long by 19.8 feet wide, with approximately 14.7-foot wide by 15.5-foot high portal openings. Alternate Name: None known. Located in Lynchburg off Main Street.

Don Spencer built Mulberry Creek Bridge, in 2001, at the entrance to his proposed recreational vehicle park. Already, a high water mark is inscribed on the concrete wing wall, at the east-southeast end, showing the water level from the flood of January 23, 2002.

The 49.9-foot long stringer bridge is supported by four steel I-beam stringers, set on poured concrete abutments with wing walls. The one-lane bridge also has a decorative double queenpost truss, one queenpost below the horizontal member of the other queenpost truss. The decorative queenpost truss is constructed with 6-inch by 6-inch timbers, the joints fastened together with bolted iron plates. The flooring consists of transverse planking with six-plank-wide wheel treads. Gray weathered vertical boarding covers the sides and the gables, the sides open under the eaves to facilitate air circulation within the bridge. The roof is constructed with red painted metal nailed to a plywood sub-roof and extends, along with the gables, beyond the entrances, to provide weather protection for the interior. Six 6-inch by 6-inch timber posts along each side support the low peaked roof. A cattle gate secures the west-northwest entrance. A covered 3.5-foot-wide walkway is on the upstream side of the bridge. The walkway is supported by thirteen steel guy wires, attached under the extended roof overhang to thirteen transverse steel angle supports, under the walkway. A wire guard-fence affixed to the guy wires runs along the outer side of the walkway, and longitudinal planking forms the floor.

Aligned west-northwest/east-southeast, the single-span bridge crosses East Fork Mulberry Creek, which joins the Elk River, south of Mulberry. A crushed stone roadway runs to and from the bridge, which is well exposed, with Main Street at the left downstream or west-northwest end; trees fill in the east-southeast corner, on the downstream side.

Directions: From Lynchburg, on SR 55 (Majors Boulevard), go southbound past the Post Office to South Elm Street, on the left; go 0.7 mile to the bridge, visible just off the roadway, on the left. South Elm Street becomes Main Street a short distance from SR 55.

Healy Covered Bridge

OVERTON COUNTY

Brief Statistics

Type: Non-authentic modern covered bridge. World Guide Covered Bridge Number: TN-67-A. Built in 1982 by Lynn Edward Patrick Healy. Three-span bridge crossing West Fork Obey River. Stringer bridge 87.8 feet long by 11.9 feet wide, with approximately 11.0-foot wide by 7.9-foot high portal openings. Alternate Name: None known. Located in Allred off Shiloh Road.

Lynn Edward Patrick Healy built Healy Bridge, in 1982, with only the center span, of the three-span stringer bridge with approach ramps, covered. Lynn Edward Patrick Healy added covers over the two end spans, in 1988. Including the ramps, the bridge has a total length of 127.1 feet; however, only the center three spans, totaling 87.8 feet, are covered.

The three center spans of the stringer bridge are supported by three 8-inch by 15-inch railroad trestle timber stringers placed on four piers, constructed with three upright wood utility poles capped with a railroad trestle timber; the upright utility poles are set in footings, except the second pier from the left downstream end is set in a poured concrete base. Three large-diameter wood utility pole stringers, spanning from the ground to the end piers, support the ramps. The flooring consists of transverse planking that extends down the ramps. The sides and portals are covered with horizontal rough-cut boarding. Dull tin, with a minor amount of rust, covers the roof, which is supported by sixteen wood posts, a combination of 4-inch by 6-inch and 4-inch by 8-inch timbers. One small, double-hung, casement glass-pane window is on each side of the center span, and one horizontal window opening is on each side of each end span. The steep approach ramps at each end have handrails. A wood sign inscribed *The Healy's* is over the left downstream entrance.

Healy Bridge uniquely curves to the right or downstream, as each covered span is slightly angled, producing an alignment from north-northeast/south-southwest to east-northeast/west-southwest. The one-lane bridge carries the graveled driveway to the residence, across the wide, rocky-bottomed swiftly flowing West Fork Obey River, which enters the Obey River, near Riverton. The site is heavily wooded, with the bridge situated 100 feet in from graveled Shiloh Road. A ford crossing is on the downstream side, and a barn is at the southwest end.

Directions: From Livingston, go east on SR 52 through Alpine to SR 85; go right to Allred, continuing 1.0 mile past the Post Office, in the white mobile home on the left, to Shiloh Road, on the right, just before the West Fork Obey River Bridge. Follow about 3.0 miles to the bridge, visible in the woods, on the right. Private property: request permission to visit.

Hidden Hollow No. 1 Covered Bridge

PUTNAM COUNTY

Brief Statistics

Type: Non-authentic modern covered footbridge. World Guide Covered Bridge Number: TN-71-a. Built in 1955 by Arda Lee. Two-span footbridge crossing tributary to Falling Water River. Stringer footbridge 35.0 feet long by 5.4 feet wide, with approximately 5.1-foot wide by 6.4-foot high portal openings. Alternate Name: None known. Located in Cookeville at Arda Lee's Hidden Hollow at 1901 Mt. Pleasant Road.

Arda Lee built Hidden Hollow No. 1 Footbridge, in 1955, the oldest of the three covered bridges he built in his Hidden Hollow recreation park. This stringer bridge is the oldest non-authentic covered footbridge in Tennessee. Wisteria entwined in the bridge rafters forms the novel roof, beautiful to behold when in full bloom.

The 35.0-foot long bridge is supported by three stringers—a wooden utility pole on the upstream side, wooden planks in the center and downstream side—that rest on concrete block and poured concrete abutments, and one concrete block pier. Pressure-treated transverse planking forms the floor; open sides have redwood-lumber framing and handrails; the roof has redwood framing for the wisteria vines and is supported by seven 2-inch by 4-inch posts along each side. There are three wooden buttress braces on each side.

The two-span footbridge crosses a tributary to the nearby Falling Water River, which joins the Caney Fork River at Center Hill Lake, before entering the Cumberland River, at Carthage. The southwest/northeast aligned bridge crosses to a man-made island in the widened stream area of 86-acre Hidden Hollow park, which is partially hidden between two steep wooded slopes. The park caters to family outings, offering playgrounds, fishing, swimming, hiking, petting zoo, sports, and shelters with grills. During the season, Christmas lights and displays abound, including a 50-foot cross erected on a hilltop.

Directions: From I-40, exit 290, US 70N/SR 24, in Cookeville; go south 0.1 mile to Poplar Grove Road; go left, then immediately bear sharp left; go 2.5 miles (road makes a sharp left turn at the bridge, at 1.7 miles) to Mt. Pleasant Road, on the left; go 0.3 mile to Hidden Hollow, at 1901 Mt. Pleasant Road, on the left, passing through Hidden Hollow No. 3 Covered Bridge, at the entrance. The bridge is in the park, not visible from the roadway. Nominal admission charge.

Hidden Hollow No. 2 Covered Bridge

PUTNAM COUNTY

Brief Statistics

Type: Non-authentic modern covered footbridge. World Guide Covered Bridge Number: TN-71-b. Built in 1975 by Arda Lee. One-span footbridge crossing tributary to Falling Water River. Stringer footbridge 10.0 feet long by 7.7 feet wide, with approximately 6.3-foot wide by 6.7-foot high portal openings. Alternate Name: None known. Located in Cookeville at Arda Lee's Hidden Hollow at 1901 Mt. Pleasant Road.

Arda Lee built Hidden Hollow No. 2 Footbridge, in 1975, the last of the three covered bridges he built in his Hidden Hollow recreation park. This 10.0-foot long bridge is the shortest non-authentic covered footbridge in Tennessee.

The 31.3-foot long stringer bridge, with the 10.0-foot long cover near the center, is supported by two wooden utility pole stringers, resting on poured concrete abutments and one concrete block support. The floor is longitudinal planking; the sides are open with a full-length bench along the upstream side and a low railing along the downstream side. The flat roof consists of diagonal boarding, covered with a plywood platform, enclosed by a low handrail, accessible by a ladder from the deck, at the concrete block support on the upstream side. Steps at both ends ascend to the bridge.

The single-span, gray weathered wood footbridge crosses a tributary to the nearby Falling Water River, which joins the Caney Fork River, at Center Hill Lake, before entering the Cumberland River, at Carthage. The southwest/northeast-aligned bridge is very open within the confines of the man-made ponds and tributary canals, in 86-acre Hidden Hollow park.

Directions: From I-40, exit 290, US 70N/SR 24, in Cookeville, go south 0.1 mile to Poplar Grove Road; go left, then immediately bear sharp left; go 2.5 miles (road makes a sharp left turn at the bridge, at 1.7 miles) to Mt. Pleasant Road, on the left; go 0.3 mile to Hidden Hollow, at 1901 Mt. Pleasant Road, on the left, passing through Hidden Hollow No. 3 Covered Bridge, at the entrance. The bridge is in the park, not visible from the roadway. Nominal admission charge.

Hidden Hollow No. 3 Covered Bridge

PUTNAM COUNTY

Brief Statistics

Type: Non-authentic modern covered bridge. World Guide Covered Bridge Number: TN-71-C. Built in 1973 by Arda Lee. One-span bridge crossing wet weather stream. Stringer bridge 37.7 feet long by 18.4 feet wide, with approximately 13.6-foot wide by 14.6-foot high portal openings. Alternate Name: None known. Located in Cookeville at Arda Lee's Hidden Hollow at 1901 Mt. Pleasant Road.

Arda Lee built Hidden Hollow No. 3 Bridge, in 1973, the second of the three covered bridges he built in his Hidden Hollow recreation park. The 37.7-foot long bridge is on the entrance road, crossing a wet weather stream that feeds the tributary to the nearby Falling Water River. The other two bridges in Hidden Hollow span the tributary.

The stringer bridge is supported by two large steel I-beam stringers, set on poured concrete abutments, and two smaller steel I-beam stringers, located to the outside of the two large I-beams. These smaller I-beams are set on three steel I-beam supports that extend beyond the sides of the bridge, to support the three timber buttress braces along each side. The flooring consists of thirteen diagonal sets of three 2-inch by 6-inch planks, on edge, and one 6-inch by 6-inch timber; five-plank-wide wheel treads run the length of the span and the asphalt roadway extends into the interior of the bridge, meeting the flooring. The sides have diagonal, 3-inch boards spaced apart to railing height, with four rows of diagonal planks, spaced apart above the railing, forming two herringbone patterns that converge toward the flooring. The portals are covered with horizontal tongue and groove boarding that wraps around about three feet inside the bridge. The low peaked roof is covered with bright tin. The gray weathered bridge has a camera surveillance system advertised by the signs at each side of the entrance portal. A white electric carriage lantern hangs in each gable.

The west-northwest/east-southeast aligned bridge sits among open fields, with a thicket along the wet weather stream. Rustic fencing and tall gateposts that hold up the *HIDDEN HOLLOW* sign, made from split tree branches, are at the entrance end of the bridge.

Directions: From I-40, exit 290, US 70N/SR 24, in Cookeville, go south 0.1 mile to Poplar Grove Road; go left, then immediately bear sharp left; go 2.5 miles (road makes a sharp left turn at the bridge, at 1.7 miles) to Mt. Pleasant Road, on the left; go 0.3 mile to Hidden Hollow, at 1901 Mt. Pleasant Road, on the left. The bridge is at the entrance, visible from the roadway.

Papa's Covered Bridge

PUTNAM COUNTY

Brief Statistics

Type: Non-authentic modern covered bridge. World Guide Covered Bridge Number: TN-71-D. Built in 1982 by John D. Stites. One-span bridge crossing East Blackburn Fork. Stringer bridge 48.8 feet long by 13.3 feet wide, with approximately 11.7-foot wide by 12.5-foot high portal openings. Alternate Name: None known. Located in Cookeville at 5395 Liberty Church Road.

The Stites' three children have built their homes on the 800-acre J & S Farms property, two on the far side of East Blackburn Fork, the only access being across a ford. One morning, following a heavy overnight rain, their daughter-in-law's vehicle stalled in the high water, her children with her in the car! The time to build a bridge had come. And build a bridge John D. Stites did; in 1982, he built a 48.8-foot long single-span covered bridge. Papa's Bridge, in addition to being sturdy, functional, and long, has an observation playroom on the roof, where the grandchildren can play and sleep out. When approaching the bridge across the open pastureland, the observation playroom on top gives the appearance of riding piggyback on the main bridge.

The stringer bridge rests on poured concrete abutments, supported by four steel I-beam stringers. Transverse planking with four-plank-wide wheel treads makes up the floor. Gray weathered vertical boarding covers the sides and gables, except for a 22-inch gap at the center of the sides and an opening under the eaves for the length of the flooring. A shiny tin roof is supported by the 2-inch by 6-inch studding. A shiny tin-roofed children's playroom, centered on top of the bridge, is also enclosed with gray weathered vertical boarding, with an observation window opening at children's eye level around the four sides. Entry is up a ladder on the bridge interior wall, through a trap door. A weathered wood sign inscribed *Papa's Bridge* is mounted on the outer gable, over the right downstream entrance. The roof and gables extend beyond the entrances, providing additional shelter from the elements for the bridge interior.

Papa's Bridge spans rocky-bottomed East Blackburn Fork, which joins West Blackburn Fork, over the Jackson County line, to form Blackburn Fork, before flowing northward into the Roaring River. The north-northeast/south-southwest aligned bridge is at the edge of open pastureland nestled in old growth woods along the stream and beyond, the old ford on the downstream side providing a little open space.

Directions: From Cookeville, at the I-40 exit at SR 135, South Willow Avenue, go north 3.1 miles to West 12th Street; go left 4.8 miles (becomes Gainesboro Grade where the 4 lanes become 2 lanes) to Shipley Road, on the right; go right, then immediately turn left on Liberty Church Road; go 0.5 mile to No. 5395, on the right. The bridge is not visible from Liberty Church Road. Private property: request permission to visit.

Brookside Motel Covered Bridge

SEVIER COUNTY

Brief Statistics

Type: Non-authentic modern covered footbridge. World Guide Covered Bridge Number: TN-78-a. Built in 1963 by unknown builder. One-span footbridge crossing Roaring Fork. Stringer footbridge 35.1 feet long by 5.2 feet wide, with approximately 4.0-foot wide by 7.0-foot high portal openings. Alternate Name: None known. Located in Gatlinburg at 463 East Parkway (US 321).

Built in 1963 by an unknown builder, Brookside Motel Footbridge is behind the Brookside Resort, formerly called Brookside Motel, providing foot traffic access to the resort units on the other side of Roaring Fork. Two steel I-beam stringers, set on a poured concrete cap on top of mortared field stone retaining walls, support the 35.1-foot long bridge. The flooring is constructed with longitudinal planking. The sides are open, with benches in alcoves on each side at the center of the bridge, railing extending along both sides. The entrance gables are closed with plywood. The weathered, moss-covered, wood-shingled roof is in two levels, the higher level extending out over the bench-lined alcoves.

The rock- and boulder-strewn Roaring Fork swiftly cascades along, between retaining walls and under the single-span bridge, on its way to the nearby West Prong of the Little Pigeon River, then into the Little Pigeon River, at Sevierville, before entering the French Broad River. The red painted bridge has resort units at both ends. Large mature trees with overhanging foliage stretch up and down the north side of the creek, keeping the north/south aligned bridge in perpetual shade during the foliage season.

Directions: In Gatlinburg, from the junction of US 441 and US 321 (East Parkway), go 0.4 mile north on US 321 to Brookside Resort, at 463 East Parkway, on the left. The bridge is to the left of the office, at the back of the resort.

Doc Prater Covered Bridge

SEVIER COUNTY

Brief Statistics

Type: Non-authentic modern covered footbridge. World Guide Covered Bridge Number: TN-78-b. Built in 1968 by unknown builder. One-span footbridge crossing Roaring Fork. Stringer footbridge 30.1 feet long by 4.6 feet wide, with approximately 3.9-foot wide by 6.2-foot high portal openings. Alternate Name: None known. Located in Gatlinburg at 463 East Parkway (US 321).

Built in 1968 by an unknown builder, Doc Prator Footbridge is behind the Brookside Resort, providing foot traffic access to Brookside Cottage No. 535, on the other side of Roaring Fork. The 30.1-foot long bridge is the narrowest non-authentic covered footbridge in Tennessee, measuring 4.6 feet wide.

Similar in construction to the earlier Brookside Motel Bridge downstream, Doc Prator Bridge is supported by two steel I-beam stringers, set on mortared rocks and boulders, on each embankment. Longitudinal planking forms the floor. The open sides have railings along the length of the bridge, with bench-lined alcoves on both sides, at the center of the span. The portal gables are covered with plywood, the roof with brown asphalt shingles. A gate closes the brown painted bridge at the north portal. A three-line sign, reading *BROOKSIDE/COTTAGE/535*, is mounted on the south gable.

The single-span bridge crosses rock- and boulder-strewn Roaring Fork, which flows into the nearby West Prong of the Little Pigeon River, at Sevierville, then enters the Little Pigeon River, on its way northward to the French Broad River. The mature-tree-lined creek has rock and retaining wall embankments; a dam is on the upstream or west side.

Directions: In Gatlinburg, from the junction of US 441 and US 321 (East Parkway), go 0.4 mile north on US 321 to Brookside Resort, at 463 East Parkway, on the left. The bridge is to the right of the office, across the parking area and behind the resort units.

Emerts Cove Covered Bridge

SEVIER COUNTY

Brief Statistics

Type: Non-authentic modern covered bridge. World Guide Covered Bridge Number: TN-78-F. Built in 2000 by Town of Pittman Center. Three-span bridge crossing Middle Prong of the Little Pigeon River. Stringer bridge 84.1 feet long by 20.0 feet wide, with approximately 11.5-foot wide by 12.5-foot high portal openings. Alternate Name: None known. Located in Pittman Center on Hills Creek Road.

A most impressive covered bridge, built in 2000 by the Town of Pittman Center, under the supervision of Steve McCarter and Garry Shultz, Emerts Cove Bridge derived its name from the area's first settler, Frederick Emert, who came here with his family in 1790. The three-span bridge crosses the Middle Prong of the Little Pigeon River, designated an Outstanding Natural Water Resource, in March 1998, by the State of Tennessee.

The 84.1-foot long stringer bridge rests on poured

A downstream view of the Middle Prong of the Little Pigeon River is seen through the triangle-pattern iron grille of Emerts Cove Bridge.

concrete abutments and two poured concrete piers, supported by four steel I-beam stringers under the roadbed and two steel I-beam stringers under the walkway. The asphalt roadway extends through the single-lane bridge; cement extends through the enclosed walkway, on the upstream side. Vertical boarding with battens covers the sides, the portals, and the partition inside the walkway, which separates the walkway from the traffic lane. Ten opposing, vertical, rectangular window openings extend down the sides and the internal partition, between the walkway and the traffic lane, all with black painted triangle-patterned iron grilles. Supported by the 2-inch by 4-inch wall studding, the steeply-peaked green-painted aluminum roof has three equally spaced, small cupolas on top. The interior traffic lane and walkway have a plywood-paneled ceiling. Steel guardrails pass through the bridge at both sides of the traffic lane, extending a considerable distance out the portals.

The east/west aligned bridge crosses a rock- and boulder-strewn river, which has densely foliated mature trees along the banks, more open on the downstream side. The Middle Prong of the Little Pigeon River feeds into the nearby Little Pigeon River, which enters the French Broad River, north of Sevierville. A bridge informational sign is near the bridge, at the right downstream or west portal, and a bridge construction credit sign is mounted on the left downstream portal.

Directions: From Gatlinburg, take US 321 north 6.3 miles to SR 416, on the left; go 0.5 mile to Hills Creek Road, on the left; go 0.1 mile to the bridge.

Harrisburg Covered Bridge

SEVIER COUNTY

Brief Statistics

Type: Authentic historic covered bridge. World Guide Covered Bridge Number: TN-78-01. Built in 1875 by Elbert Stephenson Early. One-span bridge crossing East Fork of the Little Pigeon River. Queenpost truss bridge 63.8 feet long by 14.5 feet wide, with approximately 11.8-foot wide by 10.1-foot high portal openings. Alternate Name: Pigeon Bridge. Located in Harrisburg on Old Covered Bridge Road.

Built in 1875 by Elbert Stephenson Early, a local resident who owned nearby Newport Mills, Harrisburg Bridge, named after the local community, is the oldest remaining authentic covered bridge in Tennessee. Also called Pigeon Bridge, after the Little Pigeon River, Harrisburg Bridge, along with Elizabethton Bridge, are the only two historic covered bridges in Tennessee still carrying motor traffic. The old single-span bridge had severely deteriorated and was destined for destruction, when it was saved and restored, in 1972, by the joint efforts of the Great Smokies Chapter and the Spencer Clack Chapter of the Daughters of the American Revolution, dedication ceremonies being held on June 14, 1972. Under the National Historic Covered Bridge Preservation Program, the Federal Highway Administration approved $96,392, on September 26, 2001, for current restoration work.

Harrisburg Bridge is supported by a non-modified, half-height queenpost truss, set on mortared cut-stone piers at each portal and a poured concrete pier, added at the center at a later date, likely 1972. Ramps extend between the piers and the poured concrete abutments. Bible Bridge is the only other queenpost truss covered bridge in Tennessee. Transverse 4-inch by 6-inch timbers, set on edge, form the flooring, which is laid on a combination of timbers and faced bark-covered-log floor joists. Gray weathered horizontal lapped siding covers the sides, a long horizontal window opening, consisting of six square segments, at the center of the upstream side and a long horizontal window opening, consisting of five square segments, at the center of the downstream side. The portals are uncovered, a bronze 1972 restoration dedication

This half-height queenpost truss still supports the traffic crossing the 63.8-foot long Harrisburg Bridge built in 1875.

plaque mounted above the left downstream entrance. Rusted tin covers the moderately peaked roof.

The 63.8-foot long bridge crosses high above the mud- and rock-bottomed East Fork of the Little Pigeon River, which flows westward into the nearby Little Pigeon River, before reaching the French Broad River. Harrisburg Bridge, aligned east-northeast/west-southwest, nestles among the heavy summer foliage of the dense woods along the river, with some homes nearby. An informational sign about the bridge is nearby, at the junction of Harrisburg Road and SR 339.

Harrisburg Bridge, the last historic covered bridge remaining in Sevier County, was listed on the National Register of Historic Places on June 10, 1975.

Directions: From Sevierville, the Sevier County seat, at the intersection of US 411/SR 35 and SR 66, go east 4.2 miles on US 411/SR 35 to SR 339, on the right; follow 0.9 mile to Harrisburg Road, on the right (bridge informational sign); follow to Old Covered Bridge Road; go right, to the bridge.

Hidden Mountain East Covered Bridge

Sevier County

Brief Statistics

Type: Non-authentic modern covered bridge. World Guide Covered Bridge Number: TN-78-E. Built in 1997 by Apple Valley Construction. One-span bridge crossing Serenity Brook. Stringer bridge 24.0 feet long by 12.5 feet wide, with approximately 12.5-foot wide by 12.4-foot high portal openings. Alternate Name: None known. Located in Sevierville on Walnut Way in Hidden Mountain East Resort.

Hidden Mountain East Bridge was built, in 1997, by Apple Valley Construction, in Hidden Mountain East Resort over Serenity Brook, whose waters reach the West Prong of the Little Pigeon River, a short distance downstream, continuing into the nearby Little Pigeon River and eventually emptying into the French Broad River. The single-span stringer bridge is supported by steel I-beam stringers, set on upright concrete-filled corrugated steel pipe abutments. Transverse planking makes up the floor. Horizontal, simulated-log boarding covers the sides, which have a large horizontal rectangular window opening with a window box and decorative shutters centered on each side. Vertical boarding with battens covers the portals, a colorful, oval, *Welcome to the Enchanted Forest* sign mounted on the left downstream portal, over the entrance. Tin, gleaming in the sun, covers the steeply peaked roof, which is held aloft by four wood posts along each side. At the portals, planter boxes extend out along each edge of the roadway to an electric lantern, mounted on a black metal post surmounting a cement base. Fountains are in the center of the brook adjacent to each side of the gray painted bridge.

The 24.0-foot long bridge is in an open area, in the middle of Hidden Mountain East Resort, where three roads converge, two roads running parallel to the brook, one on each side and Walnut Way crossing the brook through the bridge. Mountain resort log cabins abound around the bridge on the steep, heavily wooded hill slopes. Weeping willow trees droop over the left downstream portal, one on each side of the southeast/northwest-aligned bridge. From any vantage, the little bridge makes a very pretty sight. *See Hidden Mountain East Covered Bridge in the color photograph section: C-25.*

Directions: From Sevierville, at the junction of US 411 and US 441, go south on US 441 (Parkway) to Apple Valley Road, on the right; follow to Hidden Mountain Realty, at 475 Apple Valley Road. Pass the Realty Office and immediately take a left on Hickory Ridge Way, to a 180-degree right turn onto Walnut Way; follow as it winds its way to the bridge.

Covered Bridge

SUMNER COUNTY

Brief Statistics

Type: Non-authentic modern covered bridge. World Guide Covered Bridge Number: TN-83-A. Built in 1980 by Braxton Dixon. One-span bridge crossing Mud Hollow Creek. Stringer bridge 22.2 feet long by 14.0 feet wide, with approximately 12.3-foot wide by 8.8-foot high portal openings. Alternate Name: White House Bridge. Located in Hendersonville at 3016 New Hope Road.

Braxton Dixon built Covered Bridge on the driveway to his property, the stringer bridge completed in April 1978, with the cover added in 1980. Although a modern structure, the bridge was constructed with old timbers, deck planking and siding salvaged from old log cabins and barns. The rafter timbers (round beam with square ends) inside the portals are applewood, from racklifters salvaged from an old Michigan barn. The bridge also has been called White House Bridge, the name derived from the nearby springs, a former resort that attracted notables, including several United States Presidents, thus becoming know as the White House.

The 22.2-foot long stringer bridge rests on cement block abutments with a poured concrete cap, supported by two massive timber stringers placed outside two steel I-beam stringers. The deck or floor consists of transverse 2-inch by 12-inch planks. The sides have vertical boarding to rail height, open above; the gables also have vertical boarding, scalloped along the bottom edge. Rusted tin covers the roof, which has a small spire terminating in a ball, mounted at the center of the peak. The roof is supported by very large timber posts at the portals and a small double wood post at the center of each side. There is an ornamental roof brace at each side of both portals. The old weathered wood gives the bridge a very pleasing gray color, contrasting harmoniously with the rust red roof.

The single-span bridge crosses rocky-bottomed Mud Hollow Creek, which flows into Drakes Creek, just downstream, before entering the Cumberland River at Hendersonville. The west/east-aligned bridge is in a relatively open area at the driveway entrance, with New Hope Road at the west portal. The creek banks have light foliage, but a mature tree is near the bridge, on the west downstream side.

Directions: From White House, at the city limit sign on SR 258 southbound, go 3.4 miles to New Hope Road, on the left. The bridge is on the right, at No. 3016, just north of SR 258, and visible from the roadway. Private property: request permission to visit.

Sycamore Springs Covered Bridge

SUMNER COUNTY

Brief Statistics

Type: Non-authentic modern covered bridge. World Guide Covered Bridge Number: TN-83-B. Built in 2001 by Charles Haynes. One-span bridge crossing Bledsoe Creek. Stringer bridge 92.8 feet long by 18.9 feet wide, with approximately 11.4-foot wide by 14.3-foot high portal openings. Alternate Name: None known. Located in Bransford at 3712 US Highway 31E.

Charles Haynes built Sycamore Springs Bridge at the entrance to his Sycamore Springs property. The stringer bridge was completed on November 23, 1964, with the cover added in March 2001. The 92.8-foot long bridge is the longest non-authentic covered motor vehicle bridge in Tennessee.

The impressive stringer covered bridge is supported by four reinforced concrete slab stringers, set on poured concrete abutments and two piers, each pier constructed with three upright steel I-beams, with additional support from two diagonal steel I-beam braces, between the uprights. The flooring is the concrete slabs. Vertical boarding with battens, beginning to show signs of weathering, covers the sides and portals, the sides to rail height, boxed in. Three long rectangular openings are along each side, with vertical boarding above to the eaves. The siding on the portals wraps around into the interior. Green painted aluminum covers the steel-framed roof, which is supported by four steel girders along each side. The bridge has interior electric lighting and floodlights mounted in the gables. A security gate is at the north or left downstream portal, and a cattle gate is at the south portal.

The single-span bridge crosses wide, rocky-bottomed, normally placid Bledsoe Creek, which flows southerly into the Cumberland River, near Cairo. Open fields are on the south end, US Highway 31E is on the north end, and the creek is heavily foliated with mature trees along the embankments.

Directions: From east of Gallatin, at the junction of US 31E and US 231, go west 0,7 mile on US 31E, to the bridge, visible on the left, at the entrance to Sycamore Springs. Private property: request permission to visit.

Anderson Covered Bridge

WILLIAMSON COUNTY

Brief Statistics

Type: Non-authentic modern covered bridge. World Guide Covered Bridge Number: TN-94-C. Built in 1988 by Calvin Lehew. One-span bridge crossing Robinson Branch. Stringer bridge 16.9 feet long by 12.1 feet wide, with approximately 10.9-foot wide by 7.6-foot high portal openings. Alternate Name: None known. Located south of Boston off Johnson Hollow Road.

Calvin Lehew of Franklin, Tennessee, built Anderson Bridge, in 1988, over Robinson Branch, on a driveway that led up the hill to a cabin. *LEHEW* over *1988* is inscribed in the top of the concrete wing wall, on the downstream side, at the northeast end of the bridge. Anderson Bridge has the distinction of being the shortest non-authentic covered motor vehicle bridge, at 16.9 feet long, in Tennessee.

The stringer bridge is supported by eight 6-inch by 6-inch timber stringers, placed in two groups of four and set on poured concrete abutments with poured concrete wing walls. A transverse planked floor is fastened to the stringers. Weathered gray and black vertical 1-inch by 8-inch boards cover the sides and portals, the sides with two large window openings between the roof supports. A red-painted galvanized steel roof extends about four feet beyond the entrances, providing protection for the interior from inclement weather. Guardrails enclosed with vertical boarding extend the sides of the bridge, for the 4.7-foot length of the wing walls, at the northeast or left downstream entrance. A four-board rail fence, abutting the guardrails, leads traffic the thirty feet to the bridge from Johnson Hollow Road. A cattle guard gate secures the southwest portal.

A small unnamed creek joins gravel-bottomed Robinson Branch, just before it passes beneath the single-span bridge, on its way to Leipers Fork, at Boston, which enters the West Harpeth River, then the Harpeth River, before entering the Cumberland River, northwest of Ashland City. The bridge is situated in a lightly wooded area.

Directions: From south of Leipers Fork, at the junction of Pinewood Road (SR 46) and Old Hillsboro Road/Leipers Creek Road, go 5.6 miles south on Leipers Creek Road to Mobleys Cut Road, on the left; go 2.7 miles to the tee at Robinson Road/Johnson Hollow Road; go left 0.1 mile on Johnson Hollow Road to the bridge, visible on the right. Private property: request permission to visit.

Green Pastures Farm Covered Bridge

WILLIAMSON COUNTY

Brief Statistics

Type: Non-authentic modern covered bridge. World Guide Covered Bridge Number: TN-94-D. Built in 1995 by Eddie Dixon. One-span bridge crossing tributary to Little Harpeth River. Stringer bridge 24.7 feet long by 17.9 feet wide, with approximately 15.5-foot wide by 11.2-foot high portal openings. Alternate Name: None known. Located at Green Pastures Farm in Brentwood.

Eddie Dixon, General Contractor, built the cover for Green Pastures Farm Bridge, in 1995, over the existing abutments and stringer bridge of an earlier date, at Green Pastures Farm. The 24.7-foot long stringer bridge is supported by six steel I-beam stringers, set on a poured concrete abutment, on the south end, and rocks and concrete, on the north end. The flooring consists of longitudinal planking, fastened to timber joists placed across the stringers. Weathered vertical boarding covers the sides under the eaves and the gables, the gables arch cut at the bottom. The lower sides are open, with a handrail between the roof supports, which is constructed with 4-inch by 6-inch timbers, braced with an inverted kingpost design. Four 10-inch by 14-inch timbers support the shiny tin roof.

The bridge appears to be constructed from old wood.

The single-span bridge crosses a muddy-bottomed tributary to the Little Harpeth River, 100 yards downstream, which flows into the Harpeth River, south of Bellevue, before entering the Cumberland River, northwest of Ashland City. Open fields are at both ends of the north/south-aligned bridge. Mature trees form a wood line along the banks of the tributary.

Directions: Withheld at the owner's request.

Keeler's Covered Bridge

WILLIAMSON COUNTY

Brief Statistics

Type: Non-authentic modern covered bridge. World Guide Covered Bridge Number: TN-94-B. Built in 1997 by Joe and Linda Murphy. Five-span bridge crossing Leipers Fork. Stringer bridge 59.7 feet long by 12.4 feet wide, with approximately 10.9-foot wide by 10.0-foot high portal openings. Alternate Name: None known. Location withheld at owner's request.

In 1997, Joe and Linda Murphy added a red painted cover to an existing stringer bridge on the driveway to, at that time, their private residence, the bridge now known as Keeler's Covered Bridge. The five-span bridge crosses Leipers Fork, which flows into the West Harpeth River, then enters the Harpeth River, followed by the Cumberland River.

Five steel I-beam stringers, resting on poured concrete abutments and four poured concrete piers, support the 59.7-foot long stringer bridge. The flooring is transverse planking. Vertical boarding covers the sides, portals and interior. Five window openings, shaped like the upper half of a hexagon, are opposing along each side. Shiny tin covers the roof, which is supported along each side by six timber posts, the roof and gables extending beyond the entrances, providing shelter from the weather for the interior. Matching red painted rail fencing, at the right downstream or northwest portal, funnels traffic into the red painted bridge.

The picturesque bridge appears to be emerging from dense woods, as the southeast side of the creek has a low escarpment that is heavily wooded, enveloping the far portal of the bridge. The near portal is very open, with horse paddocks on the roadside of the long curving river-stone driveway. *See Keeler's Covered Bridge in the color photograph section: C-26.*

Directions: Withheld at the owner's request.

Leeson Covered Bridge

WILLIAMSON COUNTY

Brief Statistics

Type: Non-authentic modern covered bridge. World Guide Covered Bridge Number: TN-94-A. Built in 1968 by Jim Leeson. Two-span bridge crossing Big East Fork Creek. Stringer bridge 32.3 feet long by 11.9 feet wide, with approximately 10.3-foot wide by 7.8-foot high portal openings. Alternate Name: None known. Location withheld at owner's request.

Leeson Bridge was built, in 1968, by Jim Leeson at the entrance to, at that time, his property. The 32.3-foot long two-span bridge is supported by five wood utility poles, on the east-southeast span, and six wood utility poles, on the west-northwest span, resting on three piers, each constructed with five upright wood utility poles, cross braced, set on a poured concrete base. A ramp with low wood guard rails on each side is at the east-southeast end, extending from the mortared stone abutment to the bridge, the ramp supported in the center by a fourth pier. The pier at the west-northwest end of the bridge is placed next to the mortared stone abutment. The flooring consists of transverse planking with five-plank-wide wheel treads that extend to the end of the ramp. Vertical boarding covers the sides and gables, the sides having two opposing small rectangular window openings, with a vertical board shutter drop-hinged from the windowsill along each side. Bright tin covers the roof, which is held aloft by three wood utility pole posts along each side. An electrically operated security gate is in the center of the bridge.

The red painted bridge spans Big East Fork Creek, which flows into the South Harpeth River, thence into the Harpeth River, before entering the Cumberland River. The main road is at the east-southeast end of the bridge; open grassland is on the west-northwest end; the moderately wooded rock- and gravel-bottomed creek is in-between.

Directions: Withheld at the owner's request.

Cedarvine Manor Covered Bridge

WILSON COUNTY

Brief Statistics

Type: Non-authentic modern covered footbridge. World Guide Covered Bridge Number: TN-95-a. Built in 1995 by Jack George. One-span footbridge crossing cove to man-made pond. Stringer footbridge 37.4 feet long by 8.2 feet wide, with approximately 6.7-foot wide by 7.0-foot high portal openings. Alternate Name: None known.

Located in Lebanon at 8061 Murfreesboro Road at Cedarvine Manor.

Jack George built Cedarvine Manor Bridge, in 1995, over the cove of the man-made pond at Cedarvine Manor Events Facility & Bed and Breakfast. The 37.4-foot long stringer footbridge is supported by a two-axle flatbed trailer, set on the ground, forming the stringers for the bridge. The flooring consists of transverse boarding. The sides are open, with an old 3-inch by 4.5-inch timber handrail between the roof supports and braced by a radiating array of five small barkless logs, centered under each handrail section. The gables are enclosed with a 7-inch by 10-inch rafter below a 7-inch by 8-inch rafter, and diagonal cedar boards above. An old rusted metal roof is supported by five 4-inch by 6-inch timber posts enclosed with bark-covered cedar slabs. Old lumber is used in the construction of the bridge. Pieces of knotted wood and a woodcarving are mounted on each gable. There are two electric lights within the bridge and a large electric lamp mounted on the peak over the gable, on the east end.

The bridge is amidst the immaculately maintained grounds of Cedarvine Manor, which is used for activities such as weddings, reunions and corporate functions. The surroundings abound with well-placed antiques and vintage automobiles. The single-span footbridge is aligned east/west.

Directions: From south of Murfreesboro, at I-24, exit 81 onto US 231, go north 17.4 miles to 8061 Murfreesboro Road (US 231), at Cedarvine Manor, on the right. The bridge is not visible from the roadway. Private property: request permission to visit.

Virginia

At one time, Virginia had over one hundred covered bridges gracing its landscape and providing utility for the movement of goods and people. This number had dwindled to about fifty by 1936 and, in 1996, following the loss of Marysville Bridge in Campbell County, from a flood on September 6, only eight remained.

The covered bridges in Virginia now number twenty-six. Of this number, ten are authentic and sixteen are non-authentic, fifteen of the non-authentic being stringer type bridges and one being a post supported roof over a roadway, with all sixteen non-authentic having been built since 1977. The ten authentic bridges are comprised of eight historic bridges and two modern bridges—Bramley Bridge, in Carroll County, and Bridge at Schwabisch Hall, in James City County. These twenty-six bridges were constructed between 1857 and 2001 in eighteen of the ninty-five Virginia counties. The queenpost truss supports five of Virginia's authentic covered bridges, with the multiple kingpost truss supporting four and a Smith truss supporting one.

Virginia's eight historic covered bridges are all located along the I-81 corridor, in the following counties, in order of southbound travel: Shenandoah, Rockingham, Alleghany, Giles and Patrick. Meem's Bottom Bridge in Shenandoah County is the only historic bridge on a public highway that is open to motor traffic, and three historic bridges on private lands are open to private motor traffic.

Humpback Covered Bridge

ALLEGHANY COUNTY

Brief Statistics

Type: Authentic historic covered bridge. World Guide Covered Bridge Number: VA-03-01. Built in 1857 by Thomas McDowell Kincaid. One-span bridge crossing Dunlap Creek. Multiple kingpost truss 106.8 feet long by 15.5 feet wide, with approximately 12.2-foot wide by 11.5-foot high portal openings. Alternate Name: Old Humpback Bridge. Located in Dunlap Beach off SSR 600 in Humpback Bridge Wayside Park.

An interesting history of covered bridges precedes Humpback Bridge, the fourth in a series of covered bridges built at this site, all for the James River and Kanawha Turnpike. This turnpike had three different bridges crossing Dunlap Creek, where it "S" curves around Peters Mountain, all three along a one mile stretch of the turnpike. The site of Humpback Bridge was referred to as the lower Dunlap Creek bridge, as this bridge was the furthest downstream; the other bridges were referred to as middle Dunlap Creek bridge and upper Dunlap Creek bridge. A single-span, two-lane (double-barreled), arched kingpost truss bridge was the first bridge built at this site for the turnpike, by John Carruthers in 1824. Charles Callaghan covered this bridge later in 1824, and James Knox added a dry, river-stone pier to strengthen the bridge, in 1828. Probably due to poor construction, the bridge was rebuilt, in 1833, by David Johnson for $320. Then, on May 12, 1837, a devastating flood washed out all three bridges over Dunlap Creek. In May 1838, the second lower Dunlap Creek bridge was constructed, with improved masonry of dry, rough-cut stone for abutments and piers. However, masonry rebuilding was required in December 1841, albeit in vain, as on July 13, 1843, another disastrous flood washed out all three bridges along Dunlap Creek, a second time. A third bridge, utilizing a queenpost truss of 85-foot length, was built in 1843, at a cost of $350. This bridge collapsed in a freshet, in 1856, thus leading to the construction of the present bridge.

The oldest and the only pre-Civil War covered bridge remaining in Virginia, Humpback Bridge, the fourth bridge at this site, was built, in 1857, by Thomas McDowell Kincaid of Alleghany County, later a Captain in the Confederate Army, under contract with Mr. Venable, who was paid $1,500 for the construction. Interestingly, Mr. Venable received contracts to build the other two bridges on the James River and Kanawha Turnpike crossing Dunlap Creek; one prior to this bridge and the other during or immediately following the Civil War, both by Thomas McDowell Kincaid.

The bridge derives its name from the "humped" effect, caused by the extreme camber of the approximately four-foot rise, at the center of the bridge. The bridge has also been affectionately called Old Humpback Bridge and, at 106.8 feet long, is the longest single-span authentic covered bridge remaining in Virginia. This bridge is the only historic bridge with a multiple kingpost truss surviving in Virginia; however, a modern bridge, Bridge at Schwabisch Hall, in Busch Gardens in Williamsburg, also has a multiple kingpost truss.

Humpback Bridge was bypassed—and closed to traffic—in 1929, by an open, steel truss bridge on the far side of the Chesapeake and Ohio Railroad Bridge, just downstream, since replaced by the Interstate 64 concrete bridge. Abandoned, Humpback Bridge, for a time leased to a local farmer for hay storage, saw 24 years of neglect, until fund raising efforts of the Covington Business and Professional Women's Club and the Covington Chamber of Commerce, along with $3,000 from the State Highway Commission, achieved the $10,000 necessary to purchase the bridge and surrounding property and cover the costs of restoration. The acquisition occurred on February 28, 1953, and the bridge restoration was completed in 1954, with dedication ceremonies held on May 26, 1954.

Humpback Bridge has a ten-panel multiple kingpost truss, attached to arched upper and lower chords that create the "humped" effect. The truss and chords are the original hand-hewn oak timbers, fastened together by locust wood treenails. The single-span bridge is set on dry, rough-cut stone-block abutments, with long wing walls, mortar having been added where necessary, at a later date. The extremely cambered floor consists of transverse planks. Dark weathered horizontal lapped siding covers the sides, which have a wood-shingle-roofed skirt over the lower chord timbers, for weather protection. Weathered horizontal boarding encloses the gables, and wood shingles cover the portal sides. Black weathered, wood shingles cover the roof, which has steeply-peaked facades at the gables, an unusual feature.

Wide, rocky-bottomed Dunlap Creek glides beneath the unique Humpback Bridge, on its way to the Jackson River, at Covington. The bridge is now situated in Humpback Bridge Wayside Park, which offers parking, picnic tables, grills and trash receptacles. Mature trees are scattered about the well-maintained park. The Chesapeake and Ohio Railroad Bridge is at a comfortable distance downstream of the east-northeast/west-southwest aligned bridge. A bridge informational plaque is at the parking area; however, it gives an incorrect construction date.

The last authentic historic covered bridge remaining in Alleghany County, Humpback Bridge was the first historic covered bridge in the southeastern states to achieve recognition on the National Register of Historic Places, being listed on October 1, 1969. Covington is the county seat of Alleghany County. *See Humpback Covered Bridge in the color photograph section: C-30.*

Directions: From Covington, go west on I-64 to the US 60/SR 159 exit; go right 1.0 mile on US 60 to SSR 600; go right 0.1 mile to the bridge, in Humpback Bridge Wayside Park.

Bramley Covered Bridge

Carroll County

Brief Statistics

Type: Authentic modern covered footbridge. World Guide Covered Bridge Number: VA-18-b.* Built in 1983 by William Bondurant. One-span footbridge crossing Chances Creek. Smith truss footbridge 26.8 feet long by 9.3 feet wide, with approximately 7.7-foot wide by 7.2-foot high portal openings. Alternate Name: Bondurant Bridge. Located in Drenn at junction of Chances Creek Road and Forest Haven Drive.

William Bondurant built Bramley Bridge, in 1983, when he was the owner of the property. The bridge now carries the name of the present owner. Bramley Bridge holds the distinctions of being the shortest authentic covered footbridge, at 26.8 feet long, and the narrowest authentic covered footbridge, at 9.3 feet wide, in Virginia. Formerly called Bondurant Bridge, the footbridge is the only Smith truss covered bridge in Virginia.

Bramley Bridge is supported by an eight-panel Smith truss, fastened together and to the lower chord, upper chord, and secondary upper chord with treenails. The bridge is set on mortared natural stone abutments. The flooring consists of transverse planking with three-plank-wide wheel treads, the wheel treads to help replicate the appearance of an authentic historic bridge, as the date 1875 is carved over the southeast entrance, presumably the date of the historic bridge after which this bridge was modeled. The sides have horizontal boarding up to 20 inches above the flooring and open to the eaves, exposing the Smith truss. Weathered wood shingles cover the roof.

The single-span bridge crosses Chances Creek, which flows into nearby Crooked Creek, the latter entering the New River, near Riverhill. The northwest/southeast-aligned bridge is in the overgrown front yard among many trees. The northwest end appears to have been washed out in a flood, although it may have been only an approach ramp that was lost. In either case, at the northwest portal, two planks nailed to the truss end posts bar the entrance.

Directions: From Fancy Gap, at I-77, exit 8 SSR 775 (Chances Creek Road), go west 6.1 miles to Forest Haven Drive, on the left (0.6 mile past Misty Trail, SSR 700, on the left). The bridge is in the northeast corner of Forest Haven Drive and Chances Creek Road. Private property: request permission to visit.

**Assigned number is incorrect, as this is an authentic truss footbridge.*

Kanawha Valley Farm Covered Bridge

CARROLL COUNTY

Brief Statistics

Type: Non-authentic modern covered bridge. World Guide Covered Bridge Number: VA-18-A #2. Built in 2000 by B & H Bridge. One-span bridge crossing Burks Fork. Stringer bridge 52.1 feet long by 13.3 feet wide, with approximately 11.6-foot wide by 11.3-foot high portal openings. Alternate Name: None known. Located southeast of Dugspur off Dugspur Road (SSR 638).

The original Kanawha Valley Farm Bridge was built by J. C. Hendrix, in 1973. In 2000, faulty electrical wiring caused a fire that destroyed the bridge. The present Kanawha Valley Farm Bridge was built by B & H Bridge, in 2000, at Kanawha Valley Farm, a housing development for summer residents. Three steel I-beam stringers, set on the poured concrete abutments from the previous bridge, support the 52.1-foot long bridge. Transverse planking comprises the floor; horizontal lapped simulated-log-boarding covers the sides and the sides of the portal entrances; the gables are open. Brown painted aluminum covers the roof, which is supported along each side by 2-inch by 4-inch studding and extends beyond the entrances, providing additional shelter from the weather for the interior. Five buttress braces are evenly spaced along each side. Each side has four opposing horizontal window openings. A *B & H BRIDGE WT. LIMIT. 5 TONS* sign is mounted above the right downstream or east-southeast entrance.

Wide, deep and placid Burks Fork passes beneath the single-span bridge, on its way to nearby Big Reed Island Creek, which joins the New River, at Reed Junction. Kanawha Valley Farm Bridge is surrounded by open pasture for quarter horses, a large single-story barn in the downstream west-northwest corner.

Directions: From Hillsville, take US 221 east to Dugspur; go right 5.4 miles on Dugspur Road (SSR 638) to Kanawha Lane, on the right. The bridge, visible from the roadway, is 0.1 mile down Kanawha Lane.

Smith-Bailey Covered Bridge

CARROLL COUNTY

Brief Statistics

Type: Non-authentic modern covered bridge. World Guide Covered Bridge Number: VA-18-C. Built in 1986 by J. C. Hendrix. One-span bridge crossing Burks Fork. Stringer bridge 40.3 feet long by 12.2 feet wide, with approximately 10.0-foot wide by 11.6-foot high portal openings. Alternate Name: None known. Located south of Dugspur off Poplar Hill Road (SSR 662).

Smith-Bailey Bridge had a cover added, in 1986, over an older stringer bridge, by J. C. Hendrix, who also built the first Kanawha Valley Farm Bridge over Burks Fork, upstream of this bridge. Four steel I-beam stringers, set on poured concrete abutments that have railroad tie wing walls, support the 40.3-foot long stringer bridge. The flooring consists of transverse planking. Gray-weathered, vertical boarding covers the sides and the portals; the three buttress braces along each side are covered with gray-weathered diagonal boarding—the end buttresses flush with the portals. Each side of the bridge has two, opposing, large horizontal window openings. Shiny metal, with spots of rust, covers the low-peaked roof, which is supported by 2-inch by 6-inch studding.

The single-span bridge crosses rocky-bottomed Burks Fork, which flows into nearby Big Reed Island Creek, thence into the New River, at Reed Junction. Open fields are at each end of the north-northeast/south-southwest aligned bridge, which is a few hundred feet in from Poplar Hill Road, on the driveway to the residence. A ford crossing is upstream from the bridge. Bushes and trees line the creek banks at both sides of the bridge.

Directions: From Hillsville, go north on US 221 to Poplar Hill Road (SSR 662), on the right; go 2.9 miles to the bridge, on the left. Or, from Kanawha Valley Farm Bridge, continue 0.2 mile south on Dugspur Road to Fairhaven Road (SSR 628), on the right; go 1.0 mile to the first right, Burkes Ford Road (still SSR 628); go 2.1 miles to Poplar Hill Road (SSR 662); go right 0.4 mile to the bridge, on the right. The bridge is visible from the roadway. Private property: request permission to visit.

Cobblestone Covered Bridge

FAIRFAX COUNTY

Brief Statistics

Type: Non-authentic modern covered footbridge. World Guide Covered Bridge Number: VA-29-a. Built in 1984 by unknown builder. One-span footbridge crossing

Hidden Bridge Drive. Stringer footbridge 52.3 feet long by 6.3 feet wide, with approximately 4.6-foot wide by 7.3-foot high portal openings. Alternate Name: None known. Located north of Lorton off Pohick Road in Covered Bridge Townhouse Subdivision.

An unknown builder built Cobblestone Bridge, in 1984, in a single span over Hidden Bridge Drive, the entrance road to Covered Bridge Townhouse Subdivision. The 52.3-foot long stringer footbridge is supported by two 6-inch by 24-inch pine timber stringers, set on cement block abutments faced with mortared natural stone. Longitudinal 2-inch by 6-inch planks comprise the flooring. The sides and the portals are covered with vertical boarding, the sides open from waist height to under the eaves, for the length of the bridge, except for about four feet at each end, exposing four braced roof supports. Weathered wood shingles cover the steeply peaked roof, which is supported by six 4-inch by 6-inch wood posts down each side. The gray-green painted bridge has three electric lights in the interior. Split rail fencing flanks the walkway, at both ends of the bridge.

Cobblestone Bridge, at 6.3 feet wide, is the narrowest non-authentic covered footbridge in Virginia. Aligned north-northwest/south-southeast, the bridge sits high above the professionally landscaped entrance to the subdivision, flanked by the draping branches of two large trees. Pohick Road parallels the bridge on the west-southwest side.

Directions: From south of Fairfax, at the intersection of SR 123 and Fairfax County Parkway, go east on Fairfax County Parkway to Hooes Road; go right to Pohick Road; go left to Hidden Bridge Drive, on the left. Visible from Pohick Road, the bridge spans Hidden Bridge Drive.

Lake Anne Covered Bridge

Fairfax County

Brief Statistics

Type: Non-authentic modern covered footbridge. World Guide Covered Bridge Number: VA-29-b. Built in 1984 by unknown builder. One-span footbridge crossing wet weather stream. Stringer footbridge 28.0 feet long by 10.3 feet wide, with approximately 9.3-foot wide by 9.9-foot high portal openings. Alternate Name: None known. Located in Reston at Northgate Condominium on Northgate Square.

Named after nearby Lake Anne, Lake Anne Bridge was built, in 1984, by an unknown builder at Northgate Condominium, over a wet weather stream. Lake Anne Bridge, at 28.0 feet long, is the shortest non-authentic covered footbridge in Virginia.

The stringer footbridge is supported by seven small steel I-beam stringers, set on abutments, the west abutment constructed with a four-foot diameter concrete pillar placed horizontally on a poured concrete footing and the east abutment constructed with mortared cement blocks. The floor is constructed with transverse planks. The red painted bridge has vertical facsimile tongue and groove paneling on the sides, the portals, and the interior. The sides of the bridge extend out at the floor level, slanting back diagonally to the roof, forming wings at each end. These wings have a small vertical window opening, opposing, at each end. A large square window opening is centered on the downstream side, and a small square window opening, centered in a large vertical box inset, is at the midpoint, on the upstream side. The asphalt-shingled, low-peaked roof is supported by 2-inch by 4-inch studding along each side.

The single-span bridge sits in a densely wooded area, over a normally dry streambed amid asphalt paved walkways. The apartment complex is at the east end, the tennis court is in the west upstream sector, and an apartment building in the west downstream sector.

Directions: From Reston, on SR 267, go east to Hunter Mill Road (SSR 674); go left to Baron Cameron Avenue; go left to Wiehle Avenue; go left to North Shore Drive; go right to Northgate Square; and go left into Northgate Condominium. The bridge is on the path behind the complex, on the right. The bridge is not visible from the roadway.

Helen Lindsay Covered Bridge

Floyd County

Brief Statistics

Type: Non-authentic modern covered bridge. World Guide Covered Bridge Number: VA-31-A. Built in 1979 by unknown builder. Three-span bridge crossing Burks Fork. Stringer bridge 24.0 feet long by 12.0 feet wide, with approximately 8.5-foot wide by 6.5-foot high portal openings. Alternate Name: None known. Located southeast of Willis at 3275 Conner Grove Road (SSR 799).

An unknown builder constructed Helen Lindsay Bridge, in 1979, at the rear of the property over Burks Fork, which flows into Big Reed Island Creek, then enters the New River, at Reed Junction. Helen Lindsay Bridge, at 24.0 feet long, is the shortest non-authentic covered motor vehicle bridge in Virginia, although it is no longer open to motor vehicles.

The three-span stringer bridge is supported by two sets of triple 2-inch by 8-inch planks, placed on edge, for stringers, which are anchored to poured concrete footings at each end and two piers, constructed with three upright 4-inch by 4-inch wood posts, capped by a 4-inch by 4-inch wood crossbeam. The flooring consists of transverse planking. The sides are open, with wood benches along each side, the backs slanting outward and capped with a 1-inch by 6-inch board. The gables are enclosed with plywood, with battens spaced two feet apart to simulate vertical boarding with battens. The gray asphalt-shingled, low peaked roof is supported by four 4-inch by 4-inch wood posts along each side. All wood has weathered to a pleasing gray color.

Helen Lindsay Bridge has open pasture on the north-northeast or left downstream end, and woods and a pond on the south-southwest end. Thick foliage flanks the stream on the upstream side, and overgrown weeds flank the stream on the downstream side.

Directions: From Hillsville, take US 221 northeast to Willis to Conner Grove Road (SSR 799), on the right; go 6.2 miles to No. 3275, on the right. Or, from the Blue Ridge Parkway, take Conner Grove Road west 2.8 miles to No. 3275, on the left. The bridge is visible, at the rear of the property. Private property: request permission to visit.

Bridge Over Troubled Waters Covered Bridge

FRANKLIN COUNTY

Brief Statistics

Type: Non-authentic modern covered bridge. World Guide Covered Bridge Number: VA-33-A. Built in 1992 by T. Carmon Bennett. One-span bridge crossing tributary to Pigg River. Stringer bridge 40.2 feet long by 12.2 feet wide, with approximately 11.0-foot wide by 12.3-foot high portal openings. Alternate Name: None known. Located in Rocky Mount at 4634 Six Mile Post Road (SSR 640).

The Arrington's driveway to their residence, at the top of the hill, crossed a tributary to the Pigg River, which flowed through culverts under the asphalt-paved driveway. On three occasions, floods washed out the culverts, forcing the Arrington's to clamber up and down the slippery banks and wade across the tributary to access their home. To solve these difficult and dangerous times, Kearney W. Arrington designed a single-span covered bridge and, in June 1992, T. Carmon Bennett, a local carpenter, completed Bridge Over Troubled Waters Covered Bridge.

The 40.2-foot long stringer bridge is supported by four 18-inch steel I-beam stringers, set on poured concrete footings on the stream banks. Transverse 4-inch by 6-inch planking comprises the floor. Gray weathered vertical boarding covers the sides to rail height and under the eaves, and also covers the gables. The sides have seven openings between the roof supports, the top of each opening cut in an inverted V toward the eave. Red painted aluminum covers the low peaked roof, which is supported along each side by eight 4-inch by 6-inch wood posts. A copper and brass eagle weathervane, on top of a copper-roofed cupola, adorns the roof. The roof has gutters along each side, protecting the landscaping below from erosion. The bridge interior has electric lighting and an electronic motion sensor. A large ornate "A" for Arrington is mounted above the north or right downstream entrance.

The picturesque bridge is situated in the open, landscaped front yard, crossing a narrow, sediment- and rock-bottomed stream, flowing into the nearby Pigg River, which flows into Leesville Lake (Roanoke River), west of Pittsville. Woods are near the bridge, on the upstream side. A small plaque, inscribed *HATS OFF/TO/KEARNEY W. ARRINGTON/T.CARMON BENNETT/JUNE 1992*, is mounted on the north gable, centered over the entrance.

Directions: From Rocky Mount, go west on SR 40 to Six Mile Post Road (SSR 640); go right 2.9 miles to No. 4634, on the left. The bridge is visible from the roadway. Private property: request permission to visit.

C. K. Reynolds Covered Bridge

GILES COUNTY

Brief Statistics

Type: Authentic historic covered bridge. World Guide Covered Bridge Number: VA-35-03. Built in 1919 by James Maurice Puckett and Harvey Black Reynolds. One-span bridge crossing Sinking Creek. Queenpost truss bridge 36.0 feet long by 12.4 feet wide, with approximately 10.6-foot wide by 11.5-foot high portal openings. Alternate Names: Craig Bridge, Maple Shade Bridge, Red Maple Farm Bridge, Reynolds Bridge. Located north of Newport off SR 42 (Blue Grass Trail).

C. K. Reynolds Bridge, named after Clarence King Reynolds, is privately owned by the Reynolds Family and has also been known as Reynolds Bridge, Maple Shade Bridge, Red Maple Farm Bridge—presumably at one time the name of the farms—and Craig Bridge, the name derived from an early name for this area. James Maurice Puckett and Harvey Black Reynolds built the bridge in 1919. The 36.0-foot long bridge is the shortest authentic covered motor vehicle bridge remaining in Virginia.

C. K. Reynolds Bridge is supported by a half-height queenpost variant truss, modified with iron rods, very similar to the trusses in Clover Hollow Bridge and Link's Farm Bridge. Four steel I-beam stringers were added, in 1980, to strengthen the bridge. The bridge rests on dry natural stone abutments that have had poured concrete

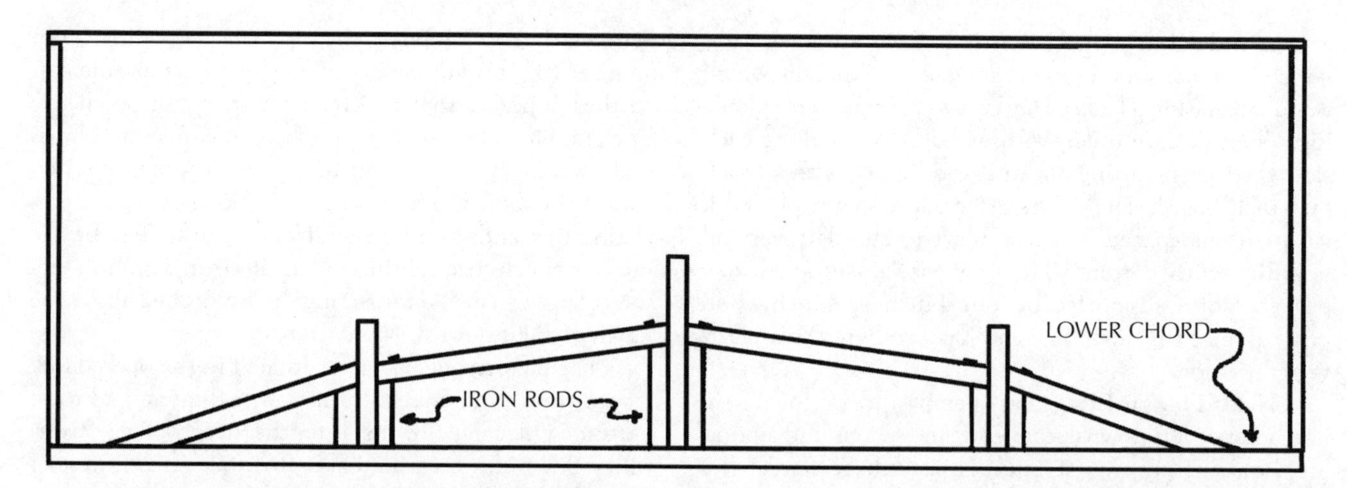

C. K. Reynolds Bridge queenpost variant truss modified with iron rods. ***See color photograph section: C-28.***

added, in 1988, in front of them. Transverse planking makes up the floor. Horizontal lapped siding covers the sides, weathered to a pleasant gray tone, and the gables, painted red. Rusted tin covers the roof, which is supported by 2-inch by 4-inch studding, as the half-height truss requires no upper chord.

The single-span bridge crosses rock-bottomed and boulder-strewn Sinking Creek, which empties into the New River, southeast of Pembroke. Open pasture runs up the hillside to State Road 42, on the northwest end of the bridge, with more open pasture and a homestead on the southeast or right downstream end. Thick foliage flanks the Sinking Creek, upstream of the bridge, and the downstream side has scattered trees. The bridge currently accommodates occasional farm vehicles and, more frequently, cows. C. K. Reynolds Bridge is one of three historic covered bridges remaining in Giles County: this bridge, Clover Hollow Bridge and Link's Farm Bridge. *See C. K. Reynolds Covered Bridge in the color photograph section: C-28.*

Directions: From Newport, take SR 42 (Blue Grass Trail) north, going 2.5 miles past the junction with Clover Hollow Road (SSR 601), on the left, to the bridge, on the right, across from No. 1581 Blue Grass Trail. The bridge is down the gated dirt road, at the far side of the pasture, not easily seen from the highway. Private property: request permission to visit.

Clover Hollow Covered Bridge

Giles County

Brief Statistics

Type: Authentic historic covered bridge. World Guide Covered Bridge Number: VA-35-01. Built in 1996 by James Maurice Puckett and Will Wingo. One-span bridge crossing Sinking Creek. Queenpost truss bridge 70.7 feet long by 14.3 feet wide, with approximately 11.8-foot wide by 10.5-foot high portal openings. Alternate Names: Newport Bridge, Sinking Creek Bridge. Located in Newport on bypassed section of Clover Hollow Road.

Built in 1916 by James Maurice Puckett and Will Wingo, Clover Hollow Bridge derived its name from its location area and Clover Hollow Road, which originally transited the bridge. The bridge has also been called Newport Bridge, as it is located in Newport, and Sinking Creek Bridge, for the creek passing beneath. Clover Hollow Bridge was bypassed, in 1949, and closed to traffic, in 1964. The bridge was restored during 1999 and 2000 and rededicated on September 23, 2000. Covered bridges from yesteryear are frequently selected as wedding sites, and weddings have been held in Clover Hollow Bridge.

The 70.7-foot long bridge is supported by a six-panel queenpost variant truss modified with iron rods, of the same design used in nearby Link's Farm Bridge, and is one of five queenpost truss bridges that have survived in Virginia. The bridge rests on mortared natural stone abutments and has transverse planking for a floor. Horizontal tongue and groove siding covers the sides and gables of the red painted bridge. Green painted tin covers the sharply peaked roof, which extends beyond the entrances, providing extra weather shelter for the interior. The construction date, 1916, in white numbers, contrasting vividly against the red gable, is over the right downstream or north-northeast entrance.

The single-span bridge is well exposed, over wide, rocky-bottomed Sinking Creek, which joins the New River, southeast of Pembroke. Dense woods are at the north-northeast end; a few trees line the south-southwest bank of the creek. Subscription bricks pave the way from Clover Hollow Road to the south-southwest portal. The

The magnificently maintained Clover Hollow Bridge has a subscription brick paved walkway leading into the bridge where you may sign the guest register. ***See Clover Hollow Covered Bridge in the color photograph section: C-29.***

one-lane bridge and immediate area is superbly maintained. The concrete bypass bridge is located a comfortable distance downstream. Clover Hollow Bridge displays a significant sag, acquired over the years of hard service. Visitors are encouraged to sign the guest register in the bridge.

Of the three historic covered bridges that have survived in Giles County, Clover Hollow Bridge is the only covered bridge open to the public. C. K. Reynolds Bridge and Link's Farm Bridge are privately owned. Pearisburg is the county seat of Giles County. *See Clover Hollow Covered Bridge in the color photograph section: C-29.*

Directions: From Newport, at the junction of US 460 and SR 42 (Blue Grass Trail), go north 1.0 mile on SR 42 to Clover Hollow Road (SSR 601), on the left, and follow 0.6 mile to the bridge, on the right, visible just off the roadway.

Link's Farm Covered Bridge

GILES COUNTY

Brief Statistics

Type: Authentic historic covered bridge. World Guide Covered Bridge Number: VA-35-02. Built in 1912 by James Maurice Puckett. One-span bridge crossing Sinking Creek. Queenpost truss bridge 49.0 feet long by 12.1 feet wide, with approximately 10.5-foot wide by 10.0-foot high portal openings. Alternate Names: Bradley Bridge, Mountain Lake Bridge. Located west of Newport on bypassed section of Mountain Lake Road (SSR 700).

James Maurice Puckett built privately owned Link's Farm Bridge in 1912. Although 1912 appears to be the accepted construction date, interestingly, the date 1913 is inscribed into the northwest or left downstream abutment. Link's, or Link, Farm Bridge, named after the present owner's farm, has also been called Bradley Bridge, after the present owner's grandfather and former owner, Samuel Alexander Bradley, and Mountain Lake Bridge, after Mountain Lake Road, which transited the bridge before it was bypassed in November 1949. Link's Farm Bridge, at 12.1 feet wide, is the narrowest historic covered motor vehicle bridge in Virginia. The old bridge still carries private motor vehicle traffic.

A six-panel queenpost variant truss, modified with iron rods, supports the 49.0-foot long bridge, the truss very similar to the queenpost trusses in C. K. Reynolds Bridge and Clover Hollow Bridge. The bridge rests on abutments constructed at the northwest end with mortared large natural stones, and at the southeast end with poured concrete, which was added, at a later date, in front of dry natural stones. The dry natural stone abutments appear to be from an earlier bridge. The flooring consists of transverse planking with two-plank-wide wheel treads. Horizontal lapped siding covers the sides and gables of the red painted bridge, the gables inverted V cut over the entrances. Green painted metal sheeting covers the roof, which is supported by six timber posts along each side, four of which are part of the truss.

All three remaining Giles County historic covered bridges cross Sinking Creek, whose waters flow under the single-span Link's Farm Bridge and join the New River, southeast of Pembroke. The rocky-bottomed creek bed is heavily wooded at the northwest end and upstream of the bridge, but open with scattered trees at the southeast end and downstream. A new bypass bridge, built in 1963, replaced the original 1949 structure and is located a very comfortable distance upstream.

Directions: From Newport, go west on US 460 to Mountain Lake Road (SSR 700), on the right; follow 0.3 mile to the bridge, on the left, at private Covered Bridge Lane. Private property: request permission to visit.

Stoney Run Covered Bridge

Henrico County

Brief Statistics

Type: Non-authentic modern covered bridge. World Guide Covered Bridge Number: VA-43-A. Built in 1979 by unknown builder. Post supported roof over roadway crossing Stoney Run. Bridge 61.3 feet long by 40.7 feet wide, with approximately 35.9-foot wide by 14.1-foot high portal openings. Alternate Name: None known. Located northwest of Richmond on Brookmont Drive.

Stoney Run Bridge was built, in 1979, by an unknown builder, as a post supported roof over asphalt-paved Brookmont Drive, over a concrete triple-square-conduit culvert, allowing the waters of Stoney Run to pass by. The cover has gray weathered 2-inch by 6-inch lumber, forming a lattice on the sides that extends from the roadbed to the eaves. Gray weathered vertical boarding covers the gables, which have long board signs, inscribed *STONEY RUN 1979*, mounted above each entrance. There is a 3.1-foot wide uncovered walkway on the downstream side. Weathered wood shingles cover the roof, which is supported by eight 8-inch by 8-inch timber posts, evenly spaced down each side. An electric light is mounted inside the bridge and on each gable.

The 61.3-foot long bridge has Stoney Run passing through the culvert beneath the roadway, on its way into nearby Deep Run, which flows into Tuckahoe Creek, before entering the James River at Bosher. At 35.9 feet wide, the bridge is the widest non-authentic covered motor vehicle bridge in Virginia. Located in a subdivision, the west/east-aligned two-lane bridge has heavily wooded empty lots in three corners, with a home in the west downstream corner.

Directions: From Richmond, take I-64 northwest to the North Parham Road exit; go left to Three Chopt Road; go right 2.4 miles to Church Road; go left 0.6 mile to Guyana Drive; go left 0.3 mile to Cox Road; go right 0.2 mile to Brookmont Drive; go right 0.3 mile to the bridge.

Bridge at Schwabisch Hall Covered Bridge

James City County

Brief Statistics

Type: Authentic modern covered footbridge. World Guide Covered Bridge Number: VA-47-01. Built in 1974 by unknown builder. One-span footbridge crossing Flume Lake. Multiple kingpost truss footbridge 105.8 feet long by 14.9 feet wide, with approximately 12.0-foot wide by 8.7-foot high portal openings. Alternate Name: Old German Bridge. Located in Busch Gardens in Williamsburg.

Built in 1974 in Busch Gardens by an unknown builder, Bridge at Schwabisch Hall Covered Bridge was modeled after a 59-foot long single-span covered footbridge, built in 1955, that crossed the meandering Kocher River, below Schwabisch Hall in Baden-Wurttemberg, in southern Germany. The footbridge provides access between Rhineland and New France, across Flume Lake, with Le Scoot log flume ride and Alpengeist ride passing beneath the bridge. The footbridge has also been called Old German Bridge. Bridge at Schwabisch Hall Footbridge has the distinctions of being the oldest authentic covered footbridge in Virginia, as well as being the longest, at 105.8 feet long, and the widest, at 14.9 feet wide.

The single-span footbridge is supported by a six-panel, single-post multiple kingpost truss, set on abutments constructed with simulated, mortared-rough-cut rock covering concrete. The west-northwest/east-southeast aligned bridge displays a significant camber. The flooring consists of longitudinal planking. Vertical facsimile tongue and groove paneling covers the sides up to rail height, open above to the eaves, exposing the kingpost truss. The gables of the brown stained bridge are also open. Weathered cedar shingles cover the roof. The bridge timbers are fastened together with the use of bolts and large steel plates. Electric lanterns are mounted at each side of the entrances and hung from the rafters, in the interior.

Directions: From Williamsburg, take I-64 south to exit 243A to Busch Gardens. The bridge is in the park, between Rhinefeld and New France, over the Alpengeist ride. Admission.

McDonald's Mills Covered Bridge

MONTGOMERY COUNTY

Brief Statistics

Type: Non-authentic modern covered bridge. World Guide Covered Bridge Number: VA-60-A. Built in 1992 by Jeff Ligon. One-span bridge crossing North Fork Roanoke River. Stringer bridge 41.4 feet long by 13.3 feet wide, with approximately 11.3-foot wide by 11.0-foot high portal openings. Alternate Name: David Kohl Bridge. Located in McDonalds Mill at 3755 Catawha Road (SSR 785).

Designed by David Kohl and built by Jeff Ligon, starting in the winter of 1991 and completing in the spring of 1992, McDonald's Mills Bridge, also known as David Kohl Bridge, is located in McDonalds Mill, over the North Fork Roanoke River. The North Fork Roanoke River merges with the South Fork Roanoke River, to give birth to the Roanoke (Staunton) River, at Lafayette. Two large steel I-beam stringers, alternating between three small I-beam stringers, set on poured concrete abutments faced with artificial natural stone, support the 41.4-foot long stringer bridge. Transverse planking makes up the flooring. Orange-brown-stained, vertical, tongue and groove boarding covers the sides and the portals, the sides open under the eaves, exposing the decorative Town lattice truss. Dark gray weathered wood shingles cover the roof, which is supported along each side by 2-inch by 4-inch studding and the decorative Town lattice truss. The roof and the gables extend beyond the entrances, to shelter the interior from the weather. Electric brass carriage lanterns are mounted over the entrances, on each side of the gables. Three electric lights are mounted inside the bridge. A sign made with black, wood cutout letters, mounted in an arc on the left downstream or northwest gable, read *McDONALD'S MILLS/COVERED BRIDGE/1991-92*. The interior of the bridge is festooned with a collection of automobile license tags from around North America.

The single-span bridge is set back from the roadway, on the graveled driveway to the residence, and is fully exposed, surrounded by open fields. There is landscaping around the bridge, with a large weeping willow tree, just downstream. A ford crossing is upstream of the bridge.

Directions: From Blacksburg, take US 460 west to Coal Bank Hollow Road (SSR 649); go right 1.6 miles to Mt. Tabor Road (SSR 624); go left 8.3 miles to Gallion Ridge Road (SSR 630); go right 1.5 miles to Catawha Road (SSR 785); go right to No. 3755 on the left. The bridge is visible from the roadway. Private property: request permission to visit.

Bob White Covered Bridge

PATRICK COUNTY

Brief Statistics

Type: Authentic historic covered bridge. World Guide Covered Bridge Number: VA-68-01. Built in 1921 by Walter G. Weaver. Two-span bridge crossing Smith River. Queenpost truss bridge 79.8 feet long by 12.5 feet wide, with approximately 10.8-foot wide by 10.0-foot high portal openings. Alternate Names: Lower Bridge, Woolwine Bridge. Located southeast of Woolwine on SSR 869.

Built in 1921 by Walter G. Weaver, Bob White Bridge acquired its name from the nearby former Bob White Post Office, which derived its name from the abundant Bobwhite quail in the area. The historic bridge has also been called Lower Bridge, presumably because it is downstream from Jack's Creek Bridge, and Woolwine Bridge, after the nearby community. The historic bridge was bypassed and closed to traffic in 1981. Bob White Bridge is one of five covered bridges, remaining in Virginia, with a queenpost truss.

The 79.8-foot long two-span bridge is supported by a white oak, half-height queenpost truss in each span and is anchored to poured concrete abutments and one centrally located poured concrete pier, with a pointed end facing upstream. The floor consists of transverse planking with two-plank-wide wheel treads. Gray weathered vertical boarding with battens covers the sides and the gables. The interior of the bridge is covered with a double layer of diagonal boarding overlaid in opposite directions, concealing the queenpost truss; however, with the assistance

Bob White Bridge was named after the nearby former Bob White Post Office that received its name from the abundance of Bobwhite Quail in the area.

of a flashlight, one may trace the truss timbers through the spaces between the boards. Dull tin, with patches of rust, covers the low peaked roof, which, along with the gables, extends beyond the entrances, providing protection for the interior from inclement weather.

The wide, rocky-bottomed Smith River passes beneath the two-span Bob White Bridge on its way to the Dan River, at Eden, North Carolina. The south-southwest/north-northeast aligned bridge is relatively open with a few scattered trees in three sectors, the downstream south-southwest sector heavily wooded, the dense woods blanketing the hillside beyond. The one-lane Bob White Bridge is now closed to traffic, the bypass bridge a considerable distance upstream. A typewritten bridge information sheet is mounted, behind glass, inside the north-northeast or right downstream portal, on the upstream side.

The Patrick County–owned bridge is one of two historic covered bridges remaining in the county. The county seat is Stuart. On May 22, 1973, Bob White Bridge and Jack's Creek Bridge were the first covered bridges in Virginia to be listed on the National Register of Historic Places.

Directions: From Woolwine, take SR 8 (Woolwine Highway) south about 1.5 miles to Elamsville Road (SSR 618), on the left; go 1.0 mile to SSR 869, on the right; go 0.1 mile to the bridge, behind the residences.

Clifford Wood Covered Bridge

Patrick County

Brief Statistics

Type: Non-authentic modern covered bridge. World Guide Covered Bridge Number: VA-68-A. Built in 1977 by Clifford Wood. One-span bridge crossing Smith River. Stringer bridge 41.9 feet long by 13.9 feet wide, with approximately 12.0-foot wide by 10.8-foot high portal openings. Alternate Name: None known. Located south of Woolwine at 8079 Woolwine Highway (SR 8).

Clifford Wood built Clifford Wood Bridge, in 1977, to gain access to his fields on the far side of the Smith River, whose waters pass under historic Jack's Creek Bridge, not too far upstream, and historic Bob White

Bridge, a little further away downstream. Clifford Wood Bridge is the oldest non-authentic covered motor vehicle bridge in Virginia. Closed to traffic by secured doors on the west-northwest portal, the bridge is presently being used for automobile storage.

The 41.9-foot long stringer bridge is supported by three large steel I-beam stringers, set on two piers. The east-southeast pier is constructed with two cement-filled, 55-gallon steel drums placed on a poured concrete base and capped with two steel I-beam crossbeams; the west-northwest or right downstream pier is constructed with two steel I-beam crossbeams placed on top of a poured concrete pier. Transverse-planked railless ramps extend from the poured concrete abutments to the piers. The flooring consists of transverse planking. Weathered vertical boarding with battens covers the sides and the portals, including the doors on the west-northwest end. Rusty-patched corrugated metal covers the roof.

Wide, rocky-bottomed Smith River passes beneath the single-span bridge, on its way through Philpott Reservoir to the Dan River, at Eden, North Carolina. The east-southeast riverbank is clear, with a few mature trees near the bridge, and lawn and fields extending to Woolwine Highway. The west-northwest riverbank is lightly wooded with thickets; open fields beyond extend to densely wooded hills. *See Clifford Wood Covered Bridge in the color photograph section: C-29.*

Directions: From Woolwine, go south on Woolwine Highway (SR 8) to No. 8079, on the right. The bridge is visible from the highway, toward the north rear of the property. Private property: request permission to visit.

Jack's Creek Covered Bridge

PATRICK COUNTY

Brief Statistics

Type: Authentic historic covered bridge. World Guide Covered Bridge Number: VA-68-02. Built in 1914 by Charlie Vaughan. One-span bridge crossing Smith River. Queenpost truss bridge 48.3 feet long by 13.8 feet wide, with approximately 11.0-foot wide by 11.5-foot high portal openings. Alternate Name: Upper Bridge. Located south of Woolwine on bypassed section of Jacks Creek Road (SSR 615).

Charlie Vaughan, a local carpenter from nearby Buffalo Ridge, built Jack's Creek Bridge in 1914. Jack's Creek Bridge derived its name from adjacent Jack's Creek Primitive Baptist Church and also has been known as Upper Bridge, presumably because it is located upstream from Bob White Bridge. The bridge is one of five queenpost truss covered bridges remaining in Virginia. The bridge was bypassed, in 1932, by a steel I-beam stringer bridge with a wood planked deck and subsequently closed to motor traffic. The present bypass bridge, with an asphalt covered concrete deck, was built in 1979, by the Virginia Department of Highways and Transportation. Jack's Creek Bridge had a new roof installed in 1969 and a major restoration by the Virginia Department of Highways completed in November 1974, at a cost of $4,554.14.

Jack's Creek Bridge is supported by a white oak, half height queenpost truss, anchored to poured concrete abutments. The flooring consists of transverse planking. The sides are covered with vertical boarding with full width board battens, and the gables are covered with vertical boarding without battens. The interior of the bridge is covered with a single layer of diagonal boarding, concealing the queenpost truss; however, with the assistance of a flashlight, one may trace the truss timbers through the spaces between the boards. The brown painted bridge has a shiny tin roof, which, with the gables, extends beyond the entrances, providing weather protection for the interior.

The 48.3-foot long single-span bridge crosses the steep-banked Smith River, which flows through Philpott Reservoir, before joining the Dan River, at Eden, North Carolina. The bypass bridge is too close to the upstream side of the covered bridge. The south/north-aligned bridge has the Baptist Church, up on the hill, and a grassed area, cleared to the river, on the north end; and large uplifted rock strata in the riverbed, on the downstream or east side. The south downstream riverbank has dense foliage, with an open field extending to State Road 8 beyond.

Jack's Creek Bridge has been owned by the Patrick County Historical Society since 1974. Only two historic covered bridges have survived in Patrick County, Jack's Creek Bridge and Bob White Bridge; these two bridges were the first covered bridges in Virginia listed on the National Register of Historic Places on May 22, 1973. Stuart is the county seat of Patrick County.

Directions: From Woolwine, go south about 2.3 miles on SR 8 (Woolwine Highway) to Jacks Creek Road (SSR 615), on the right; go 0.1 mile to the bridge, on the right side of the roadway.

Brumfield Covered Bridge

PITTSYLVANIA COUNTY

Brief Statistics

Type: Non-authentic modern covered footbridge. World Guide Covered Bridge Number: VA-69-b. Built in 2001 by Wayne Brumfield. Three-span footbridge crossing unnamed creek. Stringer footbridge 33.3 feet long by 6.6 feet wide, with approximately 3.8-foot wide by 6.8-foot high portal openings. Alternate Name: None known. Located west of Gretna off SR 40.

In 2001, Wayne Brumfield built Brumfield Bridge, in three spans, over the headwaters of a small unnamed creek, where the water is backed up from the creation of a small pond by a downstream earthen dam. The west-southwest/east-northeast aligned bridge is very exposed amid a few scattered mature trees.

The 33.3-foot long stringer footbridge is supported by two 2-inch by 8-inch wood stringers, attached to upright wood utility poles driven into the ground, two at each end of the bridge and two sets between, which act as piers. Earthen ramps are at each end of the bridge. The flooring consists of longitudinal 2-inch by 6-inch planking. The sides and the portals are open, a bench with a backrest running between the roof support posts, on the upstream side. White painted aluminum covers the low peaked roof, which is supported by the four sets of wood utility poles.

Directions: From Gretna, go west on SR 40 to just before SSR 1332, on the left, and SSR 672, just beyond SSR 1332. The bridge is visible, on the right, across from No. 1997 SR 40. Private property: request permission to visit at Compton & Nichols Plumbing, on the left just beyond the bridge.

Wards Farm Covered Bridge

Pittsylvania County

Brief Statistics

Type: Non-authentic modern covered bridge. World Guide Covered Bridge Number: VA-69-C. Built c. 1987 by Isaac (Ike) W. Ward. One-span bridge crossing tributary to Old Womans Creek. Stringer bridge 36.7 feet long by 14.5 feet wide, with approximately 13.1-foot wide by 10.7-foot high portal openings. Alternate Name: Dewberry Bridge. Located north of Gretna at 2632 Paisley Road.

Circa 1987, Isaac (Ike) W. Ward added a cover to an existing stringer bridge—the former state-owned stringer bridge was part of an old bypassed section of Secondary State Road 756, built sometime prior to 1950—creating Wards Farm Covered Bridge. The cover was added after the road was bypassed and transferred to Ike Ward, the road becoming an extension of the driveway to a residence on Ike Ward's property. The bridge has also been called Dewberry Road Bridge.

The 36.7-foot long stringer bridge is supported by eight steel I-beam stringers set on abutments. Each abutment is constructed with three upright wood utility poles capped with a 12-inch by 12-inch timber; the whole holding back stacked 12-inch by 12-inch timbers that form a retaining wall, mortared rocks forming wing walls. Outside the cover along each side is a guardrail, constructed with steel angle uprights capped with a steel I-beam and bolted to the steel I-beam stringers. This is all part of the original state bridge. The flooring consists of transverse planking with three-plank-wide wheel treads, added on top of the 3-inch by 4-inch decking from the original state bridge. Weathered vertical boarding with battens covers the sides and the portals, the sides below the guardrails covered with horizontal lapped siding. Red-painted galvanized steel covers the roof, which is supported by wood posts at each side of the entrances. The roof and the portals slant slightly outward, forming weather panels for the entrances.

The single-span bridge crosses a narrow rock- and sand-bottomed tributary to nearby Old Womans Creek, which flows northward about two miles into Leesville Lake (Roanoke River). The southeast/northwest-aligned bridge is fully exposed, with pastures on both sides of the curving dirt driveway. Wood rail and wire fencing extends along the driveway from the northwest portal to Paisley Road and along the northeast side of the driveway, on the southeast end. *See Wards Farm Covered Bridge in the color photograph section: C-30.*

Directions: From Danville, go north on US 29 to Gretna, continuing 6.5 miles past the intersection with SR 40 to Dewberry Road (SSR 756), on the left; follow 0.6 mile to the crossroads, continuing 1.8 miles straight through on Paisley Road (SSR 756) to No. 2632. The bridge is visible, on the left. Private property: request permission across the road to visit.

George A. Anderson Covered Bridge

Prince Edward County

Brief Statistics

Type: Non-authentic modern covered bridge. World Guide Covered Bridge Number: VA-71-A. Built in 1987 by Dr. Charles Anderson. One-span bridge crossing Dry Creek. Stringer bridge 38.1 feet long by 11.0 feet wide, with approximately 10.0-foot wide by 9.6-foot high portal openings. Alternate Name: None known. Located north of Farmville at 285 South Airport Road (SSR 637).

In August 1987, Dr. Charles Anderson completed the construction of George A. Anderson Bridge, that he named in memory of his father. The covered bridge is on a dirt entrance road to Dr. Anderson's Sandy Ford Farm, which has been in the family since 1743. The 11.0-foot wide bridge is the narrowest non-authentic covered motor vehicle bridge in Virginia.

The 38.1-foot long stringer bridge is supported by five steel I-beam stringers, designed by Dr. Anderson, with the

George A. Anderson Bridge was built in 1987 by Dr. Charles Anderson and named in memory of his father.

inner three stringers supporting the flooring and the outer two stringers supporting the cover, so that no vibrations from high-speed vehicles crossing the bridge passes to the walls. The stringers are set on poured concrete abutments, anchored by four 35-foot deep concrete pilings at each abutment, to secure in place, due to soft, wet ground. The abutments are faced with mortared red brick and have mortared red brick wing walls. Transverse planking forms the floor, and red-stained vertical boarding covers the sides and the portals. The sides have two adjoining, nearly square window openings on each side. Shiny galvanized steel covers the roof, which is supported by eleven 4-inch by 4-inch wood posts along each side. The roof and gables extend beyond the entrances, to provide weather protection for the interior.

Sandy bottomed and usually dry, Dry Creek passes beneath the single-span bridge, on its way to join the nearby Appomattox River, which enters the James River, at Hopewell. The west-southwest/east-northeast aligned bridge stands solitary along the straight dirt roadway, flanked by encroaching thick foliage of the creek banks, and surrounded by woods stretching to the horizon. Small open grassed areas are at each side of the roadway, on the east-northeast or right downstream end. A wood sign, inscribed *SANDY FORD/FARM/EST. 1743*, and a bronze plaque, inscribed *THE/GEORGE A. ANDERSON/ BRIDGE/By His Children/and Grandchildren/August 1987*, are mounted on the east-northeast gable, centered over the entrance.

Directions: From Farmville, at the intersection of SR 45, Business US 15 and Business US 460, go north 2.8 miles on SR 45 to South Airport Road (SSR 637), on the left; go 1.6 miles to No. 285 at Sandy Ford Farm, on the left. The bridge is down the dirt driveway, not visible from the roadway. Private property: request permission to visit.

Tacketts Mill Covered Bridge

PRINCE WILLIAM COUNTY

Brief Statistics

Type: Non-authentic modern covered footbridge. World Guide Covered Bridge Number: VA-73-b. Built in 1984 by unknown builder. Five-span footbridge crossing Tackett Creek. Stringer footbridge 71.9 feet long by 12.9 feet wide, with approximately 11.3-foot wide by 8.9-foot

The old restored grist mill viewed through the lower opening in the side of Brown's Bridge.

high portal openings. Alternate Name: Tackett Village Bridge. Located west of Occoquan in Tacketts Mill Shopping Center off Harbor Drive.

Tacketts Mill Bridge, also known as Tackett Village Bridge, was built in Tacketts Mill Shopping Center, in 1984, by an unknown builder. At 71.9 feet long and 12.9 feet wide, the stringer footbridge has the distinctions of being the longest and the widest non-authentic covered footbridge in Virginia.

The stringer bridge is supported by framed panels of 2-inch by 12-inch planks, set on abutments of timbers laid on rocks, and four piers, constructed with two cross-braced, upright wood utility poles capped with a large timber. The flooring consists of transverse planking. Vertical facsimile tongue and groove paneling covers the sides to rail height, open above to the eaves. The gables are also open. Weathered cedar shingles cover the roof, which is supported by eight 6-inch by 6-inch wood posts. The footbridge is painted light gray, on the exterior and the interior, and has three electric interior lights.

The five-span footbridge crosses Tackett Creek, dammed to form ponds at each side of the southwest/northeast-aligned bridge. Landscaped at both ends, the bridge provides access between the shopping center and office buildings at the far side of the ponds.

Directions: From Woodbridge, at the junction of US 1 and SR 123 (Gordon Boulevard), go north on SR 123 to Old Bridge Road, on the left; follow 1.5 miles to Harbor Drive, on the left; go 1 block to Tacketts Mill (Shopping Center), on the right. Go into the center, bearing left to the lower parking lot. The bridge is beyond the left corner of the parking lot.

Brown's Covered Bridge

ROANOKE COUNTY

Brief Statistics

Type: Non-authentic modern covered bridge. World Guide Covered Bridge Number: VA-77-A. Built in 1979 by H. Morton Brown. One-span bridge crossing Turner Branch. Stringer bridge 69.5 feet long by 13.7 feet wide, with approximately 10.9-foot wide by 10.3-foot high portal openings. Alternate Name: Pine Lake Bridge. Located southeast of Roanoke off Sterling Road (SSR 663).

H. Morton Brown added the cover to Brown's Bridge, on the long driveway to his residence, starting in 1977 and completing in 1979. The steel I-beam single-lane stringer bridge, with a wood-planked floor, was built in 1927, by the Virginia Department of Highways, on nearby Bandy Road (SSR 666), where it faithfully served local traffic for 44 years, until it was replaced by a two-lane bridge in 1971. The single-lane stringer bridge was, at that time, acquired by a local farmer, who set the bridge aside until 1977, when, after six years of haggling, he sold the bridge to H. Morton Brown. Mr. Brown contracted with Hawkins & Cox, of nearby Vinton, to move the stringer bridge and set it in place at its present location, in 1977.

The 69.5-foot long stringer bridge is supported by five steel I-beam stringers, set on the asphalt roadbed of an earlier driveway that serves as abutments. The flooring consists of transverse planking. Vertical boarding covers the sides on the exterior and the interior and the gables, the lower part of the sides open, exposing the timbers that support the roof. Vertical-board-covered skirting runs down each side, at floor level, protecting the flooring and stringers from the weather. Shiny galvanized corrugated steel covers the roof, which, along with the gables, extends beyond the entrances, affording protection for the interior from inclement weather. The brown stained bridge has a sign mounted over the southwest entrance inscribed, *Built by:/H. M. Brown/1979.*

Brown's Bridge, also known as Pine Lake Bridge, crosses Turner Branch in a single span, at the downstream side of the mill spillway dam. Turner Branch flows into nearby Back Creek, which enters the Roanoke (Staunton) River, at the Bedford County line. The southwest/northeast-aligned bridge has dense foliation along the driveway, on the northeast end, and open grassed areas, at the southwest stream banks.

A small pond is held back by the dam, originally built in 1911, washed out two years later, and rebuilt in 1914. The millpond supplied the waterpower for the grist mill, just downstream below the bridge, which operated until the 1940s, but now, in a restored condition, enhances the picturesque site. At 69.5 feet long, Brown's Bridge is the longest non-authentic covered motor vehicle bridge in Virginia.

Directions: From Roanoke, at the intersection of I-581/US 220 and SR 116, go south on SR 116 (becomes Jae Valley Road), just past the Blue Ridge Parkway (stone arch bridge) to Sterling Road (SSR 663), on the right; follow 0.7 mile to the first fork (actually, a paved private road, on the left), bearing left to the bridge. The bridge is not visible from the roadway. Private property: request permission to visit at the residence up the hill, beyond the bridge.

Biedler Farm Covered Bridge

ROCKINGHAM COUNTY

Brief Statistics

Type: Authentic historic covered bridge. World

Guide Covered Bridge Number: VA-79-01. Built in 1896 by Daniel Ulrich Biedler. One-span bridge crossing Smith Creek. Multiple kingpost truss with Burr arch bridge 92.3 feet long by 15.9 feet wide, with approximately 12.4-foot wide by 12.1-foot high portal openings. Alternate Name: Biedler Bridge. Located north of Tenth Legion at 13134 Valley Pike.

Daniel Ulrich Biedler built Biedler Farm Bridge, also called Biedler Bridge, in 1896, on Biedler Farm. The bridge carried old Secondary State Road 796 (now a private road) over Smith Creek and remained in the Biedler family until 1969. Biedler Farm Bridge and nearby Meem's Bottom Bridge are the sole surviving covered bridges in Virginia supported by a multiple kingpost truss with a Burr arch.

Biedler Farm Bridge has an eight-panel, single-post, multiple kingpost truss, sandwiched between a double half-height Burr arch, set on mortared, rough-cut basaltic stone abutments with wing walls. The floor pattern is unusual, with each end having a few transverse planks going into the bridge, then a few longitudinal planks along each side and diagonal planks filling in the center. Four-plank-wide wheel treads run the length of the diagonal planking. This flooring was installed in April 1990. Creosote-coated, horizontal lapped siding covers the sides and the portals, the siding enclosing the lower part of the walls inside the entrances. Bright tin covers the roof.

Smith Creek passes well below the single-span bridge, on its way into the North Fork Shenandoah River, at South Jackson. Dense, deciduous-wooded banks enshroud Smith Creek, and open pastures extend beyond the woods. The 92.3-foot long bridge aligns south-southeast/north-northwest along the tree-lined private road. Biedler Farm Bridge is the only historic covered bridge remaining in Rockingham County. *See Biedler Farm Covered Bridge in the color photograph section: C-28.*

Directions: From Mauzy, at the intersection of I-81 and US 11, go north on US 11 through Tenth Legion to SSR 796 (Newdale School Drive), on the left. At this point, take the private road, on the right, across from Newdale School Drive. The bridge is not visible from US 11. Private property: request permission to visit at the first house on the right.

Meem's Bottom Covered Bridge

Shenandoah County

Brief Statistics

Type: Authentic historic covered bridge. World Guide Covered Bridge Number: VA-82-01. Built in 1894 by John W. B. Woods. Four-span bridge crossing North Fork Shenandoah River. Multiple kingpost truss with Burr arch bridge 204.0 feet long by 19.3 feet wide, with approximately 15.8-foot wide by 13.5-foot high portal openings. Alternate Name: None known. Located in Mt. Jackson on Wissler Road (SSR 720).

John W. B. Woods, a master bridge builder of Shenandoah County, under contract from Franklin Hiser Wissler, who owned the property, built Meem's Bottom Bridge in 1894. The date 1894 is inscribed in a limestone block on the east-southeast abutment. The bridge obtained its name from the local area, called Meems Bottom, which derived its name from John G. Meem, who purchased 3,000 acres of the area in 1841. Originally built as a single-span truss bridge, privately owned by the Wissler family until the 1920s, the bridge was first acquired by the county and, then, in 1932, passed on to the state, which retains ownership today. Meem's Bottom Bridge has the distinctions of being the longest, at 204.0 feet long, and the widest, at 19.3 feet wide, authentic historic covered bridge remaining in Virginia.

Meem's Bottom Bridge was set on fire by vandals, on October 28, 1976, as a Halloween prank. The bridge was reconstructed, salvaging the majority of the truss timbers, by the Virginia Department of Transportation, using the badly damaged decking as a work platform, with completion of the reconstruction and reopening of the bridge in September of 1979. The cost of this reconstruction was $240,000.00. In 1982, a floor beam broke, causing the closing of the bridge again. It remained closed until 1983. During this time, three piers and steel I-beam stringers were added, technically converting the bridge from a single-span truss bridge to a four-span stringer bridge. The cost of these repairs was $140,000.00. Thankfully, the bridge has remained open to traffic since.

Meem's Bottom Bridge is one of two covered bridges remaining in Virginia, having a multiple kingpost truss with a Burr arch—the other being nearby Biedler Farm Bridge. The truss in Meem's Bottom Bridge is a sixteen-panel, single-post, multiple kingpost, sandwiched between a full-height laminated Burr arch, all timbers of local pine. The bridge is now supported by three large steel I-beam stringers, added in 1983, set on three poured concrete piers, also added in 1983, and poured concrete stringer piers, added in 1983, in front of the original, dry, rough-cut limestone block abutments. The original abutments have long, stepped, limestone-block, curved wing walls to retain the raised roadway to the entrances. The limestone blocks were quarried from the river bluff, about one mile upstream, at Rude's Hill. The flooring is constructed with laminated 2-inch by 4-inch pine lumber, set on edge and placed transversely, with four-plank-wide oak wheel treads. Treated orange-brown horizontal lapped pine siding covers the sides and the portals, the siding wrapped around

Built in 1894 by John W. B. Woods, 204.0-foot long Meem's Bottom Bridge is the longest covered bridge in Virginia.

inside the portals and the sides open under the eaves, to provide air circulation within. Two 2-inch by 12-inch planks run the length of the inside of the bridge along each side, forming wheel guards to protect the timbers from vehicles. Bright terne-coated stainless steel—terne is a lead and tin alloy—covers the roof.

The North Fork Shenandoah River passes well below Meem's Bottom Bridge, on its way to merge with the South Fork Shenandoah River, at Riverton Junction, to form the Shenandoah River, which enters the Potomac River, at Harpers Ferry. The swiftly flowing, wide, stone-bottomed river has dense woods on the west-northwest side and a wooded riverbank with agricultural fields extending beyond, on the east-southeast end. A parking area with picnic tables is nestled among some of the trees that form a canopy over Wissler Road, on the east-southeast end. A bronze bridge informational plaque is mounted inside the east-southeast portal, on the downstream side. A bridge informational sign is on the right side of US Highway 11 (Old Valley Pike), south of Wissler Road.

Meem's Bottom Bridge is the sole surviving historic covered bridge remaining in Shenandoah County. Woodstock is the county seat of Shenandoah County. Meem's Bottom Bridge was listed on the National Register of Historic Places on June 10, 1975.

Directions: From north of New Market, on I-81, take exit 269, SSR 730, east 0.5 mile to US 11 (Old Valley Pike); go left 0.9 mile to SSR 730 (Wissler Road); go left 0.4 mile to the bridge. Parking is available, on the left.

Village Motel Covered Bridge

SMYTH COUNTY

Brief Statistics

Type: Non-authentic modern covered footbridge. World Guide Covered Bridge Number: VA-83-a. Built in 1979 by unknown builder. One-span footbridge crossing Middle Fork Holston River. Stringer footbridge 36.4 feet long by 8.1 feet wide, with approximately 7.0-foot wide by 7.5-foot high portal openings. Alternate Name: None known. Located in Groseclose at Village Motel at 7253 Lee Highway (US11).

Village Motel Bridge was built in 1979 by an unknown builder, at the Village Motel in Groseclose. Surrounded by maintained grassed areas, the fully exposed footbridge has a clump of evergreen trees a short distance downstream and a cedar tree at the east-northeast end, near the bridge. The Village Motel and Restaurant are on the west-southwest side of the narrow, rocky-bottomed river, with the bridge providing access to a house and the Village Truck Stop, on the east-northeast side. An asphalt-paved walkway, lined with a white-painted split-rail fence, adding a nice touch to the bridge, leads to the west-southwest entrance; in contrast, a dirt path lined with an old weathered and lichen covered split-rail fence, partially fallen down and sections missing, adding a little character to the bridge, leads to the east-northeast entrance.

Two steel I-beam stringers, resting on cement block abutments, support the 36.4-foot long stringer footbridge. Longitudinal 2-inch by 6-inch planks constitute the floor, and plywood encloses the sides and the gables, the sides with three square window openings, opposing along each side. The red painted bridge has matching red asphalt shingles on the roof and *WELCOME,* in very large, yellow cutout letters, extending the length of the bridge, affixed on top of the shingles. The roof is supported by 2-inch by 4-inch studding, the roof and gables extending beyond the entrances, to provide protection from inclement weather for the interior. An electric light illuminates the interior, and a large electric lamp, mounted on each gable, illuminates the exterior, around the entrances.

The narrow, twisting Middle Fork Holston River passes beneath the single-span Village Motel Bridge, on its way to join the South Fork Holston River, near Alvarado, then merging with the North Fork Holston River, west of Kingsport, Tennessee, to give birth to the Holston River. The Village Motel Bridge is the oldest non-authentic covered footbridge in Virginia.

Directions: From Marion, go north on I-81 to exit 54, SSR 683 for Groseclose; go right to US 11 (Lee Highway); go right 0.1 mile to the Village Motel, on the right, at 7253 Lee Highway. The bridge is visible from the highway, between the motel and the Village Truck Stop.

Berea Church Road Covered Bridge

STAFFORD COUNTY

Brief Statistics

Type: Non-authentic modern covered bridge. World Guide Covered Bridge Number: VA-86-A. Built in 1987 by Ricky Carpenter. One-span bridge crossing Falls Run. Stringer bridge 44.9 feet long by 14.2 feet wide, with approximately 11.8-foot wide by 10.2-foot high portal openings. Alternate Name: None known. Located in Hartwood at 169 Berea Church Road.

Ricky Carpenter built Berea Church Road Bridge, in 1987, along the dirt driveway to the residence, at that time his property. Although still open to motor traffic, the bridge has not been used for many years.

The 44.9-foot long stringer bridge is supported by two large steel I-beam stringers, set on the ground. The flooring consists of rough-cut oak planking, abutted to poured concrete ramps at both entrances. The red painted bridge is enclosed on the sides with vertical particleboard paneling and, on the uniquely arced gables, with vertical facsimile tongue and groove paneling, flared out to fill in under the eaves. Each side has two opposing vertical window openings. The unusual roof, semicircular rather than peaked, is covered with galvanized corrugated steel and is supported by 2-inch by 4-inch studding. Two electric lights illuminate the interior.

Falls Run flows under the single-span bridge, southward bound for the Rappahannock River, at Fredericksburg. The south-southwest/north-northeast aligned bridge is just inside the margin of dense deciduous woods to the west. A large open field is off to the northeast. The bridge is semi-open on the downstream side, which faces Berea Church Road.

Directions: From the intersection of I-95 and US 17 (exit 133), north of Fredericksburg, go north 2.0 miles on US 17 to Berea Church Road, on the right; go 0.8 mile to the bridge, visible on the left, at No. 169. Private property: request permission to visit.

West Virginia

West Virginia had hundreds of covered bridges scattered over its picturesque landscape, many along its early turnpike systems and verdant valleys. This number had dwindled to eighty-nine, by 1947, and, ten years later, only fifty-eight remained. In the mid-1980s, seventeen historic covered bridges remained. The Civil War and the devastating flood of 1888 took a toll on covered bridges and progress in the mid-twentieth century replaced many with larger and stronger concrete and steel bridges. Fortunately, spectacular Philippi Bridge has survived, the only "double-barreled" covered bridge remaining in the southeastern states.

Of the twenty-four covered bridges in West Virginia, seventeen are authentic and seven are non-authentic modern stringer type bridges, with fifteen of the seventeen authentic being historic. These 24 bridges were constructed between 1852 and 2002 in seventeen of West Virginia's fifty-five counties. West Virginia's seventeen authentic covered bridges comprise a diversity of truss types, including Howe with Burr arch, kingpost, multiple kingpost, multiple kingpost with Burr arch, Long, Long with Burr arch, queenpost, queenpost/Long combination and double Warren.

West Virginia's historic covered bridges were built between 1852 and 1911. These bridges are spread around the state, nine in the north-central area, five in the south-central area, and three in the western part of the state. All are easy to locate and offer splendid scenery along West Virginia's byways and backroads, especially during the foliage season.

Carrollton Covered Bridge

BARBOUR COUNTY

Brief Statistics

Type: Authentic historic covered bridge. World Guide Covered Bridge Number: WV-01-02. Built in 1856 by Emmett J. O'Brien and Daniel O'Brien. Three-span bridge crossing Buckhannon River. Multiple kingpost truss with Burr arch bridge 156.5 feet long by 19.7 feet wide, with approximately 15.4-foot wide by 12.5-foot high portal openings. Alternate Name: Buckhannon River Bridge. Located in Carrollton on Carrollton Road (CR 36).

Deriving its names from the town in which it is located and the river it spans, Carrollton Bridge, also known as Buckhannon River Bridge, was originally built in 1856 by Emmett J. O'Brien and Daniel O'Brien, at a cost of $4,819. As originally built, the pre–Civil War bridge carried traffic of the day along the Middle Fork Turnpike, across the Buckhannon River, in a single span. In 1962, heavy vehicular traffic dictated a major change, and a new concrete slab bridge, supported by steel I-beams on two concrete piers and the original abutments, was constructed. The old wooden truss cover was then anchored upon this new bridge, thus preserving the historic Carrollton Covered Bridge for two generations and many more to come. The Highway Departments of the State of West Virginia and Barbour County built this new three-span bridge, at a cost of $50,000.

The new three-span bridge is set on dry, large sandstone-block abutments, with poured concrete, added in front of the abutments to anchor the Burr arch, and two piers, added in 1962, each constructed with two poured concrete pillars on a poured concrete base topped by a large concrete cap. The original plank floor was replaced with three reinforced concrete slabs, in 1962. The bridge has a fourteen-panel, single-post, multiple kingpost truss, sandwiched between Burr arches; the only other multiple kingpost truss with a Burr arch remaining in West Virginia is Barrackville Bridge, in Marion County. The red painted sides and abutment skirts are covered with vertical boarding with battens; the red painted portals are enclosed with horizontal boarding in the gables, the portals having ver-

Philippi Bridge is a two-lane or "double-barreled" bridge having an entrance lane and an exit lane. This is the right downstream portal view. ***See Philippi Covered Bridge in the color photograph section: C-32.***

tical boarding at the sides of the entrances, with three short diagonal boards in the upper corners. Shiny metal covers the roof, replacing the old wood shakes. A four-foot wide, internal, raised concrete sidewalk runs the length of the bridge, along the downstream side. A white-painted, four-bladed propeller-like cross is mounted above each entrance, contrasting against the red gable

Nine of the seventeen authentic covered bridges in West Virginia are still carrying motorized traffic, and Carrollton Bridge is one of these. Owned by the West Virginia Department of Transportation, the 156.5-foot long bridge is the second longest authentic covered bridge in West Virginia. Crossing the wide, rocky-bottomed Buckhannon River, which joins the Tygart Valley River three-quarters of a mile downstream, Carrollton Bridge has railroad tracks that follow the river, very near the left downstream or east-southeast portal, with dense woods running up the steep hillside, away from the bridge. Several homes nestled among trees are near the west-northwest portal, and many mature trees line the riverbanks.

Carrollton Bridge was listed on the National Register of Historic Places on June 4, 1981, simultaneously with ten other historic covered bridges in West Virginia. Carrollton Bridge and Philippi Bridge are the last remaining of the many covered bridges that once spanned the creeks and rivers of Barbour County. Philippi is the county seat of Barbour County.

Directions: From Philippi, at the junction of US 119 and US 250, go south 6.0 miles on US 119 to CR 36 (Carrollton Road), on the left, following 0.7 mile to the bridge, in Carrollton.

Philippi Covered Bridge

Barbour County

Brief Statistics

Type: Authentic historic covered bridge. World Guide Covered Bridge Number: WV-01-01. Built in 1852 by Lemuel Chenoweth and Eli Chenoweth. Four-span bridge crossing Tygart Valley River. Long truss with Burr arch bridge 301.1 feet long by 27.5 feet wide, with approximately 23.2-foot wide by 12.4-foot high portal openings. Alternate Name: Tygart River Bridge. Located in Philippi on US 250.

Representing the best in covered bridge architecture and design, Philippi Bridge was built, in 1852, by Lemuel

Chenoweth and Eli Chenoweth, at a cost of $12,181.24. Emmett J. O'Brien built the abutments and center pier. Originally built as a two-span toll bridge, carrying the Staunton-Parkersburg Turnpike across the Tygart Valley River, at Philippi, whence the bridge derives its name, the bridge, also called Tygart River Bridge, now carries US 250 across the river, the only covered bridge in the United States carrying a Federal Highway. Philippi Bridge also has the distinctions of being the oldest (1852), the longest, at 301.1 feet long, and the widest, at 27.5 feet wide, authentic covered bridge remaining in West Virginia. Philippi Bridge is the second longest and the widest authentic covered bridge in the southeastern United States.

During the Civil War, both Union and Confederate troops, at varying times, used the bridge. Philippi Bridge was the site of the first land engagement of the Civil War, when, on June 3, 1861, Union soldiers of the 7th Indiana Volunteers surprised the Confederate soldiers, under the command of Colonel George Porterfield, forcing a retreat. The Union forces then took command of the bridge and used it for their barracks.

Philippi Bridge suffered severe fire damage on February 2, 1989, and was rebuilt, retaining its original appearance. The bridge was reopened to traffic on September 16, 1991. This two-lane or "double-barreled" bridge has a fourteen-panel, single-post Long truss, sandwiched between a double Burr arch, for each of the original two spans, the trusses running along the outside walls as usual, with a third truss running down the center of the bridge, separating the two lanes. The Burr arches positioned along the outer sides of the Long trusses are visible along each side of the bridge, as the arches protrude beyond the siding. The bridge timbers were cut from native yellow poplar. The original two-span bridge rested on mortared, large sandstone-block abutments and a similarly constructed central pier. In 1934, four large steel I-beam stringers and two poured concrete support piers, centered under each of the original spans, were added, creating a four-span bridge, the additional piers necessary to accommodate the shorter length of the I-beam stringers. In 1938, concrete slabs replaced the wood plank flooring. These alterations became necessary to carry the heavier weight of the motorized vehicles of the period. An external aluminum-railed, cement floored, uncovered walkway was added, on the upstream side, during the 1934 alterations. The white painted bridge has horizontal lapped siding, leaving the arches exposed on the sides, and the same siding on the gables, the gables with arched cutouts above each side-by-side traffic lane. Red poplar wood shingles cover the roof, which, with the gables, extends beyond the entrances and exits, to provide a weather shelter for the portals. The bridge features interior lane lighting, a smoke detection system and a sprinkler system.

West Virginia has four Long truss bridges remaining, two of these with Burr arches, from the many that dotted the countryside in the past.

Owned by the West Virginia Department of Transportation, the pre–Civil War bridge crosses the wide, rock-bottomed Tygart Valley River, which merges with the West Fork River, at Fairmont, giving birth to the Monongahela River. The picturesque bridge is in a fully exposed setting, with small parks on both ends on the upstream side, which offer visitor parking. The downstream side is wooded, on the right downstream or west-southwest end, and open, on the east-northeast end.

On September 14, 1972, Philippi Bridge became the first historic covered bridge in West Virginia to be listed on the National Register of Historic Places. Only two historic covered bridges remain in Barbour County: Philippi Bridge and Carrollton Bridge. *See Philippi Covered Bridge in the color photograph section: C-32.*

Directions: The bridge is located in Philippi, the county seat, on US 250, at the junction with US 119.

Milton Covered Bridge

CABELL COUNTY

Brief Statistics

Type: Authentic historic covered bridge. World Guide Covered Bridge Number: WV-06-01. Built in 1876 by R. K. Baker. One-span bridge crossing pond. Howe truss with Burr arch bridge 114.1 feet long by 18.9 feet wide, with approximately 11.2-foot wide by 10.5-foot high portal openings. Alternate Names: Mud River Bridge, Sink's Mill Bridge. Located in Milton Pumpkin Festival Park at One Pumpkin Way.

Originally built in 1876 by R. K. Baker, on County Road 25 in Milton, over Mud River, Milton Bridge was disassembled in February 1998 and moved into storage at its new site, Milton Pumpkin Festival Park, less than one mile away, until bids could be solicited and a contract let for its reconstruction. The old bridge was rebuilt, in November 2001, by Ahern and Associates of South Charleston, West Virginia, utilizing new materials in the reconstruction, except for the truss and abutment sandstone blocks, which were salvaged from the original bridge and site. Milton Bridge, at its original site, was also known as Sink's Mill Bridge, after a nearby mill, and Mud River Bridge, after the waters passing beneath.

The present bridge is supported by a ten-panel Howe truss that had a four-ply laminated Burr arch added, during 1971 repairs, by the West Virginia Division of Highways. The Howe truss is constructed with double member braces, single member counterbraces, and double iron tension rods. The 114.1-foot long bridge is set on poured concrete abutments, faced with mortared large rectangular

Center Point Bridge, viewed from downstream, was built in 1890 and was privately owned until 1981.

sandstone blocks from the original site. Guard walls at the portals are constructed with leftover rectangular sandstone blocks, with the upper row placed on edge. Broken rock spreads out from the base of the guard walls, to prevent erosion. The flooring consists of transverse planking laid on top of diagonal planking. Vertical boarding with battens covers the sides and the gables, the sides open under the eaves to promote air circulation in the bridge. The red painted bridge has a bright and shiny new metal roof, which, along with the gables, extends beyond the entrances to form a weather panel, protecting the entrances from inclement weather.

The single-span bridge now spans a pond, the larger part with a small island on the northeast side. Restricted to foot traffic, an asphalt-paved walkway leads to the southeast portal and extends a short distance from the northwest portal. The bridge is fully exposed, with scattered mature trees nearby.

Milton Bridge is the final survivor of several historic covered bridges that, in bygone times, carried traffic across Cabell County waterways. Huntington is the county seat of Cabell County. Milton Bridge was listed on the National Register of Historic Places, under the name Mud River Bridge, on June 10, 1975.

Directions: From Milton, at I-64, exit 28, US 60, go south on Mason Road (CR 13) 0.5 mile to US 60; go right 0.5 mile to the traffic light at Fairgrounds Road, on the left; go 0.5 mile, bearing left on James River Turnpike; go 0.2 mile to the bridge, visible on the left, in Milton Pumpkin Festival Park, at One Pumpkin Way.

Center Point Covered Bridge

DODDRIDGE COUNTY

Brief Statistics

Type: Authentic historic covered bridge. World Guide Covered Bridge Number: WV-09-01. Built in 1890 by John Ash and S. H. Smith. One-span bridge crossing Pike Fork. Long truss bridge 44.0 feet long by 15.1 feet wide, with approximately 12.5-foot wide by 12.5-foot high portal openings. Alternate Name: None known. Located in Center Point off CR 10 (Pike Fork Road).

John Ash and S. H. Smith built Center Point Bridge in 1890, one of four Long truss bridges remaining in West Virginia, two of the four having Burr arches. Named after the community of Center Point, where it is located, the historic bridge carried public motor traffic until 1940, when the bridge became privately owned. In 1981, Ron and Jean Lackey, the previous owners, donated Center Point Bridge to the Doddridge County Historical Society, the present owners. Local community volunteers, in 1982, restored the 44.0-foot long bridge, which is no longer used by motorized traffic.

Center Point Bridge is supported by a four-panel Long truss, constructed with double timber braces and single timber counterbraces, fastened with iron bolts at the center, resting on dry, large sandstone-block abutments. The original underlying floor consists of diagonal planking, with an assortment of planks laid on top at a later date, most laid transversely with some longitudinal planks thrown in. Naturally gray weathered, vertical boarding covers the sides and the gables, the upstream side and the gables with battens. Uniformly rusted metal covers the roof, which, with the gables, extends a short distance beyond the entrances, affording extra weather protection for the interior. Two bridge informational plaques are mounted on the gable over the left downstream or north-northwest entrance.

The single-span bridge crosses Pike Fork, before it joins McElroy Creek, about 200 yards downstream, which then flows into Middle Island Creek, at Alma in Tyler County, thence into the Ohio River, at Middle Island above St. Marys. The fully exposed bridge has an expansive lawn area at the right downstream end and a building in the left downstream corner. A wooded hill rises behind the bridge, across Pike Fork Road.

The last historic covered bridge remaining in Doddridge County, Center Point Bridge was listed on the National Register of Historic Places on August 29, 1983. West Union is the county seat of Doddridge County.

Directions: From Salem, take SR 23 north to Center Point. Go to Pike Fork Road (CR 10), on the right. The bridge is on the right, just after the building on the right.

Red Wing Meadow Covered Bridge

Doddridge County

Brief Statistics

Type: Non-authentic modern covered bridge. World Guide Covered Bridge Number: WV-09-A. Built in 1984 by Roy Leggett. One-span bridge crossing South Fork Hughes River. Stringer bridge 43.2 feet long by 11.1 feet wide, with approximately 10.3-foot wide by 6.9-foot high portal openings. Alternate Name: None known. Located east of Oxford off South Fork Hughes River Road (CR 19-11).

Roy Leggett built Red Wing Meadow Bridge in 1984, with assistance on the abutment concrete work and I-beam stringers from Chase Petroleum Company. The 43.2-foot long and 11.1-foot wide private bridge has the distinctions of being the longest single-span and the narrowest non-authentic covered motor vehicle bridge in West Virginia.

Red Wing Meadow Bridge is supported by three steel I-beam stringers and two 12-inch diameter steel pipe stringers, set on abutments, each constructed with two upright 20-inch diameter steel pipes capped with two steel I-beam crossbeams. Three lengths of 12-inch diameter steel pipe are stacked horizontally behind each set of upright pipes, to form a retaining wall for the riverbanks. A 20.2-foot long cattle guard ramp is on the north end of the bridge. The flooring consists of 3-inch thick transverse planking with four-plank-wide wheel treads, the timbers harvested by Roy Leggett, from trees on his property. Gray weathered vertical boarding covers the sides and the portals, the west or upstream side dark gray and algae covered. The shiny-aluminum low-peaked roof is supported along each side by four 4-inch by 12-inch timbers, each flanked by 2-inch by 4-inch lumber. Each side of the bridge has three opposing horizontal window openings. Welded steel pipe handrails extend along each side of the cattle guard ramp, at the north end, and flare out from the south portal.

The South Fork Hughes River passes beneath the fully exposed bridge, to merge with the North Fork Hughes River, near Cisco, to form the Hughes River, which flows into the Little Kanawha River, at Greencastle. The south/north-aligned bridge is very close to South Fork Hughes River Road, at the south or left downstream end, with open fields on the north end. A ford crossing is about 100 feet downstream of the bridge.

Directions: From the crossroads in the center of Oxford, go east 0.3 mile on CR 19-11 to Taylor Drain Road (CR 19), on the right. Continue on CR 19-11, past Taylor Drain Road, 0.9 mile to South Fork Hughes River Road (still CR 19-11), on the right, following 0.6 mile to the bridge, visible just off the road, on the right. Private property: request permission to visit.

Herns Mill Covered Bridge

Greenbrier County

Brief Statistics

Type: Authentic historic covered bridge. World Guide Covered Bridge Number: WV-13-01. Built in 1884 by unknown builder. One-span bridge crossing Milligan Creek. Queenpost truss bridge 58.3 feet long by 13.2 feet wide, with approximately 8.9-foot wide by 11.8-foot high portal openings. Alternate Name: None known. Located northwest of Lewisburg on Herns Mill Road (CR 40).

Herns Mill Bridge was built at an approximate cost of $800 by an unknown builder, in 1884, to provide access to S. S. Herns Mill, for homesteaders on the far side of Milligan Creek. One of five queenpost truss bridges remaining in West Virginia, two of the five having a queenpost/Long combination truss, Herns Mill Bridge is owned by the West Virginia Department of Transportation. The historic bridge underwent major renovations, in 2000, that added steel I-beam stringers, concrete caps on the abutments to hold the I-beams, stone guard walls, new portal timbers, a new roof and new siding. The 13.2-foot wide bridge has the distinction of being the narrowest authentic historic covered bridge in West Virginia.

The 58.3-foot long bridge has a four-panel half-height queenpost truss, modified with iron rods at each side of the vertical tension members, and is supported by four steel I-beam stringers, added in 2000. The bridge rests on dry, large sandstone-block abutments, which have a poured concrete cap, added in 2000, at the top. Longitudinal wood planks make up the floor; brown stained vertical boarding with battens covers the sides and gables, the sides open under the eaves; shiny tin covers the roof. Slate-faced concrete guard walls, at both ends, funnel traffic into the single-lane bridge. A bridge information plaque is mounted on the right downstream guard wall. The queenpost horizontal truss member, on the upstream side, has the oldest carving found in the bridge, the initials *HEG* over *1895*, within a square.

The single-span bridge crosses rocky-bottomed and boulder-strewn Milligan Creek, which disappears into a sink, 0.4 mile southeast of the bridge. The downstream or southwest side of the bridge is heavily wooded on both sides of the creek. The creek widens out on the upstream or northeast side of the bridge, which is densely wooded. Asphalt-paved Herns Mill Road winds and curves through the woods, to and from the bridge. Parking pull-offs are at the right downstream or southeast end. Herns Mill Bridge is one of nine authentic covered bridges in West Virginia still open to motorized traffic.

Herns Mill Bridge and Hokes Mill Bridge are the only historic covered bridges remaining in Greenbrier County, known to have had over fourteen covered bridges in the past. Herns Mill Bridge was jointly listed with ten other historic covered bridges in West Virginia on the National Register of Historic Places on June 4, 1981.

Directions: From Lewisburg, the county seat, take US 60 west 3.4 miles to Bungers Mill Road (CR 40), on the left, following 1.0 mile to the bridge.

Hokes Mill Covered Bridge

GREENBRIER COUNTY

Brief Statistics

Type: Authentic historic covered bridge. World Guide Covered Bridge Number: WV-13-02. Built in 1899 by unknown builder. One-span bridge crossing Second Creek. Queenpost/Long combination truss bridge 82.0 feet long by 15.2 feet wide, with approximately 12.0-foot wide by 12.1-foot high portal openings. Alternate Name: None known. Located in Hokes Mill on bypassed section of Hokes Mill Road.

Hokes Mill Bridge was built in 1899, at a cost of $700, by an unknown builder, most likely the builders of Indian Creek Bridge, in adjacent Monroe County, the year before, as both bridges share the same unusual queenpost/Long combination truss. That bridge was built by Ray Weikel, Oscar Weikel, E. P. Smith and A. P. Smith. Hokes Mill Bridge, bypassed and closed in 1991, is in sad condition, but, fortunately, that will change, as $450,000 has been appropriated for restoration, begun in the spring of 2002 by the West Virginia Division of Highways.

The West Virginia Department of Transportation-owned bridge has a seven-panel queenpost/Long combination truss, modified with iron rods, with the queenpost horizontal member stretching across the center three panels, doing double duty as the upper chord also. The queenpost diagonal compression members extend across two panels at each end, with the Long truss under the three center panels and the first adjacent panel at each end. Vertical iron rods are placed at each side of the six vertical tension members of the Long truss, tying the queenpost to the lower chord. The truss rests on mortared, large rough-cut stone-block abutments, poured concrete having been added in front of the right downstream or south-southwest abutment. Sometime prior to 1959, two massive steel I-beams, acting as stringers, have been added inside the bridge, along each side, up against the truss, and bolted to six small steel I-beam cross-braces, placed under the lower chord to strengthen the bridge. The flooring is constructed from transverse 2-inch by 4-inch lumber, set on edge. Vertical boarding covers the sides and the gables, the sides with several boards missing here and there and the right downstream portal partially collapsed, taking part of the roof with it. The red painted bridge has a badly rusted tin roof, the roof and gables extending beyond the entrances, to provide additional protection from the weather for the interior. A metal sign, reading *HOKES MILL/BRIDGE/BUILT 1899*, in white letters on a green background, is mounted on the left downstream or north-northeast gable. Galvanized steel guardrail barriers are at each entrance.

Second Creek flows under the single-span 82.0-foot long bridge, on its way to the Greenbrier River, near Rockland, 2.0 miles downstream. The wide, rock-bottomed creek has wooded banks, upstream of the bridge, and a cliff, in the south-southwest corner, upon which a residence is built. The downstream side is open, except for a few trees near the bridge, with the concrete and steel bypass bridge a comfortable distance away. A bypassed section of narrow Hokes Mill Road leads to and from the bridge.

Only two historic covered bridges remain in Greenbrier County, soon to be restored Hokes Mill Bridge and nicely restored Herns Mill Bridge. Hokes Mill Bridge was listed on the National Register of Historic Places on June 4, 1981, simultaneously with ten other historic covered bridges in West Virginia. Lewisburg is the county seat of Greenbrier County.

Directions: From Ronceverte, go south on US 219, across Greenbrier River; then right on River Road (CR 48) 3.7 miles to a white church, on the right, where the road becomes Hokes Mill Road (CR 62); continue 1.4 miles to the bridge, visible from the highway, on the left.

Fletcher Covered Bridge

HARRISON COUNTY

Brief Statistics

Type: Authentic historic covered bridge. World Guide Covered Bridge Number: WV-17-03. Built in 1891 by Solomon Swiger and Lloyd E. Sturm. One-span bridge crossing Tenmile Creek. Multiple kingpost truss bridge 62.0 feet long by 14.8 feet wide, with approximately 12.7-foot wide by 12.6-foot high portal openings. Alternate Name: Tenmile Creek Bridge. Located north of Maken on CR 5-29.

Built in 1891, at a cost of $937.46, by Solomon Swiger and Lloyd E. Sturm, Fletcher Bridge received its name from the nearby Fletcher family. Owned by the West Virginia Department of Transportation, the 62.0-foot long bridge has also been known as Tenmile Creek Bridge, after the creek flowing beneath. Since its construction in 1891, the tin roof has been replaced with asphalt roofing paper; otherwise, the bridge has been unaltered and in continuous use. The oldest graffiti carving observed is dated 1918.

Fletcher Bridge is supported by a six-panel multiple kingpost truss, set on dry, large sandstone-block abutments, the stone obtained from a quarry south of the bridge. Diagonal planking forms the flooring; vertical boarding with battens encloses the sides and the gables, the sides open under the eaves, allowing air circulation; gray asphalt roofing paper covers the roof. The roof and gables extend beyond the entrances, forming a weather shelter panel for the entrances. The red paint on the bridge has faded, and weathered away in the more exposed areas.

The single-span bridge crosses Tenmile Creek, which runs northeastward into the West Fork River, at Haywood, and continues to the Tygart Valley River, at Fairmont. The normally placid, shallow creek has a rock-littered bottom and dense woods along both sides, near the bridge. Marshville Road (CR 5) is at the north or left downstream portal of the north/south-aligned bridge.

Fletcher Bridge, together with ten other historic covered bridges in West Virginia, was listed on the National Register of Historic Places on June 4, 1981. At one time covered bridges in Harrison County numbered in excess of sixty. Today, two remain: Fletcher Bridge and Simpson Creek Bridge.

Directions: From Clarksburg, the county seat, go west on US 50 through Maken to Marshville Road (CR 5), on the right; follow 1.5 miles to the bridge, on the left, at CR 5-29.

Simpson Creek Covered Bridge

HARRISON COUNTY

Brief Statistics

Type: Authentic historic covered bridge. World Guide Covered Bridge Number: WV-17-12. Built in 1881 by Asa S. Hugill. One-span bridge crossing Simpson Creek. Multiple kingpost truss bridge 79.5 feet long by 16.8 feet wide, with approximately 14.3-foot wide by 12.6-foot high portal openings. Alternate Names: Holland Bridge, Hollen Mill Bridge, Law Farm Bridge, W. T. Law Bridge. Located near Bridgeport on old CR 24-2 (Despard-Summit Park Road).

One of four multiple kingpost truss bridges in West Virginia, Simpson Creek Bridge was originally built, in 1881, by Asa S. Hugill, at a cost of $974, one-half mile upstream from the present site, on land owned by John Lowe. The West Virginia Department of Transportation-owned bridge survived the great flood of 1888, but was washed off its abutments in 1899. Salvaged, the bridge was rebuilt at the new location on new abutments, in 1899. The 79.5-foot long bridge has also been known as Holland Bridge, Hollen Mill Bridge, Law Farm Bridge and W. T. Law Bridge; presumably Hollen Mill and Law Farm were located nearby, and Holland may be a corruption of Hollen. The name Simpson Creek Bridge, obviously, is derived from the creek passing beneath, which obtained its name from John Simpson, a well-known peddler and Indian trader, who opened a trading post nearby, in 1764.

Simpson Creek Bridge is supported by a nine-panel multiple kingpost truss, which has a single, open center panel, and rests on dry or unmortared, large sandstone-block abutments. The floor consists of longitudinal planks, laid, sometime prior to 1988, on top of the original diagonally planked flooring. Vertical boarding with

View out the left downstream portal of Fletcher Bridge, built in 1891 and named after the nearby Fletcher family.

battens covers the sides and the gables of the red painted bridge. Shiny tin covers the roof, which, with the gables, extends a short distance beyond the portal end posts, forming a shelter panel for the entrances from inclement weather. Electric lighting has been installed inside the bridge. A bridge informational sign is at the right downstream or south portal.

Closed to motor traffic sometime prior to 1990, the single-span bridge crosses wide, rock-bottomed Simpson Creek, which joins the West Fork River, at Meadowbrook. The bypassed bridge has Meadowbrook Road at the north end and Despard-Summit Park Road at the south end, with the bridge, creek, and a small park in-between; the concrete bypass bridge, a short distance upstream, carries Despard-Summit Park Road across the creek. The creek has wooded banks downstream of the bridge, the upstream side being open, with a large sycamore tree near the bridge. The oldest graffiti in the bridge is *J.A.C./SEP 13/1905*, found carved on a truss member.

At one time, Harrison County had over sixty covered bridges, but that number has dwindled to only Simpson Creek Bridge and Fletcher Bridge, at present. Simpson Creek Bridge, along with ten other historic covered bridges in West Virginia, was listed on the National Register of Historic Places on June 4, 1981. *See Simpson Creek Covered Bridge in the color photograph section: C-31.*

Directions: From Clarksburg, the county seat, go east on US 50 to I-79; go north to exit 121, at CR 24 (Meadowbrook Road); go left on CR 24, under the overpass, about 0.3 mile to the first left, CR 24-2 (Despard-Summit Park Road). The bridge is on the right.

Sarvis Fork Covered Bridge

Jackson County

Brief Statistics

Type: Authentic modern covered bridge. World Guide Covered Bridge Number: WV-18-01 #2. Built in 2000 by unknown builder. One-span bridge crossing Left Fork Sandy Creek. Long truss with Burr arch bridge 101.8 feet long by 15.7 feet wide, with approximately 10.3-foot wide by 10.8-foot high portal openings. Alternate Names: New Era Bridge, Odaville Bridge, Sandyville Bridge. Located northeast of Sandyville on Sarvis Fork Road (CR21-15).

R. B. Cunningham originally built Sarvis Fork Bridge in Angerona, at Carnahan's Ford, in 1889, at a cost of $1,050. The bridge was disassembled and re-erected at its present site on September 10, 1924. In 1969, a truck drove across the bridge, breaking through the deck planks. As a result, structural steel supports were added under the lower chords and two steel piers were also added. The segmented single Burr arch was added in the 1980s. In 2000, Sarvis Fork Bridge was completely rebuilt, using new materials and adding the steel I-beam stringers, removing the earlier steel supports and piers. Sarvis Fork Bridge derives its name from Sarvis Fork, which joins Left Fork Sandy Creek, 0.1 mile downstream. The West Virginia Department of Transportation-owned bridge has also been called New Era Bridge, after the nearest community southwest of the bridge; Odaville Bridge, after the nearest community northeast of the bridge; and Sandyville Bridge, after the nearest large community, located next to New Era.

Sarvis Fork Bridge is one of four covered bridges in West Virginia with a Long truss, two having a Burr arch. The 101.8-foot long bridge has a thirteen-panel Long truss, constructed with double-member braces, and a segmented single Burr arch, added in the 1980s. The bridge is set on mortared, large rough-cut sandstone-block abutments with small wing walls. *F.F.R./SEP/10, 1924*, the date the bridge was relocated, is chiseled in a sandstone block, at the left downstream or northwest abutment. Four large steel I-beam stringers were added, in 2000, to support the bridge. The flooring consists of transverse 2-inch by 4-inch lumber, placed on edge, which replaced the previous flooring, around 1988. The red painted bridge has vertical boarding on the sides and the gables. Shiny tin covers the roof, which, with the gables, extends beyond the portal end posts, to provide weather protection for the entrances. A bridge data sign is mounted on the gable, on the left downstream portal.

Left Fork Sandy Creek passes beneath the single-span bridge, before merging with Right Fork Sandy Creek, just south of Sandyville, to form Sandy Creek, which flows into the Ohio River, at Ravenswood. The northwest/southeast-aligned bridge is nicely exposed, with woods at the southeast end. Clumps of trees are scattered along the creek banks near the bridge. A bridge informational sign is at Sarvis Fork Road and County Road 21. Sarvis Fork Bridge is open to traffic, as are eight other authentic covered bridges in West Virginia. Graveled Sarvis Fork Road passes through the bridge.

Only two authentic covered bridges remain in Jackson County, this bridge and Staats Mill Bridge. On June 4, 1981, this bridge and ten other historic covered bridges in West Virginia were listed on the National Register of Historic Places. The county seat of Jackson County is Ripley.

Directions: From Sandyville, go north on CR 21 to New Era, where CR 21 bears a sharp right; follow 1.2 miles to Sarvis Fork Road (CR 21-15), on the right. Go 0.3 mile to the bridge.

An old painted advertisement on the upper guardrail in the Staats Mill Bridge solicits LADIES AND GENTS FINE SHOES AT A. M. CARSON CO. STORE. The bridge was built in 1887. ***See Staats Mill Covered Bridge in the color photograph section: C-32.***

Staats Mill Covered Bridge

Jackson County

Brief Statistics

Type: Authentic historic covered bridge. World Guide Covered Bridge Number: WV-18-04. Built in 1887 by Henry T. Hartley. One-span bridge crossing pond. Long truss bridge 101.6 feet long by 15.1 feet wide, with approximately 11.1-foot wide by 11.1-foot high portal openings. Alternate Name: None known. Located south of Ripley at Cedar Lakes Conference Center.

Staats Mill Bridge was originally built over Tug Fork in Statts Mills, on County Road 36, in October 1887. The wooden bridge was built by Henry T. Hartley, a local builder, at a cost of $903.95; the abutments and stone work built by Quincy and Grimm, local masons, at a cost of $710.40; and fill work and the approaches prepared by Enoch Staats, the mill owner, at a cost of $110—the total bill amounting to $1,724.35. The bridge acquired its name from Staats Mill, owned by Enoch Staats, which was located close to the bridge. In July 1982, the bridge was moved approximately 7.2 miles to Cedar Lakes Conference Center, near Ripley, where it was restored and re-erected, at a cost in excess of $104,000. Dedication ceremonies were held on June 29, 1983.

The 101.6-foot long bridge has a twelve-panel Long truss, constructed with double timber vertical tension members, double timber braces and single timber counterbraces, modified by the addition of single iron tension rods adjacent to each vertical tension member. Owned by West Virginia Department of Education, the bridge rests on new, mortared, large sandstone-block abutments. The floor is transverse planking. The red painted bridge has vertical boarding on the sides and the gables, the sides open under the eaves, to promote air circulation within the bridge, and the gable boarding scalloped along the lower edges. A shiny tin roof completes the structure. Three old advertisements, in excellent condition, are painted on the upper guardrail, which runs the length of the interior.

The single-span bridge crosses an elongated duck pond. Now closed to motor traffic, the historic bridge is in a very exposed setting, with mature cedar, maple and catalpa trees scattered in the vicinity of the south/north-aligned

Red painted Walkersville Bridge is affectionately called Old Red Bridge. The bridge, built in 1902, spans the Right Fork of the West Fork River. This is a downstream view.

bridge. A view of Heritage House, an early nineteenth century two-story log cabin, seen across the field through the north portal, makes a great photograph. A bridge informational sign is located near the parking area for the bridge.

One of two authentic covered bridges remaining in Jackson County, Staats Mill Bridge was listed on the National Register of Historic Places on May 29, 1979. The county seat of Jackson County is Ripley. *See Staats Mill Covered Bridge in the color photograph section: C-32.*

Directions: From I-77, north of Fairplain, take exit 132, at CR 21, going north and immediately go right on Cedar Lakes Road, 0.4 mile to CR 25 (left sharp curve); go left 2.9 miles on CR 25 to Cedar Lakes Conference Center, on the right. The bridge is on the left, just inside the entrance.

Walkersville Covered Bridge

Lewis County

Brief Statistics

Type: Authentic historic covered bridge. World Guide Covered Bridge Number: WV-21-03. Built in 1902 by John G. Sprigg. One-span bridge crossing Right Fork of the West Fork River. Queenpost truss bridge 39.8 feet long by 15.4 feet wide, with approximately 12.1-foot wide by 11.0-foot high portal openings. Alternate Name: Old Red Bridge. Located in Walkersville on Big Run Road (CR 19-17).

John G. Sprigg built Walkersville Bridge, in 1902, at a cost of $567.00, using a queenpost truss, one of five queenpost truss bridges remaining in West Virginia, two of them in combination with a Long truss. Walkersville Bridge, located in Walkersville, has also been affectionately called Old Red Bridge, referring to the historic bridge's red colored paint. The West Virginia Department of Transportation-owned bridge has had $260,000 approved, on September 26, 2001, by the Federal Highway Administration, for restoration under the National Historic Covered Bridge Preservation Program. The bridge was last repaired and repainted in 1963.

The 39.8-foot long queenpost truss bridge is set on dry, rough-cut, large sandstone-block abutments with long wing walls. The flooring is constructed with transverse planking that has a six-plank-wide wheel tread on the downstream side and a five-plank-wide wheel tread on the upstream side. The red painted bridge has vertical boarding with battens on the sides and portals. The shiny tin covered roof and the gables extend beyond the entrances, to provide weather protection for the interior.

The Right Fork of the West Fork River flows under the single-span bridge, joining the West Fork River 0.75 mile downstream. Aligned east-northeast/west-northwest, the bridge is very exposed, with a few mature trees at the ends. The old bridge still carries Big Run Road traffic and is one of the nine historic covered bridges in West Virginia still open to motorized traffic.

Listed on the National Register of Historic Places, simultaneously with ten other historic covered bridges in West Virginia, on June 4, 1981, Walkersville Bridge is the only historic covered bridge remaining in Lewis County. The county seat of Lewis County is Weston.

Directions: From Walkersville, take US 19 south 1.3 miles after the green steel truss bridge, just after Crane Camp Run Road (CR 44-7), on the right, to Big Run Road (CR 19-17), on the right. Follow 0.2 mile to the bridge.

Saunders Covered Bridge

Lincoln County

Brief Statistics

Type: Non-authentic modern covered footbridge. World Guide Covered Bridge Number: WV-22-a. Built in 1979 by Fred R. Saunders. One-span footbridge crossing tributary to Middle Creek. Stringer footbridge 27.9 feet long by 4.5 feet wide, with approximately 3.9-foot wide by 6.4-foot high portal openings. Alternate Name: None known. Located east of Hamlin off SR 34.

Saunders Footbridge was built by Fred R. Saunders, in 1979, in the front yard of his residence, over a small stream running adjacent to State Road 34. The small stream is a tributary to nearby Middle Creek, near its confluence with the Mud River, which flows into the Guyandotte River, at Barboursville, before entering the Ohio River, at Huntington. The 27.9-foot long bridge is fully exposed, except for an apple tree on the downstream side. Saunders Bridge is the oldest covered footbridge in West Virginia.

The single-span footbridge is supported by four 2-inch by 8-inch wood stringers, set on rocks and earth. The flooring is transverse 2-inch by 6-inch planks. The sides have a naturally weathered, fine mesh lattice attached to fifteen 2-inch by 4-inch wood posts along each side, which support the gray asphalt-shingled roof. There is a handrail on both sides of the interior of the northwest/southeast-aligned bridge. A weathervane is mounted on the center of the low peaked roof.

Directions: From Hurricane, go south on SR 34, continuing 2.2 miles past Big Creek Road (CR 34-2), on the left, to the bridge, on the right. Or, from Hamlin, go east on SR 3 to the junction with SR 34; go left 0.5 mile on SR 34 to the bridge, visible from the roadway, on the left. Private property: request permission to visit.

Michael McDonald Covered Bridge

Logan County

Brief Statistics

Type: Non-authentic modern covered bridge. World Guide Covered Bridge Number: WV-23-A. Built in 1989 by Michael McDonald. One-span bridge crossing Mud Fork. Stringer bridge 16.4 feet long by 16.2 feet wide, with approximately 15.1-foot wide by 9.3-foot high portal openings. Alternate Name: None known. Located in Verdunville off Harts Road.

Michael McDonald built Michael McDonald Bridge, in 1989, on the cement driveway at the entrance to his property, off Harts Road. A heavily wooded steep hill rises across the road from the bridge, which is fully exposed in the front yard of the residence. The single-span bridge crosses narrow Mud Fork, which joins Copperas Mine Fork, at Ridgeview, thence flowing into the Guyandotte River nearby.

The 16.4-foot long stringer bridge is supported by four steel I-beam stringers, set on stuccoed cement-block abutments. The flooring is transverse planking; the sides and the gables have dark weathered vertical boarding with battens; the roof has reddish-brown asphalt shingles. The sides are open under the eaves to promote air circulation within the bridge. The low peaked roof is supported by three 4-inch by 4-inch wood posts along each side and extends beyond the gables, to provide weather protection for the entrances. The bridge aligns west-northwest/east-southeast.

Michael McDonald Bridge has the distinction of being the shortest non-authentic covered motor vehicle bridge in West Virginia, at 16.4 feet long.

Directions: From the junction of US 119 and SR 73, west of Logan, go east 1.5 miles on SR 73 to Mud Fork Road, on the right; go 0.2 mile and bear right (still Mud Fork Road—CR 5), continuing 5.4 miles under US 119 to Harts Road, on the right. Go 0.2 mile to the bridge, visible from the roadway, on the left. Note: There is no access to CR 5 from US 119.

Barrackville Bridge was originally built in 1853 with no siding. In 1873 R. L. Cunningham added horizontal lapped siding. This is an upstream view of the white painted bridge.

Barrackville Covered Bridge

Marion County

Brief Statistics

Type: Authentic historic covered bridge. World Guide Covered Bridge Number: WV-25-02. Built in 1853 by Lemuel Chenoweth and Eli Chenoweth. One-span bridge crossing Buffalo Creek. Multiple kingpost truss with Burr arch bridge 145.5 feet long by 19.1 feet wide, with approximately 15.3-foot wide by 12.3-foot high portal openings. Alternate Name: Buffalo Creek Bridge. Located in Barrackville on bypassed section of Pike Street (CR 21).

Barrackville Bridge was built as part of the Fairmont-Wheeling Turnpike. The contract for the bridge was granted, in April 1853, to Lemuel Chenoweth and Eli Chenoweth (brothers) of Beverly, West Virginia, approved on July 6, 1853, and required completion by December 1, 1853, at a price of $12.50 per lineal foot for the 132-foot span, totaling $1,650. John McConnell and Robert McConnell constructed the stonework and abutments, presumably for $202, as the total cost of the bridge was $1,852. As originally built, the bridge had no siding, thus a contract was let to R. L. Cunningham to install horizontal lapped siding, in 1873. In 1934, C. A. Short, employed by the State of West Virginia, strengthened the bridge by adding iron rods, tying the Burr arch to the lower chord, and added the external walkway, removing the siding between the walkway and traffic lane. In 1951, the bridge was further strengthened by adding steel under the bridge. At some time in the past, the bridge was painted red on the exterior and also on the interior, remaining so into the 1990s. The truss members inside the bridge still wear the old chipped and peeling red paint.

Toward the end of the Civil War, on April 29, 1863, Barrackville Bridge was saved from destruction by retreating Confederate soldiers. Bailes Ice, who owned a grist mill and a saw mill near the bridge, provided food to the soldiers and allowed them to rest on his property, in exchange for not burning the bridge.

Barrackville Bridge, located in Barrackville, has also been known as Buffalo Creek Bridge, after the creek flowing beneath. The 145.5-foot long bridge has the distinction of being the longest single-span authentic covered bridge remaining in West Virginia.

Barrackville Bridge has a fifteen-panel, single-post,

multiple kingpost truss, encased by a double Burr arch, the center three panels with an overlapping double kingpost and the end panels containing an X brace. The historic bridge is one of four covered bridges in West Virginia with a multiple kingpost truss, two having a Burr arch. The bridge is anchored on dry, rough-cut, large sandstone-block abutments. The floor is constructed with diagonal planking topped with transverse planking. The white painted bridge has horizontal lapped siding on the sides, leaving the Burr arch visible, and also on the gables, which are arch-cut over the entrances. The sides are open under the eaves to provide air circulation inside the bridge. A red-painted metal roof completes the structure. The roof and gables extend beyond the entrances, providing additional protection from the weather for the interior. The exterior walkway and the iron rods tying the Burr arch to the lower chord have been removed, probably during the complete restoration in 1998-1999.

Bypassed and closed to motor traffic in 1987, Barrackville Bridge is on the bypassed section of Pike Street, County Road 21, with the concrete bypass bridge very near the upstream side. The West Virginia Department of Transportation-owned bridge crosses Buffalo Creek in a single span, Buffalo Creek flowing into the nearby Monongahela River, at Fairmont. There are small parks at both ends of the bridge, with a bridge informational sign at the left downstream or south end of the south/north-aligned bridge. The downstream side of the bridge has woods on the north bank of the river and strip storage units on the south bank. The oldest graffiti carving in the bridge is dated *MAR 19 1860*, found on the Burr arch.

Barrackville Bridge was listed on the National Register of Historic Places on March 30, 1973. Barrackville Bridge became the sole surviving historic covered bridge in Marion County, after the loss of Paw Paw Bridge, in Grant Town, in the flood of August 18, 1980.

Directions: From Farmington, take US 250 east to Barrackville Road, on the left; follow 1.0 mile to the bridge, on the right (Barrackville Road becomes Pike Street, CR 21).

Rock-N-Wood Heaven Covered Bridge

MINERAL COUNTY

Brief Statistics

Type: Non-authentic modern covered bridge. World Guide Covered Bridge Number: WV-29-A. Built in 1995 by Benny Aronhalt, Sr. Five-span bridge crossing Harrisons Run. Stringer bridge 68.0 feet long by 16.4 feet wide, with approximately 15.0-foot wide by 9.9-foot high portal openings. Alternate Name: None known. Located in New Creek (Harrisons Gap) off US 50.

In 1990, Benny Aronhalt, Sr., built the abutments, piers and stringer bridge for what was eventually to become Rock-N-Wood Heaven Covered Bridge. In 1995, he added the cover. As the bridge is built with rocks and wood, he aptly named it Rock-N-Wood Heaven Bridge. At 68.0 feet long and 16.4 feet wide, the impressive bridge has the distinctions of being the longest and the widest non-authentic covered motor vehicle bridge in West Virginia.

The five-span stringer bridge is supported by nine 6-inch by 6-inch timber stringers, in spans starting from the east end of 14 feet, 12 feet, 12 feet, 12 feet and 18 feet respectively. The stringers are set on abutments, constructed of horizontal locust log and large rock cribbing, and four piers, constructed with three upright locust log posts capped with an 8-inch by 8-inch timber crossbeam and braced. Transverse 2-inch by 10-inch planking, fastened to the stringers, forms the floor. The sides and the portals are covered with gray weathered diagonal boarding, the sides open under the eaves, and the bottoms of the diagonal boards square cut, to form a sawtooth edge along the bottom of the bridge. Galvanized steel covers the roof, which is supported by 2-inch by 4-inch studding.

Rock-N-Wood Heaven Bridge tilts five and one-half feet downhill, following the terrain down to the highway. A *ROCK-N-WOOD/HEAVEN* sign, inscribed in a wood plaque, is centered on the east gable. A *BUILT TO LAST/ 100 YEARS* sign, inscribed in a wood plaque, is centered on the west gable. At the east end, under the bridge, between the abutment and the piers, is a two-room dwelling, constructed by Benny Aronhalt, Sr. This dwelling has a kitchen, bathroom and a combination living room/bedroom. Two stoves provide cooking and heat, their individual smokestacks running up through the bridge and out the roof. Mr. Aronhalt dubbed this dwelling the troll house (his youngest children still look for the troll). In addition, a covered walkway goes from the troll house to a garage downstream.

The east/west-aligned bridge crosses Harrisons Run, a boulder-strewn stream swiftly cascading down the steeply inclined streambed under the bridge. Harrisons Run flows into nearby New Creek, before entering the North Branch Potomac River, at Keyser. Rock-N-Wood Heaven Bridge is in a lightly wooded area of mature trees and very close to US 50, at the west end, and hugs the mountainside, at the east end.

Directions: From Keyser, take US 220 south to SR 972, on the right; go 2.1 miles to US 50; go left 0.2 mile to the bridge, visible on the left. Private property: request permission to visit.

Dents Run Covered Bridge

Monongalia County

Brief Statistics

Type: Authentic historic covered bridge. World Guide Covered Bridge Number: WV-31-03. Built in 1889 by William Mercer and Joseph Mercer. One-span bridge crossing Dents Run. Kingpost truss bridge 40.3 feet long by 15.0 feet wide, with approximately 12.5-foot wide by 11.1-foot high portal openings. Alternate Name: Laurel Point Bridge. Located west of Laurel Point on CR 43-6.

Dents Run Bridge was built, in 1889, at a cost of $448; with William Mercer and Joseph Mercer building the wooden bridge for $250, and W. A. Loar constructing the abutments for $198. Owned by West Virginia Department of Transportation, Dents Run Bridge is named after the stream passing beneath, which obtained its name from John Dent, a settler to the area in 1775. Located west of Laurel Point, the bridge has also been called Laurel Point Bridge. Bypassed, sometime prior to the late 1980s, by the construction of a low, unobtrusive, concrete bridge a few feet downstream from the covered bridge, although still open, traffic seldom passes through the historic structure. Restoration of the bridge took place in 1984.

The 40.3-foot long bridge is one of two kingpost truss bridges remaining in West Virginia, the other being Fish Creek Bridge in Wetzel County. This single-post kingpost truss is modified by the addition of timber supports, from part way up the vertical tension member to under the center of each diagonal brace. The bridge is set on dry, large sandstone-block abutments with wing walls. Six steel I-beam stringers were added, sometime prior to 1988, to strengthen the bridge. The floor consists of transverse planking with five-plank-wide wheel treads. The red painted bridge has horizontal lapped siding on the sides and the portals, the sides open under the eaves for air circulation. Shiny tin covers the steeply peaked roof. Two small signs are mounted on the right downstream gable: one with *DENTS RUN/COVERED BRIDGE* and the other with *BUILT 1889*.

The single-span bridge crosses small, sandy-bottomed Dents Run, which flows into the Monongahela River, at Granville. The west-northwest/east-southeast aligned bridge is nicely exposed, with the bypass bridge on the downstream side and the upstream side densely foliated. A wood picket fence has been recently erected on the downstream side, between the covered bridge and the bypass bridge, obscuring the lower part of the downstream side of the covered bridge. A bridge informational sign is at Sugar Grove Road (County Road 43), at the junction with US 19.

Dents Run Bridge was simultaneously listed with ten other historic covered bridges in West Virginia on the National Register of Historic Places on June 4, 1981. This is the only historic covered bridge remaining in Monongalia County. Morgantown is the county seat of Monongalia County.

Directions: From Laurel Point, go west on US 19 to Sugar Grove Road (CR 43), on the right; go 0.7 mile to CR 43-6, a gravel road on the left; follow 0.3 mile to the bridge.

Indian Creek Covered Bridge

Monroe County

Brief Statistics

Type: Authentic historic covered bridge. World Guide Covered Bridge Number: WV-32-02. Built circa 1898 by Ray Weikel, Oscar Weikel, W. P. Smith and A. P. Smith. One-span bridge crossing Indian Creek. Queenpost/Long combination truss bridge 49.9 feet long by 14.8 feet wide, with approximately 11.6-foot wide by 12.9-foot high portal openings. Alternate Name: Salt Sulphur Springs Bridge. Located south of Salt Sulphur Springs on bypassed section of US 219.

Ray Weikel, Oscar Weikel, E. P. Smith and A. P. Smith built Indian Creek Bridge, circa 1898, at a cost of $400. Due to the similarity with the truss in Hokes Mill Bridge, in adjacent Greenbrier County, both were most likely constructed by the same builders. Indian Creek Bridge, obviously named for the creek flowing beneath, has also been known as Salt Sulphur Springs Bridge, after the community north of the bridge. The Monroe County Historical Society-owned bridge was restored in 1965 and again in 2000.

The 49.9-foot long bridge has a four-panel queenpost/Long combination truss, modified with iron rods, with the queenpost horizontal member stretching across the center two panels, doing double duty as the upper chord. The queenpost diagonal compression members extend across one panel at each end, with the Long truss under the two center panels. Vertical iron rods are placed at each side of the three vertical tension members of the Long truss and three iron rods are placed over each queenpost diagonal compression member, the rods tying the queenpost to the lower chord. The wooden structure rests on abutments, constructed with mortared rough-cut stone, the left downstream or west end abutment on a rock outcrop. The floor is transverse planking. Dark stained vertical boarding with battens covers the sides and the gables, the sides open under the eaves to allow air to circulate within the bridge. New cedar shingles cover the roof, which, with the gables, extends beyond the entrances, to form a weather panel to protect the interior. A bridge informational plaque is mounted on the right downstream gable.

Rocky-bottomed Indian Creek passes under the single-span bridge, on its way to the New River, near Crumps Bottom. Bypassed and closed to traffic, the bridge has a bypassed section of US 219 passing through it, the east end open and a rock outcrop at the west end. The concrete bypass bridge is a very comfortable distance downstream, affording a magnificent view of the covered bridge, looking up the creek or across the small field, with a wooded hill forming a backdrop.

Indian Creek Bridge and Laurel Creek Bridge are the only surviving historic covered bridges in Monroe County. Indian Creek Bridge was listed on the National Register of Historic Places on April 1, 1975.

Directions: From Union, the county seat, at the four-way stop on US 219, go south 4.6 miles on US 219 to the bridge, on the right.

Laurel Creek Covered Bridge

Monroe County

Brief Statistics

Type: Authentic historic covered bridge. World Guide Covered Bridge Number: WV-32-01. Built in 1911 by C. Robert Arnott & Sons. One-span bridge crossing Laurel Creek. Queenpost truss bridge 24.4 feet long by 14.5 feet wide, with approximately 9.9-foot wide by 12.9-foot high portal openings. Alternate Name: Lillydale Bridge. Located in Lillydale on Laurel Creek Road (CR 219-11).

In 1911, at a cost of $365, C. Robert Arnott & Sons built Laurel Creek Bridge, in Lillydale, to carry Laurel Creek Road over Laurel Creek. The bridge also has been known as Lillydale Bridge. At 24.4 feet long, Laurel Creek Bridge has the distinction of being the shortest authentic covered bridge remaining in West Virginia. The small bridge has a queenpost truss, one of five queenpost truss bridges in West Virginia, two that are in combination with a Long truss. The old bridge was rehabilitated, in 2000, at a cost of $260,000 by Hoke Brothers Contracting and Twin Springs Company. The work included replacing the roof, adding concrete caps on top of the abutments, repairing the lower chord, repairing the gables, repainting, and repaving the roadway at both ends of the bridge.

Laurel Creek Bridge has a half-height queenpost truss, modified by three iron rods, tying the horizontal truss member to the center of the lower chord. The red painted bridge rests on mortared, large natural stone abutments, atop which a poured concrete cap was added, in 2000. The flooring consists of transverse planking; the sides are covered with horizontal lapped siding, open under the eaves for air circulation; the gables are also covered with horizontal lapped siding; the roof is covered with shiny new tin. The steeply peaked roof and the gables extend beyond the entrances, providing protection from the elements for the interior. An informational plaque is mounted on each gable.

Owned by the West Virginia Department of Transportation, the single-span bridge crosses narrow, rocky-bottomed and large rock–strewn Laurel Creek, which enters Indian Creek, at Greenville, continuing to the New River, near Crumps Bottom. Still carrying traffic, the west/east-aligned bridge is fully exposed, with sycamore trees downstream of the bridge and woods in the east upstream corner.

Laurel Creek Bridge was listed on the National Register of Historic Places on June 4, 1981, simultaneously with ten other historic covered bridges in West Virginia. Monroe County has only two historic covered bridges, Laurel Creek Bridge and Indian Creek Bridge. *See Laurel Creek Covered Bridge in the color photograph section: C-31.*

Directions: From Union, the county seat, take US 219 south through Salt Sulphur Springs to Lillydale Road (CR 219-7), on the right; go 2.8 miles to Laurel Creek Road (CR 219-11), on the right, and follow 1.3 miles to the bridge.

Locust Creek Covered Bridge

Pocahontas County

Brief Statistics

Type: Authentic historic covered bridge. World Guide Covered Bridge Number: WV-38-01. Built in 1870 by unknown builder. One-span bridge crossing Locust Creek. Warren double truss bridge 119.3 feet long by 16.4 feet wide, with approximately 11.2-foot wide by 12.1-foot high portal openings. Alternate Name: Denmar Bridge. Located west of Denmar on bypassed section of Brownstown Road (CR 31).

An unknown builder built Locust Creek Bridge, at a cost of $1,325, in 1870, over Locust Creek, located west of Denmar, whereby it was also called Denmar Bridge. Owned by the West Virginia Department of Transportation, Locust Creek Bridge is the only Warren truss historic covered bridge in the southeastern United States. The old bridge was bypassed and closed to traffic in 1990.

The 119.3-foot long bridge has a ten-panel Warren double truss, constructed with a double timber truss, sandwiching a single timber inverted truss, secured with iron bolts at every crossover and double timber end post. The bridge rests on dry, rough-cut stone slab abutments and four sets of steel I-beam supports on concrete footings, which were added long ago, plus one wood timber support, added on the southeast end. The floor is constructed with transverse planking. Old naturally weathered vertical boarding with battens covers the sides and the portals. Slightly rusted dull tin covers the steeply peaked roof. Signs reading *CLEARANCE/8 FT. 6 IN.*, in white letters on a green background, are mounted over each entrance. The oldest graffiti carving found in the bridge is dated 1914.

The single-span bridge crosses moderately wide, rock-bottomed Locust Creek, which joins the Greenbrier River, 1.1 miles downstream. Aligned northwest/southeast, the covered bridge is exposed on the ends and on the downstream side, with the bypass bridge being close on the downstream side. Dense woods are at the upstream side, beyond the southeast creek bank.

The sole surviving historic covered bridge remaining in Pocahontas County was listed on the National Register of Historic Places on June 4, 1981, simultaneously with ten other historic covered bridges in West Virginia. Marlington is the county seat of Pocahontas County.

Directions: From Renick, go north on US 219 to a little past Droop Mountain Battlefield State Park, to Locust Creek Road (CR 20), on the right. Follow 3.0 miles to the end, at Brownstown Road (CR 31). The bridge is on the right.

Cool Springs Park Covered Bridge

Preston County

Brief Statistics

Type: Non-authentic modern covered bridge. World Guide Covered Bridge Number: WV-39-A. Built in 1967 by Harland Castle. One-span bridge crossing Pleasant Run. Stringer bridge 30.3 feet long by 13.1 feet wide, with approximately 11.7-foot wide by 10.0-foot high portal openings. Alternate Name: None known. Located west of Macomber in Cool Springs Park on former Cool Springs Road.

Harland Castle built Cool Springs Park Bridge, in 1967, carrying former Cool Springs Road over Pleasant Run, in Cool Springs Park. Now closed to motor traffic, Cool Springs Park Bridge has the distinction of being the oldest non-authentic covered motor vehicle bridge in West Virginia.

Five log stringers, set on abutments of concrete and rocks, support the 30.3-foot long stringer bridge. The floor consists of transverse planking. Naturally weathered vertical boarding with battens covers the sides and the gables, the sides open the length of the bridge at eye level, exposing a decorative four-panel Long truss. The weathered cedar-shingled roof, along with the gables, extends beyond the entrances, providing weather protection for the interior. The low peaked roof is supported, along each side, by triple timbers at the ends and single timbers at the three vertical tension members between the four panels of the decorative Long truss. The left downstream or south-southwest end is gated, to keep park animals out of the bridge.

Pleasant Run is confined by a concrete-lined bottom and mortared, rock-lined stream banks, as it runs through the park, passing under the single-span bridge, flowing into Flag Run, just downstream, which merges into Buffalo Creek, at Macomber, just upstream of the Cheat River. Located in a private park, the bridge blends nicely with surrounding antique trains, steam-powered contraptions, old buggies and items of the like, amid a scattering of mature trees.

Directions: From Fellowsville, at the junction of US 50 and SR 26, go east 8.2 miles on US 50 to Cool Springs Park, on the right, just after the "Cool Springs Incorporated" road sign. The bridge, visible from the highway, is to the left and rear of the retail building.

Lady in Love Covered Bridge

Preston County

Brief Statistics

Type: Non-authentic modern covered footbridge. World Guide Covered Bridge Number: WV-39-b. Built in 2002 by David Dean. Three-span footbridge crossing tributary to Cheat River. Stringer footbridge 33.3 feet long by 6.4 feet wide, with approximately 5.4-foot wide by 7.0-foot high portal openings. Alternate Name: None known. Located east of Macomber off US 50.

A photograph of Bridge Over Troubled Waters Covered Bridge in Rocky Mount, Franklin County, Virginia, published in a magazine, caught the eye of David Dean and inspired him to build a scaled-down version, as a footbridge. David Dean set to work and built Lady in Love Footbridge, in 2002, over a stream at the front of his yard. The bridge lacked steps to enter and exit, at the time of the author's visit, but Mr. Dean expects to construct steps in the near future, being temporarily delayed by unusually persistent rains. The three-span bridge crosses a small muddy bottomed tributary to the nearby Cheat River.

The 33.3-foot long stringer bridge is supported by two 6-inch by 6-inch timber stringers fastened to four piers, consisting of double upright 6-inch by 6-inch timber posts with braces. There are no abutments, as the steps to be added will go from the ground to the end piers. The flooring consists of longitudinal 2-inch by 6-inch planking. Vertical boarding covers the sides to rail height and under the eaves. The sides have five openings between the roof support posts, the top of each opening cut in an inverted V toward the eave. The gables are also covered with vertical boarding. Galvanized steel spotted with rust covers the roof, which is supported by six 4-inch by 4-inch wood posts.

The fully exposed footbridge is aligned north-northwest/south-southeast, parallel to the driveway, which is very close to the east-northeast or upstream side. An old corn planter, just off the highway, is near the south-southeast end of the footbridge, which is visible from the highway.

Directions: From Macomber, take US 50 east, going about 1.0 mile east of the junction with SR 72. The bridge is visible, on the left, just off the highway. Private property: request permission to visit.

Brushy Lick Run Covered Bridge

TAYLOR COUNTY

Brief Statistics

Type: Non-authentic modern covered bridge. World Guide Covered Bridge Number: WV-46-A. Built in 1998 by Wilford R. Jennings. One-span bridge crossing Brushy Lick Run. Stringer bridge 36.2 feet long by 14.2 feet wide, with approximately 9.6-foot wide by 8.8-foot high portal openings. Alternate Name: Jennings Bridge. Located in Thornton off US 50.

Brushy Lick Run Bridge, also known as Jennings Bridge, was built, in December 1998, by Wilford R. Jennings, assisted by Mike Heart and David Myers, over Brushy Lick Run, located at the end of the driveway to the Jennings' property. Mr. Jennings modeled his bridge after historic Scantic River Bridge, built in 1842, in Windsor, Connecticut.

The 36.2-foot long stringer bridge is supported by a reinforced concrete slab, set on six steel I-beam stringers, which span two piers, constructed with railroad ties, stacked horizontally, and four square tubular steel stringers, between the piers and the poured concrete abutments. The railroad tie piers have wing walls extending from each side of the pier to the stream bank, which are constructed with stacked horizontal railroad ties interlocked with the ties of the pier. The concrete slab forms the floor of the brown painted bridge, which has the sides and portals covered with vertical facsimile tongue and groove paneling, the sides open under the eaves for air circulation within the bridge. Red-brown asphalt shingles cover the low peaked roof, which is supported by 2-inch by 4-inch studding. *1998* and *Brushy Lick Run* signs are mounted on the gable over the right downstream or west entrance.

Narrow, rock-bottomed Brushy Lick Run passes beneath the single-span bridge and into nearby Three Fork Creek, which enters the Tygart Valley River, at Grafton. The impressive bridge is highly visible from US 50, set at the driveway entrance and fully exposed, in the front yard of the residence.

Directions: From Grafton, at the intersection of US 119 and US 50, go 3.9 miles east on US 50. The bridge is visible, on the left side of the highway. Private property: request permission to visit.

Fish Creek Covered Bridge

WETZEL COUNTY

Brief Statistics

Type: Authentic modern covered bridge. World Guide Covered Bridge Number: WV-52-01 #2. Built in 2001 by Lone Pine Construction, Inc. One-span bridge crossing Fish Creek. Kingpost truss bridge 35.3 feet long by 14.5 feet wide, with approximately 11.5-foot wide by 11.3-foot high portal openings. Alternate Names: Hundred Bridge, Rush Run Bridge. Located south of Hundred on Rush Run Road (CR 13).

Originally built in 1881 by relatives of C. W. Critchfield, Fish Creek Bridge was totally rebuilt, salvaging only four truss timber braces, in August 2001, by Lone Pine Construction, Inc., of Bentleyville, Pennsylvania, at a cost of $275,000. The West Virginia Department of Transportation-owned bridge is one of two kingpost truss bridges remaining in the State. Named for the creek passing beneath, Fish Creek Bridge has also been called Hundred Bridge, after the nearby community, and Rush Run Bridge, after Rush Run Road, which parallels Rush Run, south of the bridge, before passing through the bridge.

The 35.3-foot long bridge has a single kingpost truss, supported by six steel I-beam stringers, added in 2001, which are set on the original, dry, rough-cut, large sandstone-block abutments. A poured concrete barrier has been added at the bottom of the abutment, on the west-southwest end, to deter erosion. The flooring is constructed with transverse 2-inch by 4-inch planks, set on edge. The red painted bridge, also painted on the interior, has vertical boarding on the sides and the gables. Bright, shiny, new metal covers the roof.

The single-span bridge crosses moderately wide,

rocky-bottomed Fish Creek, which flows into the Ohio River, at Woodlands. Open to motor traffic, the east-northeast/west-southwest aligned bridge is fully exposed, with homes on the upstream side and dense woods on the downstream side.

The only authentic covered bridge remaining in Wetzel County, Fish Creek Bridge was listed on the National Register of Historic Places on June 4, 1981, simultaneously with ten other historic covered bridges in West Virginia. New Martinsville is the county seat of Wetzel County.

Directions: From Hundred, go south on US 250 to just past the junction with SR 7; go right on Rush Run Road (CR 13) 0.2 mile to the bridge.

Glossary

abutment: a structure, usually stone or concrete, that supports the end of a bridge usually at the banks of a stream.

authentic: a wood-truss-supported covered bridge.

baluster handrail: a handrail consisting of a series of upright posts between the main supporting posts.

banister style handrail: a handrail consisting of a series of upright posts between the main supporting posts.

batten: a narrow strip of wood placed over the seams of vertical boarding.

brace: diagonal timbers in a truss or wall slanting toward the center of the truss or wall for primary structural support.

Burr arch: a wooden bowed arch patented by Theodore Burr in 1806 and 1817 used in conjunction with a truss to increase the strength of the truss.

buttress brace: an external support for a bridge wall running diagonally upward from an extended transverse floor beam to a wall post or truss member.

camber: a built-in upward curvature or arch in a bridge, highest at the center, that levels down under the weight of the structure.

C-beam: a steel beam with a cross-section resembling the letter C.

Childs truss: a truss patented by Enoch and Horace Child in 1846 that is a multiple kingpost with iron rod counterbraces added to the timber brace panels.

chin brace: a timber inside a covered bridge fastened to the upper chord slanting inward at the bottom and fastened to a transverse floor beam to stabilize the truss.

chord: timbers forming the upper or lower horizontal beam to which the truss members are fastened.

compression member: a diagonal truss member that is compressed or squeezed when a load crosses the bridge.

counterbrace: diagonal timbers in a truss or wall slanting away from the center of the truss or wall for secondary structural support.

cribbing: interlocked walls constructed with horizontal logs and rocks arranged such that one holds the other in place forming a stable wall.

decking: the floor of a covered bridge.

double-barreled bridge: a covered bridge that has a truss or partition down the center separating two traffic lanes.

dry: the stone or rock of an abutment or pier that is laid without mortar.

eaves: the underside of a roof overhang.

fascia (plural fasciae): the flat horizontal facing surface immediately below the edge of a roof.

gable: the triangular section of a wall under the roof peak.

gambrel roof: a peaked two-sloped roof that has a lower roof at a different slope on all four sides.

Haupt truss: a truss patented by Herman Haupt in 1839 consisting of a series of timber braces slanting toward the center of the bridge overlapping vertical tension members, frequently accompanied by a secondary chord.

height restrictor: a barrier near the portals of a covered bridge, usually of steel construction, that prohibits an overheight vehicle from entering the bridge.

hip roof: a roof with sloping ends as well as sloping sides.

Howe truss: a truss patented by William Howe in 1840 that is a Long truss improved by substituting iron rods for the vertical tension members.

I-beam: a steel beam with a cross-section resembling the letter I.

joist: transverse or longitudinal timbers that support the floor.

kingpost truss: the oldest known truss dating back to the Middle Ages (about 400 to 1400 AD) consisting of a kingpost (a vertical tension member) and a diagonal brace slanting out from each side at the top of the kingpost.

lapped siding: horizontal board siding that overlaps the lower board.

Long truss: a truss patented by Stephen Long in 1830 consisting of X panels comprised of a timber brace and a timber counterbrace between vertical timber tension members.

lower chord: timbers forming the lower horizontal beam to which the bottom of the truss members are fastened.

mortise and tenon joint: the joining of two timbers by cutting a notch out of one timber (mortise) and an exact fit projection out of the other timber (tenon).

multiple kingpost truss: a kingpost truss that has multiple brace panels extending evenly out in each direction from the center panels.

non-authentic bridge: a covered bridge supported by stringers, usually wood or steel, or a post supported roof over a roadway provided a waterway passes beneath the roadway through a culvert or pipe.

panel: the individual sections of a truss that are separated by a vertical post.

pier: a vertical structural support between two spans of a bridge.

pilaster: a facsimile ornamental column at the sides of an entrance.

pile: a vertical support for a bridge that is driven into the ground.

portal: the end of a covered bridge that contains the entrance.

post supported roof: a roof supported by studding or posts along each side that has no attached flooring.

Pratt truss: a truss patented by Thomas Pratt in 1844 that is essentially an inverted multiple kingpost.

purlin: horizontal slats placed across the roof rafters to which is attached the roof covering, such as wood shingles.

queenpost truss: this truss is an extended kingpost truss formed by adding a second vertical tension member separated at the top by the addition of a horizontal timber between the two vertical tension members.

ridgepole: the longitudinal beam running along the peak of a roof.

secondary chord: timbers forming an additional horizontal beam between the upper chord and the lower chord to which the truss members are fastened.

shakes: large rough wooden shingles usually cut by hand.

Smith Type 4 truss: a truss patented by Robert W. Smith in the late 1860s consisting of long closely-spaced counterbraces extending out from the center of the bridge with a shorter brace crossing one counterbrace and attached to the two adjacent counterbraces.

stanchion: a vertical pole or beam support.

stringer: a longitudinal timber or beam supporting a bridge over a span.

tension member: a vertical member of a truss that is stretched when a load crosses the bridge.

tension rod: an iron or steel vertical member of a truss that is stretched when a load crosses the bridge.

terne: an alloy of lead and tin with antimony used as a protective coating on iron or steel.

tongue and groove: a joint between two wooden boards consisting of a projecting strip or tongue along the edge of one board and a mating groove along the edge of the other board.

Town lattice truss: a truss patented by Ithiel Town in 1820 consisting of closely-spaced crisscrossed diagonal planks fastened at every intersection and to the upper and lower chords.

transverse: crosswise or side-to-side.

treenails: wooden pegs that hold truss members together, also called trunnels.

trunnels: see treenails.

truss: a network of timbers in a series of triangular sections that support a bridge, for covered bridges in the walls above the floor.

upper chord: timbers forming the upper horizontal beam to which the tops of the truss members are fastened.

Warren truss: a truss patented by James Warren in 1838 consisting of a series of counterbraces and braces forming inverted V's extending out from the center of the bridge.

weather panel: a panel or partial panel at the end of a covered bridge that is under an extended roof and gable adding weather protection to the entrances.

Wheeler truss: a truss consisting of a series of X panels extending out from a vertical tension member at the center of the bridge, the X panels constructed with one-piece counterbraces and two-piece braces butted to the counterbraces, a secondary chord connecting all intersections.

wheel treads: one or more longitudinal planks laid side-by-side on the floor forming smooth runways for the wheels of a vehicle, also called runners.

wing walls: the lateral extensions of an abutment usually angled into the stream bank to prevent erosion behind the abutment.

Appendix: World Guide Covered Bridge Numbers (WGCB)

WGCB	Page
KY-09-03 #2	72
KY-19-A	74
KY-19-b	73
KY-19-c	74
KY-22-a	75
KY-35-04	77
KY-35-05	77
KY-35-06	75
KY-37-01 #2	78
KY-45-01	79
KY-45-02 #2	79
KY-47-A	80
KY-56-A	81
KY-68-03	81
KY-68-A	82
KY-68-B	82
KY-78-A	83
KY-79-a	83
KY-81-01	84
KY-81-02	85
KY-81-a	84
KY-94-01	85
KY-94-A	85
KY-101-01	86
KY-102-A #2	87
KY-105-A	87
KY-106-A	87
KY-115-01	88
KY-115-a	89
MD-01-A	90
MD-03-02	90
MD-07-01	93
MD-07-02	93
MD-10-01	99
MD-10-02	98
MD-10-03 #2	97
MD-10-a	95
MD-10-c	96
MD-10-d	96
MD-10-e	97
MD-10-f	99
MD-11-b	100
MD-12-01	90
MD-12-A	101
MD-12-b	101
MD-13-A #2	101
MD-13-B	102
MD-15-a	104
MD-15-B	103
MD-15-C	103
MD-15-D	103
MD-15-E	102
MD-16-a	104
NC-05-A	107
NC-06-a	108
NC-06-b	108
NC-08-B	108
NC-08-c	110
NC-11-A	110
NC-11-b	111
NC-18-01	111
NC-20-A	112
NC-23-A	113
NC-28-A	113
NC-34-01	114
NC-41-A	115
NC-41-B	115
NC-41-c	117
NC-41-d	116
NC-44-A	117
NC-45-a	118
NC-57-b	118
NC-58-a	119
NC-62-A #2	120
NC-65-A	120
NC-68-A	121
NC-72-A	121
NC-75-A #2	121
NC-75-C	122
NC-76-01	122
NC-81-C	123
NC-81-D	123
NC-85-A	124
NC-88-a #2	125
NC-88-B	124
NC-95-b	125
NC-95-C	126
NC-95-d	126
NC-95-E	107
SC-04-a	128
SC-04-b	128
SC-10-a	129
SC-22-a	130
SC-22-b	130
SC-23-02	131
SC-23-A	133
SC-23-b	132
SC-23-d	134
SC-23-E	132
SC-30-A	135
SC-32-A	135
SC-39-b	138
SC-39-C	136
SC-39-D	135
SC-39-E	137
SC-39-f	136
SC-39-g	137
SC-42-A	139
SC-42-B	138
SC-42-c	140
SC-43-a	140
SC-43-b	140
TN-02-A	142
TN-10-01	143
TN-11-A	144
TN-18-A	144
TN-23-01	145
TN-30-01	146
TN-30-a	146
TN-32-A	147
TN-33-A	148
TN-33-C	148
TN-33-D	148
TN-33-E	149
TN-50-A #2	149
TN-56-A	151
TN-56-B	150
TN-57-A	151
TN-62-A	151
TN-63-01 #1	152
TN-63-01 #2	152
TN-64-A	152
TN-66-01	145
TN-67-A	153
TN-71-a	154
TN-71-b	154
TN-71-C	154
TN-71-D	155
TN-78-01	157
TN-78-a	156
TN-78-b	156
TN-78-E	158
TN-78-F	156
TN-83-A	159
TN-83-B	159
TN-94-A	161
TN-94-B	161
TN-94-C	160
TN-94-D	160
TN-95-a	161
VA-03-01	163
VA-18-A #2	165
VA-18-b	164
VA-18-C	165
VA-29-a	165
VA-29-b	166
VA-31-A	166
VA-33-A	167
VA-35-01	168
VA-35-02	169
VA-35-03	167
VA-43-A	170
VA-47-01	170
VA-60-A	171
VA-68-01	171
VA-68-02	173
VA-68-A	172
VA-69-b	173
VA-69-C	174
VA-71-A	174
VA-73-b	175
VA-77-A	177
VA-79-01	177
VA-82-01	178
VA-83-a	179
VA-86-A	180
WV-01-01	182
WV-01-02	181
WV-06-01	183
WV-09-01	184
WV-09-A	185
WV-13-01	185
WV-13-02	186
WV-17-03	187
WV-17-12	187
WV-18-01 #2	189
WV-18-04	190
WV-21-03	191
WV-22-a	192
WV-23-A	192
WV-25-02	193
WV-29-A	194
WV-31-03	195
WV-32-01	196
WV-32-02	195
WV-38-01	196
WV-39-A	197
WV-39-b	197
WV-46-A	198
WV-52-01 #2	198

Index